EUROPEAN PHILOSOPHY OF SCIENCE – PHILOSOPHY OF SCIENCE IN EUROPE AND THE VIENNESE HERITAGE

VIENNA CIRCLE INSTITUTE YEARBOOK

17

VIENNA CIRCLE INSTITUTE YEARBOOK
17

Institut Wiener Kreis

Society for the Advancement of the Scientific World Conception

Series-Editor:
Friedrich Stadler
University of Vienna, Austria and Director, Institut Wiener Kreis

Editorial Address:

Institut Wiener Kreis
Universitätscampus, Hof 1
Spitalgasse 2-4, A–1090 Wien, Austria
Tel.: +431/4277 46501 (international)
or 01/4277 46501 (national)
Fax.: +431/4277 41297 (international)
or 01/4277 41297 (national)
Email: ivc@univie.ac.at
Homepage: http://univie.ac.at/ivc/

EUROPEAN SCIENCE FOUNDATION

www.pse-esf.org

Maria Carla Galavotti • Elisabeth Nemeth
Friedrich Stadler
Editors

European Philosophy of Science – Philosophy of Science in Europe and the Viennese Heritage

Springer

Editors
Maria Carla Galavotti
Dipto. Filosofia
Università di Bologna
Bologna, Italy

Elisabeth Nemeth
Inst. Philosophie
Universität Wien
Wien, Austria

Friedrich Stadler
Inst. Wiener Kreis
Inst. Philosophie
Inst. Zeitgeschichte
Universität Wien
Wien, Austria

ISSN 0929-6328
ISBN 978-3-319-01898-0 ISBN 978-3-319-01899-7 (eBook)
DOI 10.1007/978-3-319-01899-7
Springer Cham Dordrecht Heidelberg New York London

Library of Congress Control Number: 2013947192

Printed on acid-free paper

Springer is part of Springer Science+Business Media (www.springer.com)

Table of Contents

Editorial

The Research Networking Programme "The Philosophy of Science in a European Perspective" (PSE) of the European Science Foundation (ESF) dealt with the philosophies, foundations and methodologies of the sciences. The international symposium "Philosophy of Science in Europe – European Philosophy of Science and the Viennese Heritage" (Vienna, December 5–7, 2011), combined the theoretical and historical perspective focusing on the specific features of a European philosophy of science. On the occasion of the 20th anniversary of the Institute Vienna Circle the Viennese roots and influences were addressed, in addition.

There is no doubt that contemporary philosophy of science originated mainly in Europe beginning in the 19th century and has influenced decisively the subsequent development of globalized philosophy of science, esp. in North America. Recent research in this field documents some specific characteristics of philosophy of science covering the natural, social, and also cultural sciences in the European context up to the destruction and forced migration caused by Fascism and National Socialism.

The proceedings of the opening plenary conference of the Networking Programme PSE, held in Vienna, from December 18–20, 2008, were published in 2010 as The Present Situation in the Philosophy of Science.[1] They document the flourishing topicality of contemporary philosophy of science in Europe. The volume covers foundational and methodological debates, formal methods and their applications, the place of the life sciences and physical sciences in the foundations of science, and the present situation of the philosophy of the cultural and social sciences on the one hand, and some specific European manifestations, on the other hand, which can be generally identified with historical, pragmatic and interdisciplinary approaches bridging the absolute dualism of "analytic" and "continental" philosophy (of science). Therefore, also more general philosophical topics in the sciences are accompanied by a naturalistic approach, taking into account the aims and values of philosophy of science in itself and the consequences for the related

---

1 *The Present Situation in the Philosophy of Science*. Ed. by Friedrich Stadler, together with Dennis Dieks, Wenceslao J. González. Stephan Hartmann, Thomas Uebel, Marcel Weber. Further volumes in this series "The Philosophy of Science in a European Perspective", ed. by Maria Carla Galavotti and Friedrich Stadler (Dordrecht–Heidelberg–London–New York : Springer 2010f). Vol. 2: Explanation, Prediction and Confirmation. Ed. by Dennis Dieks, Wenceslao J. González, Stephan Hartmann, Thomas Uebel, Marcel Weber (2011). Vol. 3: Probabilities, Laws, and Structures. Ed. by Dennis Dieks, Wenceslao J. González, Stephan Hartmann, Michael Stöltzner, Marcel Weber (2012). Vol. 4: New Challenges to Philosophy of Science. Ed. by Hanne Andersen, Dennis Dieks, Wenceslao J. González, Thomas Uebel, Gregory Wheeler (2013).

methodology (since the *Methodenstreit*) and historiography, obviously within the frame of a theoretical pluralism.

This European perspective with the integration of history and philosophy of science and the current situation in the philosophy of science after the transatlantic interaction and transformation, and the "return" after World War II raised the question of contemporary European characteristics in the philosophy of science. The conference referred to this opening conference and its results aiming at topical issues and open questions between philosophy of science in Europe and European philosophy of science.[2]

On the occasion of the 20th anniversary of the Institute Vienna Circle, and its establishment as Department of the University of Vienna (Faculty of Philosophy and Education) in 2011, the role and function of the renowned Vienna Circle of Logical Empiricism and its impact and influence on contemporary philosophy of science was on the agenda, too. Accordingly, the general topic was addressed in two parallel sessions representing systematic-formal as well as genetic-historical perspectives on philosophy of science in an European context up to the present. The present volume largely contains the English-language contributions to this symposium. The German-language contributions will appear in a parallel volume *Die europäische Wissenschaftsphilosophie und das Wiener Erbe*, Elisabeth Nemeth und Friedrich Stadler (Hrsg.), Dordrecht–Heidelberg–London–New York: Springer, 2013 (= Veröffentlichungen des Instituts Wiener Kreis, Bd. 18).

Thanks go to Maria Carla Galavotti on behalf of the ESF-PSE programme for the joint organization and to the members of the staff of the Institute Vienna Circle – Sabine Koch, Robert Kaller, and Karoly Kokai – for their help regarding the anniversary conference and the publication of the proceedings. The University of Vienna enabled the establishment of the Institute Vienna Circle as a Department in the Faculty of Philosophy and Education, which was pleasingly reinforced on the occasion of the opening of the conference by Vice-Rector Susanne Weigelin-Schwiedrzik and the then Vice-Dean Konrad Paul Liessmann.

Vienna, April 2013

Friedrich Stadler
(Institute Vienna Circle, Head and Director)

2 Review of Stathis Psillos, in: *Metascience*, vol.20, No.2.

Friedrich Stadler

From the Vienna Circle to the Institute Vienna Circle: On the Viennese Heritage in Contemporary Philosophy of Science

The Vienna Circle as part of the intellectual movement of Central European philosophy of science is certainly one of the most important currents for the emergence of modern philosophy of science. Independent from this uncontested historical fact there remains the question of the direct and indirect influence, reception and topicality of this scientific community in contemporary general philosophy of science as well as in the philosophy of the individual sciences, including the social sciences and humanities.

In my account I will focus on the "present situation in the philosophy of science"[1] by identifying some relevant impacts, results, and unfinished projects since the classical Vienna Circle, by dealing with specific European features of this globalized philosophical tradition up to the present, and by exemplifying some future perspectives after the linguistic, historical and pragmatic turns. This reconstruction is partly linked to the history of the Institute Vienna Circle which was established in 1991 in Vienna, and which was a supporting institution of the ESF Research Network Program the "Philosophy of Science in a European Perspective" (PSE) from 2008 to 2013.

1 Analytic philosophy (of science) – the context of modernity

The Vienna Circle, which was part of the intellectual movement of Central European philosophy of science, is certainly one of the most important currents in the emergence of modern philosophy of science. Apart from this uncontested historical fact there remains the question of the direct and indirect influence, reception and topicality of this scientific community in contemporary philosophy of science in general as well as in the philosophy of the individual sciences, including the social sciences and humanities.

First, I will characterize the road from the Schlick Circle to contemporary philosophy of science. Second, I will refer to "the present situation in the philosophy of science" by identifying relevant impacts, findings, and unfinished projects since

1 Friedrich Stadler (Ed.) (2010) *The Present Situation in the Philosophy of Science*, Dordrecht–London–New York: Springer and Friedrich Stadler (2010) "On the Present Situation in the Philosophy of Science", in that volume, p. 7-10.

M.C. Galavotti et al. (eds.), *European Philosophy of Science – Philosophy of Science in Europe and the Viennese Heritage*, Vienna Circle Institute Yearbook 17,
DOI 10.1007/978-3-319-01899-7_1,

the classical Vienna Circle. Third, I will address some specific European features of this globalized philosophical tradition up to the present, and outline some future perspectives after the linguistic, historical and pragmatic turns – looking back to the "received view".

2 Viennese and European contexts

In 2011, Edward Timms referred to the Viennese cultural circles entitled "Dynamik der Kreise, Resonanz der Räume. Das Denken der Wiener Moderne" in his Opening Lecture to the anniversary conference as the broader cultural context of the Vienna Circle since Ernst Mach's pioneering role in *Fin de Siécle* Vienna.[2] Herein it becomes clear how the Vienna Circle was part of a modernist movement transgressing the boundaries of philosophy proper.

At the same time this means that the origins and development of Logical Empiricism cannot be equated with the current of analytic philosophy, as it was reconstructed critically already by Georg Henrik von Wright (1993)[3]: this specific branch of philosophy since Frege, Moore, Russell, and Wittgenstein was more or less intertwined with the establishment of Logical Empiricism based mainly on the philosophy of the natural sciences. In parallel, the idea of analytic philosophy can be described as a tradition between a school or as a research field.[4]

Already in 1991, on the occasion of the centenary conference of Carnap, Reichenbach, and Zilsel in Vienna, which was at the same time the opening conference of the Institute Vienna Circle, Carl G. Hempel dealt with the historical, pragmatic, and sociological turns ending up with the inclusion of the pragmatic component according to Neurath's option: [5]

> Thus, under the influence if internal and external critical and constructive reflection, the original ideas of the Vienna Circle evolved into a new and very different empiricist con-

2 Publication in: Hubert Christian Ehalt, Friedrich Stadler, Edward Timms, Heidemarie Uhl, *Schorskes Wien: Eine Neuerfindung*. Wien: Picus 2012. His overlapping circles cover poets, musicians, artists, philosophers, and scientists, but also politicians and social reformers. (p. 44). www.vorlesungen.wien.at. In addition: Edward Timms, *Dynamik der Kreise, Resonanz der Räume. Die schöpferischen Impulse der Wiener Moderne*. Weitra 2013.

3 Georg Henrik von Wright (1991) "Analytic Philosophy – A Historical-Critical Account", in: von Wright *The Tree of Knowledge and Other Essays*, Leiden: E. J. Brill.

4 Friedrich Stadler (2011). "The Idea of Analytic Philosophy – A School or Research Field?", Paper, delivered at the University of Innsbruck, Nov.15, 2011 in honor of Edmund Runggaldier. (forthcoming).

5 Carl G. Hempel (1993) "Empiricism in the Vienna Circle", in: Friedrich Stadler, *Scientific Philosophy: Origins and Developments*, Dordrecht-Boston-Londont: Kluwer, p. 8. See also: Rudolf Haller and Friedrich Stadler *Wien – Berlin –Prag. Der Aufstieg der wissenschaftlichen Philosophie*, Vienna: Verlag Hölder-Pichler-Tempsky.

strual of scientific knowledge and of the methodology of scientific inquiry. While surely not initially intended or anticipated, this evolution was in accord with a basic view widely held in the Circle as to how a 'scientific world view' was to be developed: not as a monolithic *a priori* system, but, like empirical science itself, under constant critical appraisal and revision.

The interaction of analytic philosophy and philosophy of science in the "Origins of Analytic Philosophy" remains on the agenda as became apparent from a controversial discussion on the comparison of "analytic" and "continental" proponents of this philosophical journey[6] determined by the distinction as described by Michael Friedman.[7]

Just in this case study, the rhetoric of science becomes apparent: alone the employed terminology and history of concepts indicate the need for systematic clarification and "rational reconstruction" with reference to terms like Scientific Philosophy, Logic of Science, Positivism, Neo-positivism, Logical Empiricism, Encyclopedia of Unified Science, Philosophy of Science, History and Philosophy of Science – as labels for overlapping conceptions between philosophy and the individual sciences.[8]

3 Vienna – Berlin – Prague: Central European Communication

After more than two decades of research on the "Rise of Scientific Philosophy" (Hans Reichenbach) between Vienna and Berlin it became also clear that Logical Empiricism was certainly not a homogeneous school or discipline in philosophy of science[9], where epistemology, probability and induction was linked to the discovery-justification distinction and different approaches to probabilism and realism emerged already before World War II. Therefore, it is not surprising that Reichenbach in his "Logistic Empiricism in Germany and the Present State of its Problems" (1936) presents a sort of a counter manifesto to that Vienna Circle manifesto of 1929.[10] In any case, it is a significant fact that Hans Reichenbach and

6 As a pre-history: Paul Kruntorad (Ed.) *Jour fixe der Vernunft* (1991), containing contributions by Ludovico Geymonat and Wolfgang Stegmüller dealing with "modern analytic philosophy" and "modern philosophy of science". The discussion took place at the list of the International Society for the History of Philosophy of Science (HOPOS-L) from 9/15/2011 to 9/22/2011.

7 Michael Friedman, (2000) *Parting of the Ways.* Chicago: Open Court.

8 See Thomas Uebel, "'Logical Positivism' – 'Logical Empiricism': What's in a Name?", in: *Perspectives on Science*, 2013, vol. 21, no. 1, pp.58-99.

9 Friedrich Stadler (2011) "The Road to *Experience and Prediction* from within: Hans Reichenbach's Scientific Correspondence from Berlin to Istanbul" in: *Synthese*, 181, Issue 1, p. 137-55.

10 Hans Reichenbach, (1936) "Logistic Empiricism in Germany and the Present State of ist Problems", *The Journal of Philosophy*, Vol. 33, No. 6, p. 141-160. See the reprint

Moritz Schlick, the main proponents and opponents of Logical Empiricism are honoured in the long run by two historical-critical edition projects, in addition to ongoing contributions in the philosophy of science.[11]

Already here we find the contested topics of (philosophical, critical, structural) realism, causality, (subjective and objective) probability and inductivism and the relativized *a priori* with reference to German Neo-Kantianism.

4 Edgar Zilsel – Import of History and Sociology of Science

Besides pure philosophy of science, the history of science and the sociology of science and scientific knowledge were present from the beginning. One representative case study for this long neglected research field is the lifework of Edgar Zilsel. Following his investigation in the problem of induction (*Das Anwendungsproblem. Ein philosophischer Versuch über das Gesetz der großen Zahlen und die Induktion* (1916), his special contributions are documented with two monographs on the history and analysis of the phenomenon and concept of „genius“:

– *Die Geniereligion. Ein kritischer Versuch über das moderne Persönlichkeitsideal, mit einer historischen Begründung* (1918).

– *Die Entstehung des Geniebegriffes. Ein Beitrag zur Ideengeschichte der Antike und des Frühkapitalismus* (1926).

Pleasingly, some of his pioneering studies in the history and philosophy of science are collected in German in the edited volume on *Die sozialen Ursprünge der neuzeitlichen Wissenschaft* (posthum, 1976), contextualizing the famous "Zilsel thesis":

> From the period from the end of the Middle Ages until 1600 the university scholars and the humanistic literati are rationally trained but they do not experiment as they despise manual labor. … About 1600, with the progress of technology, the experimental method is adopted by rationally trained scholars of the educated upper class. So the two components of scientific search are united at last: modern science is born … (Zilsel 1939).

The translation of his collected articles into English, accompanied by new research contributions, seems to be a late recognition and appraisal of this unique scholar, who committed suicide in 1944 in US exile after a tragic track of life following his forced migration from his home city of "Red Vienna". In this volume entitled

---

of the Vienna Circle Manifesto: *Wissenschaftliche Weltauffassung. Der Wiener Kreis*. Friedrich Stadler and Thomas Uebel (eds.), (2012) Wien–New York: Springer.

11 *Moritz Schlick Gesamtausgabe*: www.moritz-schlick.de, www.univie.ac.at/ivc/Schlick-Project and Reichenbach, *Gesammelte Werke*, as well as the special issue of *Synthese* (2011) 181. Reichenbach Edition, Marie Reichenbach and Andreas Kamlah (eds.), Springer.

The Social Origins of Modern Science (2000), the renowned historian of science Joseph Needham wrote on the academic outsider: [12]

> All we can be sure of, and this is where Zilsel's work is a veritable torch to light the darkness, is that we have to look for the 'sociological roots' as well as the purely intellectual ones, of science and technology, whether it be in the West or in the East. Fiat lux, we all cry, and Edgar Zilsel's life and work put him among most notable taperers in the procession of those who seek to understand.

5 Logical empiricism re-evaluated

With a critical re-evaluation of Logical Empiricism in 2001 on the occasion of the 10th anniversary of the Institute Vienna Circle[13] the origins, history and reception, the methods, scientific communication , (self-)organization, sociology of science and knowledge were addressed based on the growing research. The uprooting and alienation from (Central) Europe, the subsequent forced migration to US/UK and transformation of Logical Empiricism with mutual theory dynamics are documented in the proceedings on *The Vienna Circle and Logical Empiricism. Re-evaluation and Future Perspectives* complemented by a special volume on Carnap's role in this respect: *Language, Truth and Knowledge: Contributions the Philosophy of Rudolf Carnap* (ed. Th. Bonk 2003) Again, the three main figures Schlick, Reichenbach, and Carnap came into focus: the question of one or two circles in Berlin and/or Vienna as well as the idea of a common denomination (Logical Positivism, Logical Empiricism, Scientific World Conception, Encyclopedia of Unified Science etc.) challenging the dualism of philosophy and the sciences up to Reichenbach's *Experience and Prediction* (1938) showed the inherent theoretical pluralism. Recurring topics were present with the continuing relationship of "Positivism and Realism" (Schlick 1932), the theoretical pluralism with regard to probability theory, unity and plurality, context of discovery and context of justification, emotivism and meta-ethical non-cognitivism in the broader frame of philosophy of science. New perspectives emerged on women of Logical Empiricism (Rose Rand, Susan Stebbing, Else Frenkel-Brunswik, Olga Hahn, Olga Taussky) and the third generation in philosophy of science, esp. the Schlick's students and guests from abroad born between 1900 and 1912! (We are speaking of some 20 philosophers from Ayer to Tscha Hung = Hong Qian, who succeeded scholarly in the further academic life in their home countries.) This sociology of philosophy is to be still researched with a collective intellectual biography.

---

12 Joseph Needham (2000) "Foreword" in: Edgar Zilsel *The Social Origins of Modern Science*, (Boston Studies in the Philosophy of Science), p. xiv.

13 Friedrich Stadler (2003), "What is the Vienna Circle? Some Methodological and Histiriographical Answers", in: Friedrich Stadler (ed.) *The Vienna Circle and Logical Empiricism*, Wien–New York: Springer, p. xi-xxiii.

6 Viennese origins – European networks

The influence and impact of this Viennese culture has to be connected with European networks proper. Already Kasimir Twardowski studied in Vienna for many years before he acted as the founder of the Polish Lwow-Warsaw School of "logistic anti-irrationalism" leading up to Alfred Tarski and his contemporaries. Anna Brożek (2011) reconstructed the Viennese roots of Twardowski, and opened in her book Tarski's intellectual background. [14]

At the same time, the Brentano tradition and the phenomenological movement in philosophy and philosophy of science becomes manifest: it was above all Felix Kaufmann, the mediator between Husserl, Kelsen and the Vienna Circle, who bridged the gap between these three currents and tried to approach American pragmatism (esp. John Dewey) in his later US exile years. This occurred with the English edition of his *Methodology of the Social Sciences* (1944), which was not directly a translation of his *Methodenlehre der Sozialwissenschaften* (1936), in which he tried to solve the virulent *Methodenstreit* between the social and natural sciences. Independently, the strong presence of Husserl, from Mach to Wittgenstein and the Vienna Circle is another long neglected subject of research.[15]

There is now doubt about Ernst Mach's role and function as one precursor and pioneer of the Vienna Circle and Logical Empiricism, especially as a model for history and philosophy of science (HPS), which is currently discussed with reference to the context of discovery and context of justification distinction.[16] With his "Chair for Philosophy, particularly for the History and Theory of the Inductive Sciences" (1895) he succeeded as one of the few scientists who took over a chair in philosophy and introduced the "historical-critical method" into philosophy (of science). His impact can hardly be overrated, even if there were modifications, selective reception, and also hidden manifestations, which can be demonstrated by his influence on Wittgenstein[17] and Paul Feyerabend,[18] who in his later years came back to Mach and fostered the historical tradition in the philosophy of science – besides the uncontested Boltzmann tradition. These new contextualizations

14 Anna Brożek (2011) *Kazimierz Twardowski. Die Wiener Jahre*, Wien–New York: Springer.

15 As one exception see the Vienna Circle Lecture of Dagfinn Foellesdal on Husserl and Gödel in this volume.

16 Friedrich Stadler (2012), "History and Philosophy of Science: Between Description and Construction", in: *New Directions in the Philosophy of Science*. Ed. by Maria Carla Galavotti, Dennis Dieks, Wenceslao J. Gonzalez, Stephan Hartmann, Thomas Uebel, Marcel Weber, Springer, to appear in 2014.

17 J. Hintikka (2001) "Ernst Mach at the Crossroads of Twentieth-Century Philosophy", in: Juliet Floyd and Sanford Shieh (eds.) *Future Pasts. The Analytic Tradition in Twentieth-Century Philosophy*, Oxford University Press, p. 81-100.

18 Friedrich Stadler and Kurt R. Fischer (Eds.) (2006) *Paul Feyerabend – Ein Philosoph aus Wien*, Wien–New York: Springer.

are accompanied by the running *Ernst Mach Studienausgabe* (2008 f.), which will be completed by the forthcoming anniversary in 2016.[19]

The "French connection" was already at work with Mach at the heydays of the Vienna Circle, where the networking with the English-speaking community and the French scholars was at par before the forced migration and transformation of the philosophers of science.[20]

Similarly, even if somewhat stronger, the Austro-British Interaction since the Russell-Meinong exchange in 1905 emerged thereby challenging the analytic-continental split in the philosophy of science, which flourished till the outbreak of WW II, e.g. with the "Fourth International Congress for Unity of Science" in Cambridge (Girton College) in 1938.[21]

Only in the last decade there was a special focus on the traditional interaction between the community in the Nordic Countries and the Central European philosophy of science, as presented by Eino Kaila, Georg Henrik von Wright, Arne Naess, up to Jaakko Hintikka. This was the main topic of a conference in Helsinki and its proceedings on *The Vienna Circle in the Nordic Countries*[22], referring once more to realism, empiricism, psychologism and paving the way to experimental philosophy (of science), represented by Arne Naess.

7 Moritz Schlick – between realism and empiricism

The rediscovery and re-evaluation of Moritz Schlick's life and work is going on with the running historical-critical *Schlick Edition Project* accompanied by the *Schlick Studies* and *Schlickiana,* since 2006 a cooperation between the Institute Vienna Circle with the University of Rostock, the Moritz Schlick Research Unit (Moritz Schlick-Forschungsstelle).[23]

The first part of the edition is nearly finished with the publication of Schlick's *Allgemeine Erkenntnislehre, Reflexion des Lichtes/Raum und Zeit, Lebensweisheit/Fragen der Ethik,* and the collection of articles of his *Die Wiener Zeit.* In addition, the research oriented *Schlick Studien* (2008f.), offer most recent studies in the philosophy of the founder of the Vienna Circle, contributed also by a young-

19 *Ernst Mach Studienausgabe*, Ed. Friedrich Stadler, Berlin: Xenomoi, 2008f.

20 Elisabeth Nemeth/NicolasRoudet (Eds.), *Paris – Wien. Enzyklopädien im Vergleich*, Wien–New York: Springer 2005.

21 On the Austro-British exchange: Timms (Ed.), and Maria Carla Galavotti (Ed.) (2004) *Cambridge and Vienna. Frank P. Ramsey and the Vienna* Circle, Wien–New York: Springer.

22 Manninen/Stadler (Eds.) (2010) *The Vienna Circle in the Nordic Countries*, Wien–New York: Springer.

23 www.univie.ac.at/ivc/Schlick-Project and www.moritz-schlick.de: *Moritz Schlick-Gesamtausgabe* (MSGA): Wien–New York: Springer. 2006f.

er generation from the editorial team.[24] A intellectual biography by Massimo Ferrari will accomplish this long term project, which is funded by the Hamburgische Akademie in the meantime. There was , and is a lot of investigation on the relation between Einstein and Schlick, who exerted a decisive influence on the former as the first acknowledged philosophical interpreter of relativity theory although from the middle of the 1920s the occurred a certain alienation due to Schlick's linguistic turn under the influence of Wittgenstein. Despite the modification of his earlier critical realism as "turning point in philosophy" there is also evidence for continuity between the young Schlick in Germany and the later one of his Vienna Circle period, as is reconstructed by Hans Jürgen Wendel (2013) in „Moritz Schlick and the Metaphysics" claiming this red thread with reference to his relation to (neo-Kantian) metaphysics.[25] Besides this controversial topic the difference between Schlick and Neurath stays on the agenda: theoretical pluralism, naturalism and the vision of a unity of science is the topical frame leading up to the dualism of philosophy and the sciences with correspondence theory on the one hand, and the empiricist encyclopedia project with a coherentist variant, on the other. This unfinished internal dispute indicates the variety of Logical Empiricism, especially regarding philosophical and structural realism, "consistent empiricism" (Schlick) and the inclusion of values and ethic into the concept of scientific philosophy in general.[26]

8 Rudolf Carnap – Philosophy of Science Today

It's not a surprise that the most systematic philosopher of science is a permanent research subject in contemporary philosophy of science: besides the *The Cambridge Companion to Carnap*[27] this focus was enriched by a conference on *Carnap and the Legacy of Logical Empiricism* (R. Creath 2012) published as proceedings of a conference organized by the Institute Vienna Circle. Based on the running edition

24 Friedrich Stadler, Hans Jürgen Wendel (Eds.) *Stationen. Dem Philosophen und Physiker Schlick zum 125. Geburtstag (2009), Moritz Schlick Studien,* Band 1. Wien–New York: Springer. M. Neuber (2011) *Die Grenzen des Revisionismus. Schlick, Cassirer und das „Raumproblem", Schlick Studien,* Band 2, Wien–New York: Springer; Johannes Friedl (2013) *Konsequenter Empirismus. Die Entwicklung von Moritz Schlicks Erkenntnistheorie im Wiener Kreis, Moritz Schlick Studien Band 3*, Wien–New York: Springer.

25 This Vienna Circle Lecture 2011 is published in the German volume of the proceedings: Elisabeth Nemeth/Friedrich Stadler (Hrsg.), *Die europäische Wissenschaftsphilosophie und das Wiener Erbe*. Dordrecht: Springer (forthcoming).

26 An updated discussion in Anne Siegetsleitner (Hrsg.), *Logischer Empirismus, Werte und Moral*, Wien–New York: Springer 2010.

27 Richard Creath and Michael Friedman (2007), Cambridge University Press.

project *Collected Works of Carnap*[28] we experience a remarkable renaissance on the *Aufbau* project (1928) since Alan Richardson's reconstruction.[29]

From a specific point of view, it is remarkable to see the continuity and renaissance of research on Carnap's life and work: first, the *Collected Works of Rudolf Carnap*, the *Cambridge Companion to Carnap*, together with the intellectual biography by A. W. Carus, *Carnap in Twentieth Century Thought: Explication as Enlightenment* (2007). Not least the *Aufbau* project is reconstructed and reformulated promising as a still viable model in the philosophy of science. e.g., Hannes Leitgeb wrote (in his "New Life for Carnap's *Aufbau*?"):[30]

> Carnap's *Der logische Aufbau der Welt* ... is generally conceived of as being the failed manifesto of logical positivism. ... How much of the *Aufbau* can actually be saved? We will argue that there is an adaption of an old system which satisfies many demands of the original pogramme. In oder to defend this thesis, we have to show how a new ‚Aufbau-like' programme may solve or circumvent the problems that affected the original *Aufbau* project. In particular, we are going to focus on how a new system may address the well-known difficulties in Carnap's *Aufbau* concerning abstraction, dimensionality, and theoretical terms.

Recently, David J. Chalmers (Introduction to *Constructing the World*? – „A structuralist response to skepticism") undertook a huge (re)construction of Carnap's vision in his recent book via the concept of "scrutability":[31]

> In many ways, Carnap is the hero of this book. Like the other twentieth century logical empiricists, he is often dismissed as a proponent of a failed research program. But I am inclined to think that Carnap was fundametally right more often than he was wrong. I do not think that he was right about everything, but I think that many of his ideas have been underappreciated. So I might see this project, in part, as aiming for a sort of vindication. The title of this book is a homage to Carnap's 1928 book *Der logische Aufbau der Welt*, usually translated as either *The Logical Construction of the World* or *The Logical Structure of the World*.

Within the Vienna Circle Institute this perspective was already manifest in the conferences and publications, like the edited volumes of Thomas Bonk (*Language, Truth and Knowledge,* 2003), and Richard Creath (*Rudolf Carnap and the Legacy of Logical Empiricism,* 2012). All these elaborations are not really surprising, given Carnap's own late conviction that the unfinished *Aufbau* program principally could work out after some revisions.[32]

---

28 *Collected Works of Rudolf Carnap*, Chicago: Open Court (www.opencourtbooks.com).

29 Alan Richardson(1998) *Carnap's Construction of the World. The* Aufbau *and the Emergence of Logical Empiricism*, Cambridge University Press.

30 Hannes Leitgeb (2011) "New Life for Carnap's *Aufbau*?", *Synthese* 2011/180.

31 David J. Chalmers (2012) *Constructing the World*, Oxford University Press, p. 17.

32 Rudolf Carnap (1961), *Der logische Aufbau der Welt. Scheinprobleme in der Philosophie*. Hamburg: Meiner. Vorwort zur zweiten Auflage, S. IX.

9 Neurath's Boat Rediscovered – The "Visual Turn"

In the long run, Neurath's Boat was re-discovered, in parallel with the pictorial turn: W. V. O. Quine, since his influential "Two Dogmas of Empiricism" (1951) refers in his *Word and Object* (1960) (dedicated "to Rudolf Carnap. Teacher and Friend") to Neurath in German: "Wie Schiffer sind wir, die ihr Schiff auf offener See umbauen müssen, ohne es jemals in einem Dock zerlegen und aus besten Bestandteilen neu errichten zu können", which confirms the development and criticism of Logical Empiricism from within leading up to further so called dogmas as a product of the transfer and transformation of the philosophy of science from the "Vienna Circle to Harvard Square" (Holton)[33]. With this process we can see the tension between Carnap and Neurath regarding semantics and the correspondence theory of truth as well as the emergence of the Duhem-Neurath-Quine thesis comprising methodological holism and non-reductive naturalism. The pioneering work of multi-faceted life and work of Neurath[34] between *Encyclopedia and Utopia*[35] was dealt with, also in the context of the formation of Logical Empiricism already before World War I as the "first Vienna Circle" before the arrival of Schlick in 1922.[36]

Neurath's oeuvre covers so far neglected topics like sociology, social sciences and economy, besides his long term project of the interdisciplinary *International Encyclopedia of Unity of Science* in the midst of the global economic crisis, there is also the idea of an ecological economy (in kind) criticizing the exclusive dominating monetary market (it is the alternative of Neurath vs. Hayek between plan and market).

In addition, the trendy "visual turn" in the cultural sciences and museology was the second important innovation of Neurath's "Vienna Method of Pictorial statistics" and Isotype (International System of Typographic Picture Education) as tools for popularisation and humanisation of knowledge and science, complementing the argumentation by written language. [37]

A quick reconsideration of Neurath's lifework uncovers easily the topicality in the history and philosophy of science: a methodological and reflexive relativism, empiricism and naturalism embedded in the unity of science model vs. any

33 Gerald Holton, "From the Vienna Circle to the Harvard Square: The Americanization of a European World Conception", in: Friedrich Stadler (Ed.), *Scientific Philosophy: Origins and Developments*. Dordrecht–Boston–London 1993, pp. 47-74.

34 Jordi Cat, Nancy Cartwright, Lola Fleck and Thomas Uebel (1996) *Otto Neurath: Philosophy Between Science and Politics*, Cambridge University Press.

35 Elisabeth Nemeth and Friedrich Stadler (Eds.) (1996) *Encyclopedia and Utopia. Otto Neurath (1882-1945)*, Wien–New York: Springer.

36 Thomas Uebel (2000) *Vernunftkritik und Wissenschaft. Otto Neurath und der Erste Wiener Kreis*, Vienna/ New York: Springer.

37 Richard Heinrich, Elisabeth Nemeth, Wolfram Pichler, David Wagner (Eds.) (2012) *Image and Imaging in Philosophy, Science and the Arts*, Heusenstamm: Ontos.

"system" with foundational intention. We realize a pluralism covering the natural and social sciences and fighting any form of „pseudo-rationalism" (verification and falsification).

10 Arne Naess – a road to empirical semantics and "experimental philosophy"

One of the many guests in the Schlick-Circle from abroad was the young Arne Naess (1912–2009) from Oslo, who elaborated a radical empiricism by overcoming Carnap and Neurath. His plea for "models instead of theories and systems", "research vs. science", and empirical vs. logical semantics (according to Neurath's "Gelehrtenbehavioristik") is paving the way for contemporary experimental philosophy,[38] already elaborated in his Vienna Years in the 1930s, e.g., in his *Wie fördert man die empirische Bewegung? Eine Auseinandersetzung mit dem Empirismus von Otto Neurath und Rudolph Carnap* (1937–39, published 1956), especially in his dissertation *Erkenntnis und wissenschaftliches Verhalten* (1936), followed by *Truth as Conceived by Those who are not Professional Philosophers* (1938) and the late monograph *From an Empirical Point of View* (1992) – obviously contrasting Quine's *From a Logical Point of View* (1953). All these publications seem to be anticipations of the sociological turn in epistemology and philosophy of science, drawing on contextual meaning based on epistemological scepticism.

Together with Egon Brunswik's *Wahrnehmung und Gegenstandswelt* (1934) and *The Conceptual Framework of Psychology* (1952), Naess had planned to contribute to Neurath's *Encyclopedia* project.

This biography is only one example for the flourishing networking between the Vienna Circle and the philosophers in the Nordic countries, like Eino Kaila (who, by the way, coined the term "logical empiricism") and G. H. von Wright between the wars dealing with critical and structural realism, inductivism and probabilism, the semantic turn and the relation between psychology and "scientific philosophy".[39]

---

38 K. A. Appiah "Experimental Philosophy," Presidential Address to the Eastern Division of the American Philosophical Association, 2007. Proceedings and Addresses of the American Philosophical Association Vol. LXXXII No. 2.(November 2008): 7-22.

39 Manninen, Juha; Stadler, Friedrich (Eds.) (2010) *The Vienna Circle in the Nordic Countries*, Wien–New York: Springer.

11 Friedrich Waismann between Schlick and Wittgenstein: Vienna-Cambridge-Oxford

Addressing the "Wittgenstein-connection" we have to reconsider the triangle Schlick, Waismann and Wittgenstein, especially the tragic and dramatic relation of Waismann and Wittgenstein in Vienna and in British exile.[40] This recent research on the interaction of the Vienna Circle and Wittgenstein shows the conflict-laden communication and the difficult personal relations of the latter with some members like Carnap, including with priority disputes (surprisingly just on physicalism) and the role and function of philosophy vis a vis the sciences. A reconstruction of these intellectual and personal relations results in a confirmation of the Anti-Wittgensteinian formation with Carnap, Menger and Neurath, the continuing differences of Schlick and Wittgenstein despite of Schlick's linguistic "turn in philosophy" (1930), but also the relative independent philosophical development of Waismann (especially in Oxford), even if strongly influenced by Wittgenstein till his forced emigration in 1938 to Cambridge. After the refusal of Wittgenstein to stay in contact with his former collaborator and interpreter, Waismann further developed analytic philosophy, e.g., language strata, providing the concepts of porosity, vagueness and speech act contexts. He also contributed to philosophy of science with reformulations of probability, causality, verifiability and language analysis, open texture, argumentation theory, as becomes manifest in his *The Principles of Linguistic Philosophy* (posthumous 1965), or in his *Introduction to Mathematical Thinking* (1936/1951, with a preface by the mathematician Karl Menger).

One may ask, whether Waismann is one of the forgotten pioneers of analytic philosophy with two origins as referred to by G. H. von Wright, who cites Waismann's, "Was ist logische Analyse?" (*Erkenntnis* 8, 1939/40) and Max Black's, "Relations between logical positivism and the Cambridge School of Analysis" (*Erkenntnis* 8, 1939/40) for this early manifestation of the "analytic" (as opposed to "continental") tradition in philosophy.

One episode is worth mentioning with regard to the emergence of "historical epistemology" by Ludwik Fleck, who contacted Schlick in order to find a publisher for his path breaking book *Entstehung und Entwicklung einer wissenschaftlichen Tatsache* (1935). The published correspondence between the two is a nice document to re-evaluate the image of Fleck as an exclusive alternative to Logical Empiricism between context of discovery and context of justification.[41]

40 Brian McGuinness (Ed.) (2011) *Friedrich Waismann – Causality and Logical Empiricism*, Wien–New York: Springer.

41 On the relation between Fleck und Schlick: "Briefwechsel mit Moritz Schlick (1933–1934)" in: Sylwia Werner and Claus Zittel (Eds.) *Ludwik Fleck, Denkstile und Tatsachen*, Frankfurt a. Main: Suhrkamp, p. 561-565.

12 The 'Third Vienna Circle': Arthur Pap and the Renaissance of Analytic Philosophy (of Science)

Another initiative in the history of Logical Empiricism is to be mentioned in the context of the post-war developments in Europe: the "third Vienna Cicle" around the late Viktor Kraft from 1949–1953, who as the last member of the former Schlick circle organized a permanent discussion group, attended by Elizabeth Anscombe, Bela Juhos, Walter Hollitscher, Ernst Topitsch, Georg Henrik von Wright, Wolfgang Stegmüller, Arthur Pap, and Paul Feyerabend, inter alia. Especially, the young Pap as a Fulbright visiting professor in 1953/54 engaged Paul Feyerabend as his assistant and exerted a remarkable influence during his stay, at least with his book on *Analytische Erkenntnistheorie* (compiled by Feyerabend) as an attempt to revive the heritage of the Vienna Circle at the University of Vienna after the forced migration and expulsion of its members latest with the "Anschluss" in 1938.

The failed attempt to achieve an appointment in Vienna and the early death of Arthur Pap obscured his innovative contributions to the renaissance of analytic philosophy.[42]

Specific elements of his research were the adherence to empiricism with intuitive knowledge, a model-theoretic account of logical consequence, and dispositional and modal concepts which anticipate Kripke's later work. Pap's functional a priori between conventionalism and pragmatism can be regarded as another version of the relativized a priori anticipating the research of Michael Friedman.[43]

Viktor Kraft in his "Third Vienna Circle" promoted a critical or constructive realism, rejected meta-ethical non-cognitivism, and accepted theoretical entities by overcoming the "two dogmas", which Wolfgang Stegmüller extended with some further critique of Logical Empiricism with the establishment of analytic philosophy of science (Analytische Wissenschaftstheorie).[44]

---

42 *The Limits of Logical Empiricism. Selected Papers of Arthur Pap.* (2006) Alfons Keupink and Sanford Shieh (Eds.), Dordrecht: Springer..The Institute Vienna Circle received the Pap collection from his family, which will bet the basis for a research project on the renaissance of analytic philosophy.

43 David J. Stump (2011) "Arthur Pap's Functional Theory of the A Priori", *HOPOS Journal*, Vol. 1, No. 2, p. 273-290.

44 On these postwar developments see the contributions in: Friedrich Stadler (Hrsg.), *Vertreibung, Transformation und Rückkehr der Wissenschaftstheorie. Am Beispiel von Rudolf Carnap und Wolfgang Stegmüller.* Wien–Berlin: LIT Verlag 2010.

13 Continental interactions – Finno-Ugrian traditions

As to the international relations, we are aware of the well known interactions in "scientific philosophy" between Vienna, Berlin, and Prague between the two World Wars.[45]

Less known is the fate of the Vienna Circle in Hungary before and after WW II with the ideological rejection of "positivism" under the influence of communism. Nevertheless, we can discover a remarkable influence before and even after 1945: first, via the Brentano-School and experimental psychology, and especially by the so called "Galileo Circle" under the influence of Mach (with its members Karl and Michael Polanyi, Karl Mannheim, Georg Lukacs, and conducted by Julius Pikler). In addition, after WW I the Hungarian tenets of Bela Juhos, Egon Brunswik, Eugen Lukacs, John von Neumann, John Harsanyi becomes manifest with the reception of modern logic, quantum physics and game theory (L. Kalmár), the latter currently active with the "Budapest school" on probabilistic causality, esp. on Reichenbach's Common Cause Principle.[46]

On the other side, the emergence of scientific philosophy was crucially enriched by the interactions with philosophers in the Nordic countries, as can be drawn from the permanent scholarly exchange with the community between Vienna, Copenhagen, Helsinki, Gotheburg, Uppsala, and Oslo.[47] An important role played the journal *Theoria*, ed. by Ake Petzäll and others in Sweden (with Ernst Cassirer in exile) and addressing besides Logical Empiricism also Hans Kelsen's pure theory of law (with Alf Ross as central proponent).[48]

One culmination of this "Nordic connection" since Kaila and the Vienna Circle was the organisation of the 2nd International Congress for the Unity of Science in Copenhagen on the main topic causality (Niels Bohr, Joergen Joergensen, Arne Naess et al.).

After 1945 just these contacts (cautiously continued by Viktor Kraft in his Viennese inner exile) where continued by Paul Feyerabend before his departure from Vienna in the late 1950s.[49] It seems that the dominant transfer and transformation of the former Vienna Circle members to the Anglo-Saxon world after World War

45 See the proceedings of the opening conference of the Vienna Circle: Rudolf Haller and Friedrich Stadler (Ed.) *Wien–Berlin–Prag. Der Aufstieg der wissenschaftlichen Philosophie*, Vienna: Hölder-Pichler-Tempsky; and Friedrich Stadler (Ed.) *Scientific Philosophy. Origins and Developments*, Wien–New York: Springer.

46 See the proceedings of a long term bilateral project *The Vienna Circle in Hungary*: András Máté, Miklós Rédei, Friedrich Stadler (Eds.) *Der Wiener Kreis in Ungarn*, Wien–New York: Springer.

47 Manninen, Juha; Stadler, Friedrich (Eds.) (2010) *The Vienna Circle in the Nordic Countries*, Wien–New York: Springer.

48 Johan Strang, (2010) *History, Transfer, Politics. Five Studies on the Legacy of Uppsala Philosophy*. Tampere: Juvenes Print..

49 Friedrich Stadler and Kurt R. Fischer (Eds.) *Paul Feyerabend – Ein Philosoph aus Wien*, Wien–New York: Springer.

II obscured this productive communication in the long run, reinforced by the Cold War developments.[50]

14 Intra-continental networking between East and West

The "French Connection": Since the early reception and critique of conventionalism before World War II in the "first Vienna Circle" following the exchange between Mach, Pierre Duhem and Henri Poincaré, with Friedrich Adler as mediator, this remarkable networking with the French philosophers like Marcel Boll lasted till the outbreak of the war, with the two highlights in Paris 1935 (the 1st Congress for the Unity of Science) and in 1937 (3rd Congress for the Unity of Science and Congrès Descartes).[51]

A reconsideration of this nearly forgotten "French connection" confirms the impression of the importance of these events and mutual influences for the formation of the *International Encyclopedia of Unified Science* edited by Rudolf Carnap, Charles Morris, and Otto Neurath 1938ff. under most difficult academic and political circumstances.[52] With this theoretical initiative the tension between a preferred empiricism as a "mosaic of the sciences" (Neurath) and the semiotic conception of the sciences (Carnap and Morris) became apparent, which faded out after the unexpected early death of Neurath in British exile in 1945.

The liberalization of methodological and epistemological rules in the philosophy of science was already set in Vienna with the individual contributions by Alfred Tarski, Karl Menger, Carnap and Kurt Gödel: the emergence of the "principle of tolerance" and the semantic turn is only one indicator for the theoretical pluralism, which was apparent in the (philosophy of) mathematics and logic dealt with in Menger's "Mathematical Colloquium" in parallel to the Schlick Circle.[53] The theories of truth and confirmation, philosophical and mathematical logic, proof theory and meta-languages appeared on the agenda of the circles, latest with Tarski's input in the correspondence-theoretic tradition since Boltzmann, followed by

50 Ilkka Niiniluoto / Sami Pihlström (eds.) (2012) *Reappraisals of Eino Kaila's Philosophy*. Helsinki: Philosophical Society of Finland.

51 Philipp Frank (1949) "Introduction: Historical Background", in: Frank, *Modern* Science *and its Philosophy*, Cambridge, Mass.

52 *Paris–Wien. Enzyklopädien im Vergleich*. Ed. by Elisabeth Nemeth und Nicolas Roudet. Wien–New York: 2005.

53 Eckehart Köhler, Werner DePauli-Schimanovich and Friedrich Stadler (Eds.) (1995) *The Foundational Debate: Constructivity and Complexity in Mathematics and Physics*, Wien–New York: Springer. Jan Wolenski (Ed.) (1998) *Alfred Tarski and the Vienna Circle. Austro-Polish Connections in Logical Empiricism*, Wien–New York: Springer. On the "Mathematischen Kolloquium": Karl Sigmund and E. Dierker (Eds.) *Karl Menger – Ergebnisse eines Mathematischen Colloquiums*, Springer Wien–New York, 1998.

Feigl, and Kraft leading up to Popper's *Logic of Scientific Discovery* in a realistic epistemology.

The tragic split between the close friends Neurath and Carnap stayed unresolved, as became manifest with the latter's "Empiricism, Semantics and Ontology" (1950). It is a puzzling fact that the young Arne Naess accused both of them of having lost the consequent empiricism.[54]

15 The Austro-British Interaction Since 1900

Since the debate between Bertrand Russell and Alexius Meinong "On Denoting" (1905) and the subsequent first encounters of Wittgenstein in Cambridge, the story of „The *Wiener Kreis* in Great Britain" between the two world wars is a preceding counterpart of "The Wiener Kreis in North America" as first described by Herbert Feigl in 1968.[55]

Further on, the meetings of Frank P. Ramsey, Schlick, Waismann and Wittgenstein in Vienna,[56] and the continuation of Neurath and Waismann in Oxford are manifestations of the most important role of British (analytic) philosophy for the development of Logical Empiricism proper. See for exampe the translations of Carnap's work by Max Black 1934f. and the re-establishing of the Isotype-movement in Britain, with A. J. Ayer's pivotal book *Language, Truth and Logic* (1936) enabled the internationalization of the Vienna Circle which culminated with the 4th Congress for the Unity of Science in Cambridge, mainly supported by Susan Stebbing as the leading figure of an import of the "Logical Positivism" in the context of analytic philosophy as developed in England independently. By the way, the the myth of the "Poker-Story" between Wittgenstein and Popper is only one more symptom for the divergences within the domain between linguistic and scientific philosophy.[57]

After the war, the third Vienna Circle around Viktor Kraft with Lakatos and Feyerabend paved the way to the historical-sociological turn at the LSE Sym-

54 See my account on Naess, "Arne Naess: Dogmas and Problems of Empiricism" in: Manninen/Stadler (eds.) (2010) *The Vienna Circle and the Nordic Countries*, Wien–New York: Springer, p. 11-31.

55 Herbert Feigl, "The 'Wiener Kreis' in America", in: D. Fleming/B. Baylin (Eds.) *The Intellectual Migration 1930–1960,* Cambridge, Mass, p. 630-73 Friedrich Stadler, "The Wiener Kreis in Great Britain: Emigration and Interaction in the Philosophy of Science", in: Edward Timms/Jon Hughes (Eds.), *Intellectual Migration and Cultural Transformation. Refugees from National-Socialism in the English-Speaking World.* Wien–New York: Springer 2003, pp. 155-180.

56 Maria Carla Galavotti (Ed.) (2004) *Cambridge and Vienna. Frank P. Ramsey and the Vienna* Circle, Wien–New York: Springer.

57 On the so called Poker Story: David Edmonds and John Eidinow (2001) *Wittgenstein's Poker: The Story of a Ten-Minute Argument Between Two Great Philosophers*, New York: Harper Collins.

posium 1965 with Feyerabend, Kuhn, Popper, Lakatos, Toulmin, Watkins and others,[58] and the attempt to bridge the gap of the Stegmüller School adhering the structuralist view of scientific theories.[59]

16 Transatlantic interactions: Europe and America

The story of this dramatic move was described by Gerald Holton "From the Vienna Circle to the Harvard Square"[60] – a history of the philosophy of science in the context of the forced migration, as the transfer and transformation of Logical Empiricism from Europe to America 1930ff.[61] Recent research has uncovered both the *Origins of Logical Empiricism* (Giere and Richardson 1996) and the *Logical Empiricism in North America* (Hardcastle/Richardson 2004), mainly with reference to the convergence and divergence with (Neo-)Pragmatism represented by P. W. Bridgman, C. I. Lewis, J. Dewey, W. V. O. Quine and others. The re-transfer after World War II is now reconstructed with the come back of a modified normative analytic philosophy of science (Logik und Analytische Wissenschaftstheorie).[62]

Besides this often presented success story only in the last years some significant Latin American Relations came in the focus of the scholarship in the exile studies on the Nazi-period: the Jewish emigration to Latin and South America[63] unearthed an early inter-continental dialogue since the reception of Mach in Columbia to the more direct influence exerted by the former Schlick student Hans Lindemann in Buenos Aires, most likely followed by the work of Mario Bunge and Alberto Coffa till their academic presence in the US and Canada.[64] A first account

58 Imre Lakatos and Alan Musgrave (Eds.) (1970) *Criticism and the Growth of Knowledge*, Cambridge University Press.

59 Friedrich Stadler (Ed.) (2010) *Vertreibung, Transformation und Rückkehr der Wissenschaftstheorie: Am Beispiel von Rudolf Carnap und Wolfgang Stegmüller*, Münster: LIT Verlag as a documentation of a research project, funded by the FWF at the Institute Vienna Circle on "History and/or Philosophy of Science" with the project investigators Christian Damböck, Hans-Joachim Dahms, Christoph Limbeck-Lilienau, Michael Schorner.

60 Gerald Holton (1993) "From the Vienna Circle to Harvard Square: The Americanization of a European World Conception", in: Friedrich Stadler (Ed.) *Scientific Philosophy: Origins and Developments*, Wien–New York: Springer 1993.

61 Friedrich Stadler (2012), "History and Philosophy of Science: Zwischen Deskription und Konstruktion", in: *Berichte zur Wissenschaftsgeschichte* 35 (2012), 217-238.

62 Michael Schorner "Comeback auf Umwegen: Die Rückkehr der Wissenschaftstheorie in Österreich" and Hans-Joachim Dahms "Stegmüller und das Comeback der Wissenschaftstheorie in Deutschland", in: Friedrich Stadler (Ed.) (2010) *Vertreibung, Transformation und Rückkehr der Wissenschaftstheorie: Am Beispiel von Rudolf Carnap und Wolfgang Stegmüller*, Münster: LIT Verlag, p. 189-252 and p. 271-340..

63 Alexander and Barbara Litsauer (Eds.) (2010) *„Verlorene Nachbarschaft". Jüdische Emigration von der Donau an den Rio d la Plata*, Vienna: Mandelbaum Verlag.

64 Friedrich Stadler, "Zum Umgang mit der Vergangenheit in Österreich nach 1945", in:

of R. Campis and E. Bermudez Barrera on "Intercultural Dialogue in Philosophy: Julio Enrique Blanco, Hans Lindemann, Wittgenstein and the Austrian Tradition" (2006) is just the beginning of an investigation into this still unresearched field.

This can be said likewise of the oeuvre of another student of Schlick's (1935/36), the Portuguese Delfim Santos in Portugal who made an important academic and political career after his return to Lisbon before and during World War II.[65]

17 Emotivism and Meta-Ethical Noncognitivism: Norms and Values Revisited

Despite the most influential characterization of Logical Empiricism as a movement of emotivism based on the dualism of is-ought and facts-values, there is a gradual reconstruction and re-evaluation showing a pluralist standpoint with no consensus in the logical and epistemological status of value statements. Although ethics and moral philosophy was certainly not in the center of Vienna Circle's philosophizing, there is a surprising evidence for more or less explicit treatment of values instead of a replacement of ethics by meta-ethics already from the 1920s on: explicit ethical writings of Moritz Schlick (*Fragen der Ethik*, 1930), Otto Neurath (1931/1944), Rudolf Carnap ("Theoretische Fragen und praktische Entscheidungen", 1934), Karl Menger (*Moral, Wille, Weltgestaltung*, 1934), Felix Kaufmann (*Methodenlehre der Sozialwissenschaften*,1936/1944), Viktor Kraft (*Grundlagen einer wissenschaftlichen Wertlehre*, 1937), Friedrich Waismann and Josef Schächter (1938), Richard von Mises (*Kleines Lehrbuch des Positivismus/ Positivism*, 1939/1951) reveal the attitude of dealing with norms and values rationally and logically within the frame of "rational reconstruction".[66] Additionally, the inclusion of Dewey's "Theory of Valuation" and Abraham Edel's *Science and the Structure of Ethics* (1961) document the willingness of taking the issue of the descriptive-normative distinction seriously, leading up to different solutions. Not least, the emerging game and decision theory facilitated the overcoming of an early verificationism claiming the exclusion of value statements without any meaning. Even the late Carnap, with his research into probabilism, inductivism and theory choice (together with Richard Jeffrey) admitted value based reasoning in the long run – despite the principally well accepted dualism of normative and descriptive statements.

Alexander Litsauer/Barbara Litsauer (Hrsg.), *Verlorene Nachbarschaft. Jüdische Emigration von der Donau an den Rio de la Plata*. Wien: Mandelbaum 2010, pp. 223-231.

65 Cf. the running Santos-project in Lisbon: www.delfimsantos.org/VIENNA_PAPERS.htm.

66 A first overview and analysis in: Friedrich Stadler "Wissenschaftliche Weltauffassung und Kunst. Zur werttheoretischen Dimension im Wiener Kreis", in: *Deutsche Zeitschrift für Philosophie* 4, 1995, 635-651.

A re-evaluation of this complex matter – of course, rejecting any value realism and the Kantian categorical imperative – provides a stronger differentiated image of the is-ought question of the whole logical-empirical movement.[67]

18 Logical empiricism and Pure theory of law – Family resemblance

The dualism of is-ought is one feature in common between the Vienna Circle and the school of Pure Theory of Law of Hans Kelsen, who in his later life in the USA converged with the Unity of Science movement and contributed with a volume on *Vergeltung und Kausalität* (Retribution and Causality) 1940 (*Society and Nature* 1943) in the series "Library of Unified Science".[68]

This family resemblance and overlapping intentions in the context of Viennese modernity and afterwards are a becoming closer in exile as in Vienna till 1931, where Kelsen developed his law theory inspired by Kant, Freud, and partly Husserl in the frame of a anti-metaphysical enlightened world view fighting against any variant of totalitarian ideology and politics like most members of the Vienna Circle, as Gustav Bergmann has formulated already in 1938.[69]

Although Kelsen in his posthumous published *Allgemeine Theorie der Normen* did not share Schlick's moral philosophy from a legal point of view, the problem of the foundation of last normative principles from a neo-Kantian perspective, the relation of law and morality viz. normative and descriptive statements stayed on the agenda of a common attitude towards science and society as expressed also in Kelsen's posthumous edited volume on *Secular Religion.*[70]

---

67 Recently published by Anne Siegetsleitner (Ed.) (2010) *Logischer Empirismus, Werte und Moral. Eine Neubewertung*, Wien–New York: Springer.

68 Clemens Jabloner (Ed.) *Logischer Empirismus und Reine Rechtslehre*, Wien–New York: Springer.

69 As a source and research basis see the publications of the Vienna based Hans Kelsen Stiftung: www.hans-kelsen.org. Regarding the Bergmann reference: Gustav Bergmann, "Erinnerungen an den Wiener Kreis. Brief an Otto Neurath", in: Friedrich Stadler (Hrsg.), *Vertriebene Vernunft II. Emigration und Exil österreichischer Wissenschaft 1930–1940*. Münster: LIT Verlag 2004, S. 171-180.

70 Hans Kelsen, *Secular Religion. A Polemic against the Misinterpretation of Modern Social Philosophy, Science and Politics as "New Religions"*. Ed. from the estate of Hans Kelsen by Robert Walter, Clemens Jabloner and Klaus Zeleny. Wien–New York 2012.

19 Felix Kaufmann's Mediating Schools and Methods – Liberalism and Pluralism

Felix Kaufmann (1895–1949) was a unique mediator between some important Viennese intellectual circles: based on a lifelong commitment to Husserl's phenomenology, he commuted between Kelsen's theory of law, the Ludwig von Mises seminar linked to the so called "Geist-Kreis" of the Austrian School in economics (F. A. Hayek), and the the Schlick-Circle bridging the gap between iurisprudence, the social and cultural sciences (reconciling *Erklären* und *Verstehen*) mainly via methodological studies: his main work on *Methodenlehre der Sozialwissenschaften* (1936) is a convincing attempt to overcome the ongoing *Methodenstreit* in the social sciences since the end of the 19th century.

In addition, he contributed to the unresolved foundational debate in mathematics, e.g., with his book *Das Unendliche in der Mathematik und seine Ausschaltung* (1930), published one year before Kurt Gödel's proofs, arguing for an intuitionist conception inspired by Husserl's philosophy of mathematics.[71]

Although not a philosopher at the University (he had to earn a living as manager of the Anglo-Iranian Oil Company), Kaufmann only in exile after 1938 succeeded at the New School for Social Research in New York – together with his Viennese friend Alfred Schütz – with an academic career till the end of his too short life. In the US Kaufmann tried to approach American (neo)pragmatism, especially Dewey, who was not really convinced of a theoretical alliance between Logical Empiricism and (neo)pragmatism – even if himself a contributed to the International *Encyclopedia of Unified Science* (with "Theory of Valuation"). Therefore, Kaufmann did not only translate his 1936 book, but wrote a new one, although under the same title *Methodology of the Social Sciences* (1944), the former one pleasingly to be reprinted soon in the "Vienna Circle Collection".[72]

Generally, Kaufmann provided a merger of liberalism und pluralism in a phenomenological (later also pragmatist) perspective on scientific philosophy, even if the fusion between Logical Empiricism and pragmatism failed because of the reluctance of Dewey, Nagel and others. From a present day point of view his attempts, especially as a co-editor of the journal *Philosophy and Phenomenological Research,* are significant building blocks for a methodological approach to different philosophical positions overcoming the dualisms of "continental" and "analytic" schools – by the way, defending the Kantian heritage against the ac-

71 Felix Kaufmann, *Das Unendliche in der Mathematik und seine Ausschaltung*. Leipzig-Wien 1930. (Darmstadt 1968). Expanded English Edition: *The Infinite in Mathematics*. With an Introduction by Ernest Nagel. Ed. by Brian McGuinness. Dordrecht-Boston-London: Reidel 1978.

72 Felix Kaufmann (1944) *Methodology of the Social Sciences*, Oxford (New York 1958). Reprint of the *Methodenlehre der Sozialwissenschaften* (1936): *Theory and Method in the Social Sciences*. Ed. by Robert S. Cohen and Ingeborg Helling. With an Introductory Essay by Ingeborg K. Helling. Springer (forthcoming).

cusations of Dewey's critique in his *German Philosophy and Politics*[73]. His premature death prevented an adequate appreciation and critical evaluation of his impressive work, which is complemented by his humour and (self)irony as documented in his posthumously published songs on economy and philosophy.[74] His methodological strategies similar to P. W. Bridgman's operationalism with decision theoretical rules are developed further by the Isaac Levi and Henry Kyburg.[75] Not to forget, that Karl Popper succeeded with his *Poverty of Historicism* in the Cold War period, in which he took over directly Kaufmann's pro-naturalistic and anti-naturalistic doctrines applied to his concept of historicism, without mentioning him explicitly in this regard.[76]

20 Philosophical foundations of quantum physics and mathematics

From the beginning in the first Vienna Circle modern physics was the challenge for the conceptualization of a "scientific philosophy": Mach, Einstein, Russell and Bohr were the most appropriate research subjects as addressed by Philipp Frank, Hans Reichenbach and Moritz Schlick and the philosophical interpretation of modern (quantum) physics and relativity theory.[77] Especially Schlick's *Raum und Zeit in der gegenwärtigen Physik* (Space and Time in Contemporary Physics 1919/2006) – together with Schlick's main work *Allgemeine Erkenntnislehre* (*General Theory of Knowledge* 1918/1925) was appreciated by Einstein as the best philosophical interpretation of his relativity theory, even if since the middle of

73 Peter M. Ruttkoff/William B. Scott, *New School. A History of the New School for Social Research*. New York 1986, 137f., with reference to John Dewey's, *German Philosophy and Politics*. (1915). German translation: *Deutsche Philosophie und Deutsche Politik.* With an introduction by Axel Honneth. Berlin 2000.

74 Felix Kaufmann, *Wiener Lieder zu Philosophie und Ökonomie*. Mit einer Einführung von J. Herbert Fürth. Hrsg. von Gottfried Haberler und Ernst Helmstädter. Stuttgart–Jena–New York: Gustav Fischer Verlag 1992. On the life and work of Felix Kaufmann: Harry P. Reeder, *The Work of Felix Kaufmann*. Washington, D.C. 1991; *Klarheit und Methode: Felix Kaufmanns Wissenschaftstheorie*. Amsterdam–Atlanta 1990; Friedrich Stadler (Ed.) (1997) *Phänomenologie und Logischer Empirismus. Zentenarium Felix Kaufmann*. Wien–New York.

75 H. G. Zilian (1990) *Klarheit und Methode: Felix Kaufmanns Wissenschaftstheorie*, Amsterdam: Rodopi, p. 11.

76 Malachi Hacohen, *Karl Popper – The Formative Years*, 1902–1945. *Politics and Philosophy in Interwar Vienna*. Cambridge University Press 1997, 362ff.

77 See the related volumes of the Schlick edition: *Raum und Zeit in der gegenwärtigen Physik*, Moritz Schlick Gesamtausgabe, Abt. I, vol. 2, p. 121-345 and the Reichenbach Edition, Springer. Most important the writings of Philipp Frank, besides his biography of Einstein: *Das Kausalgesetz und seine Grenzen,* Vienna, 1932. *Between Physics and Philosophy, Science and its Philosophy*, Cambridge, Mass., 1941; *Foundation of Physics*, Chicago, 1946 and *Relativity – A Richer Truth*, Boston, 1950.

the 1920s he did not like Schlick's linguistic turn under the influence of Wittgenstein and his departure from a realist epistemology. This divergence came up again in a dispute with his teacher Max Planck, who contested strongly the "positivism" of his former favourite disciple.[78]

The issue of *Erkenntnis* (6/1936) contains the proceedings of the Copenhagen congress on the problem of causality, which was one of the crucial topics caused by quantum physics and relativity theory, to which physicists and philosophers contributed substantially in the local context of the Copenhagen interpretation. Frank's writings – from his *Kausalgesetz* (1932) to *Philosophy of Science: the Link between Science and Philosophy* (1957) formed the thematic thread of this debate, which was presented by him as a convergence of Mach, Boltzmann, Russell and Wittgenstein, the French conventionalism (Duhem and Poincaré), the Vienna Circle and American operationalism and pragmatism (Bridgman).[79]

In mathematics and formal logic, the foundational debate in mathematics emerged around 1930, which divided the community between logicism (Russell and Carnap), formalism (Hilbert), and intuitionism (Brouwer and Gödel), with an open end after the challenging results of Kurt Gödel.[80]

One decisive background for all these (re)constructions of modern science, logic and mathematics was the famous "Mathematical Kolloquium" (1928–1936) founded and conducted by Karl Menger, with the participation of such experts like Kurt Gödel, Abraham Wald, John von Neumann, Alfred Tarski, Hans Thirring, Werner Heisenberg, Karl Popper, Olga Taussky, Herman Mark, Hans Hahn and many others.[81] This parallel action vis a vis the Schlick circle was certainly the necessary intellectual environment of Gödel's achievements in the foundations of mathematics and logic – as Einstein in physics.

21 European philosophy of science – philosophy of science in Europe

Some common features of the intellectual scene as described above are the historical approach with a pragmatic tendency towards history and philosophy of science (including the sociology of science), a methodological sensitivity towards all branches of sciences, and before World War II a multilingual manifestation of philosophical schools and communities. The networking transcended national

78 Schlick, "Positivismus und Realismus" (1932) vs. Planck "Positivismus und reale Außenwelt" (1931).

79 Frank (1949) *Modern* Science *and its Philosophy*, Cambridge, Mass.

80 Eckehart Köhler, Werner DePauli-Schimanovich and Friedrich Stadler (Eds.) (1995) *The Foundational Debate: Constructivity and Complexity in Mathematics and Physics*, Wien–New York: Springer.

81 Edited with introductions by Karl Sigmund/E. Dierker, *Ergebnisse eines Mathematischen Colloquium* Springer Wien–New York, 1998.

boundaries and was mostly a manifestation of urban cultures of science in circles not restricted to universities. With these cognitive and social properties the co-existing "analytic" and „continental" philosophy of science covered the natural sciences, social and cultural sciences and humanities, even if the focus in the formal and natural sciences was certainly dominating. Furthermore, the inter- and transdisciplinarity was partly practised and intentionally implemented in the late enlightened project of the unfinished *International Encyclopedia of Unified of Science.* After the wave of forced migration and transformation of this implicit research program the academic setting and societal environment led to a narrowing towards a normative analytic philosophy (of science) with the abstention from norms and values in the Cold War period.

Summarizing, it seems to be reasonable to speak of *an* European philosophy of science and philosophy of science in Europe, which presents itself as international and global without any ethnocentric and nationalist features.[82]

But when trying to answer these questions, our attention gets drawn to the still generally discussed problems of an imagined European symbolic space (what is Europe, Central Europe, Middle-Europe, Continental Europe?), which is not identical with the area of the EU, the European Council, or the range of the ESF. This "spatial turn" as subject and method reminds us of the inherent plurality and dynamic force of an intellectual enterprise with all European scholars participating in the global village, but still challenges by a well known saying coming from the environmental sciences: "think globally, act locally". Accordingly, the preview and future perspectives are European networking with a global interaction, and the review of the "received view" in the philosophy of science – embedded in the two major initiatives in the last years, the PSE-network (2008–2013), hopefully continued within the European Philosophy of Science Association (EPSA) as a permanent and sustainable society.[83] If there is a genuine *European* philosophy of science today seems to me questionable, but there is certainly a flourishing philosophy of science *in Europe* in a cosmopolitan context.[84]

82 Examples of the annual topics of the Vienna International Summer University (VISU/SWC) since 2001: www.unuvie.ac.at/ivc/VISU and the annual topics and proceedings of the ESF Research Network Program "The Philosophy of Science in a European Perspective" (PSE): www.pse-esf.org.
- The present situation in the philosophy of science
- Explanation, prediction, and confirmation
- Probabilities, laws, and structures
- New challenges to philosophy of science
- New directions in the philosophy of science.

83 www.epsa.ac.at

84 It is worth mentioning here that the global IUHPS/DLMPST was founded by European philosophers of science (*inter alia* Alfred Tarski, Everet Beth).

Friedrich Stadler
University of Vienna
Institute Vienna Circle
Spitalgasse 2-4, Hof 1
A-1090 Wien
Austria
Friedrich.Stadler@univie.ac.at

Cristina Chimisso

A matter of substance? Gaston Bachelard on chemistry's philosophical lessons

1. Introduction

Philosophers have paid far less attention to chemistry than they have to physics.[1] It is only in the last twenty years or so that the philosophy of chemistry has gained an important place in the philosophy of science. However, before then, there have been important exceptions to the neglect of chemistry. Notably, chemistry has been very important in the French tradition: Bernadette Bensaude-Vincent has argued that the attention that Pierre Duhem, Emile Meyerson, Hélène Metzger and Gaston Bachelard paid to chemistry had a profound impact on their philosophies, which as a consequence share concerns and views of science, despite their differences. She has maintained in particular that the controversy around the role of Lavoisier in the history of chemistry 'drew attention to the scientific revolutions'.[2] Moreover, she has regarded the attention that Duhem, Meyerson and Bachelard paid to matter theories as central to the development of their respective philosophies of science.[3] I aim further to investigate the philosophical importance of chemistry within the French tradition by focussing on the significance of chemistry for some crucial aspects of the philosophy of Gaston Bachelard. Bachelard did not focus exclusively on chemistry. Indeed, physics is so central in his view of history and knowledge, that he indicated Einstein's theory of relativity as the beginning of a new era not just for physics, not even just for science, but indeed for the mind, that is to say for the way human beings think and interpret the world.[4] Nevertheless, chemistry plays a prominent role in his writing. This is very noticeable even just in terms of quantity: not only did he dedicate an entire book to the philosophy of chemistry, *Le pluralisme cohérent de la chimie moderne*, but this science also occupies a central place in several other books, notably *Le matérialisme rationnel* and *La philosophie du non*.

1 Joachim Schummer has provided a survey of the philosophy of chemistry and proposed reasons for its marginality: J. Schummer, "The philosophy of chemistry", *Endeavour* 27, no. 1 (2003).

2 Bernadette Bensaude-Vincent, "Chemistry in the French Tradition of Philosophy of Science: Duhem, Meyerson, Metzger and Bachelard", *Studies in History and Philosophy of Science* 36 (2005), p. 635.

3 Ibid.

4 Gaston Bachelard, *La formation de l'esprit scientifique: contribution à une psychanalyse de la connaissance objective* (Paris: Vrin, 1993 [1938]), p. 7.

M.C. Galavotti et al. (eds.), *European Philosophy of Science – Philosophy of Science in Europe and the Viennese Heritage*, Vienna Circle Institute Yearbook 17,
DOI 10.1007/978-3-319-01899-7_2,

Bachelard believed that science is a model of how we should reason. He insisted that 'reason must obey science', 'science in effect creates philosophy' and that 'the major lesson that the philosopher should learn from the evolution of science is that philosophy itself should be altered'.[5] These are general statements, and there is a general story to be told about how Bachelard thought that philosophy should follow science. This is a story I have told elsewhere;[6] here I shall rather focus on the specific lessons that chemistry afford philosophy according to Bachelard. For him some philosophical notions should be radically revised following chemistry's lessons. I shall concentrate on three of them that are of crucial importance: the role of analysis and synthesis in the development of knowledge, the concept of substance, and the concept of scientific object. This discussion on chemistry's philosophical lessons will form the basis of my answer to the question of whether such lessons in Bachelard's philosophy are only afforded by chemistry. In other words, was it (only) chemistry, or more specifically history of chemistry, that led Bachelard to develop his views on the above notions, and more in general, his theory of knowledge? I shall conclude by reflecting on what 'chemistry' and 'history of chemistry' are in Bachelard's philosophy. Bachelard shared his focus on chemistry with other philosophers, and yet he 'learned' quite different lessons from it than for instance Emile Meyerson. I shall argue that philosophers, just like scientists, construct their own object (e.g. history of chemistry) and that they do so differently. I shall briefly outline some of the philosophical ideas that contributed to Bachelard's construction of his own philosophical objects.

2. Analysis and synthesis

Bachelard was often polemical with philosophers, who in his view were attached to obsolete forms of knowledge. Modern science, he believed, would cure them of many bad habits and misplaced notions.[7] One of the most general lessons that chemistry could teach philosophers is for Bachelard the role of analysis and synthesis in the development of knowledge. Philosophers, Bachelard argued, are very attached to the notion of analysis as a method. They believe that if they analyse a problem, and understand its components, then they will understand the whole. This for Bachelard is highly misguided: the whole is not simply the sum of its

5 Gaston Bachelard, *The New Scientific Spirit* (Boston: Beacon Press, 1984), p. 3; Gaston Bachelard, *The Philosophy of No: A Philosophy of the New Scientific Mind* (New York: Orion Press, 1968 [1940]), p. 122; Bachelard, *Le matérialisme rationnel*, p. 135.

6 Cristina Chimisso, *Gaston Bachelard: Critic of Science and the Imagination*, (London: Routledge, 2001); Cristina Chimisso, *Writing the History of the Mind: Philosophy and Science in France, 1900 to 1960s*, (Aldershot: Ashgate, 2008), Chapter 6.

7 See his remarks about Sartre: Gaston Bachelard, *L'activité rationaliste de la physique contemporaine* (Paris: Presses Universitaires de France, 1951), p. 92; and about Bergson: Gaston Bachelard, *L'intuition de l'instant* (Paris: Stock, 1992 [1931]), p. 30.

parts. He employed a quotation from Jean-Baptiste Biot's introduction to Auguste Laurent's *Méthode de la chimie* to the effect that analysis only judges bodies after they no longer exist.[8] For Bachelard, an engagement with the way chemistry produces knowledge demonstrates that analysis alone cannot yield the results that philosophers wish for. A simple example that can be found in any chemistry handbook will show what Bachelard means. If we know sodium (Na) as a metallic and shiny element, and chlorine (Cl) as a toxic, pale green, gaseous element, we can hardly say that we know common salt; no matter that it is sodium chloride (NaCl), formed by sodium and chlorine. Here the single components are poisonous, whereas the compound is an essential part of the human diet. The 'whole', this being a chemical substance or a philosophical problem, is not reducible to its parts. Bachelard's revision of the role of analysis in philosophy on the basis of the teachings of chemistry does not mean at all that analysis has no place in the acquisition of knowledge. Again chemistry teaches us that analysis is necessary, but only as a stage in the process of knowledge. In chemistry, Bachelard pointed out, analysis is aimed at synthesis, at creating new substances, and investigating how new syntheses of elements would behave. Chemistry does consider single elements in order to understand complex compounds, but it does so in order to characterise the elements 'by their role in the synthesis of the compounds [*corps composés*]'.[9] What is important is their role in the compound rather than what they ontologically are, independently of the compound.

Synthesis means creation of substances: since the chemist cannot infer the properties of a complex substance from the known properties of its parts, she creates a new substance, and then studies it. Discovery, Bachelard argued, depends on synthesis rather than analysis.[10] As a consequence, in chemistry the technical production of objects in the laboratory takes central stage in the cognitive process. Chemistry is 'factitious',[11] it produces its own objects. When Bachelard wondered whether one should wish for 'knowledge by reduction or by production',[12] the answer is once again provided by chemistry: it is production that enables us to understand the world. As Bachelard put it, '[o]ne has more chance of knowing sugar by making sugars than by analyzing a particular sugar'; this is because 'chemical substances only get to be truly defined at the moment of their reconstruction'.[13] A post-Bachelardian example is very fitting here: this is how chemists have studied TRH (thyrotropin releasing hormone) – or TRF (thyrotropin releasing factor), depending on the team who studied it. What makes it a good example for what

8 Bachelard, *Le matérialisme rationnel*, p. 147.

9 Gaston Bachelard, *Le pluralisme cohérent de la chimie moderne* (Paris: Vrin, 1973 [1932]), p. 57.

10 Ibid., p. 57.

11 Ibid., pp. 69-71.

12 Bachelard, *Le matérialisme rationnel*, p. 119.

13 Bachelard, *The Philosophy of No*, p. 47. On this point, also see Bensaude-Vincent, "Chemistry in the French Tradition of Philosophy of Science", p. 643.

Bachelard aimed to show is that this hormone exists in too tiny quantities for chemists to analyze it. As a consequence, as Ian Hacking has narrated, in order to discover the chemical structure of TRH, the two laboratories involved synthesized several similar substances, until one of the compounds behaved just like TRH. Reportedly, one of the reasons for the Nobel Prize assigned to the heads of the two laboratories was that the idea of analyzing by synthesis was deemed to be very innovative.[14] This story fits perfectly with what Bachelard said about the importance of synthesis as a means of discovery; he went further and maintained that not only chemistry, but knowledge in general is not only, or principally, about observing and analyzing, but rather about doing and creating.

3. The scientific object

By commenting on chemistry's use of analysis and synthesis, Bachelard also told us a great deal about his own conception of the scientific object. For him scientific objects are not natural objects, and indeed one of the differences he saw between common knowledge and scientific knowledge is that their objects are different. Science does not concern itself with every-day objects, but rather with 'rectified' objects. Scientific reality is a 'transformed reality, a rectified reality, [...] a reality that has received the human mark *par excellence*, the mark of rationalism'.[15] He placed a lot of emphasis on the rationalization of objects, in other words on the rational and therefore human nature of the objects of science. For Bachelard, the rationality of knowledge is a historical process, achieved by the constant overcoming of epistemological obstacles, which originate in the natural tendency of the mind to dream and imagine.[16] This is why epistemology for Bachelard can only be historical; there is no definitive state of science that epistemology can either analyse or prescribe. In order to understand science, it is therefore necessary to examine not only its present state, but also its history. He outlined a philosophical history of the objects of chemistry, namely chemical substances: for him they presented themselves at first in a boundless variety, without any order. Later, at the beginning of the eighteenth century, the impulse was to find a general principle for all substances, and ultimately to deny their fundamental differences.[17] Modern chemistry for Bachelard results from the dialectic between diversity, or pluralism, and generality. He called the attitude of modern chemistry a 'coherent pluralism': on the one hand, there is recognition of the diversity of substances, on the other these are organized in a coherent whole. The paramount example of this organi-

14 Ian Hacking, "The Participant Irrealist at Large in the Laboratory", *British Journal for the Philosophy of Science* 39 (1988).

15 Gaston Bachelard, *Le rationalisme appliqué* (Paris: Presses Universitaires de France, 1986 [1949]), p. 8. See also ibid., p. 123.

16 Bachelard, *La formation de l'esprit scientifique*.

17 Bachelard, *Le pluralisme cohérent de la chimie moderne*, Chapter 1.

zation is of course Mendeleev's table of elements: thanks to it for Bachelard the various substances become rationalized,[18] or, as he put it elsewhere, 'diversity … loses [its] irrationality'.[19]

However, Bachelard never thought that the work of creation of scientific objects was done purely by the mind without a material process of production. For him chemistry shows that knowledge is both rational and materially productive; its 'technical materialism' has nothing to do with philosophical realism.[20] For Bachelard modern chemistry shows us that substances are the product of technology, rather than bodies found in nature.[21] The creative capacity of the chemist goes beyond nature, as chemistry is 'the science of the radical transformation of natural substances'; unlike living beings, who give birth to beings similar to themselves, chemical elements may create new substances that not resemble them at all.[22] For Bachelard, chemistry shows that the object of knowledge is made, which means that it is made in a moment in time: it is therefore historical. The creation of new objects impacts on the understanding of other objects and their relations. Bachelard argued that, because chemistry is 'eminently constructive', it cannot but refuse a 'definitive intuition of reality': in chemistry there is a continuous 're-intuition' that enables chemists to see their constructions in a new light.[23] For him the objects of chemistry bear the 'human mark' in a double and connected manner: they are both rationally rectified and technically produced. Chemistry for him is the exemplar of knowledge as a dialectical synthesis of theoretical reflection and experimental activity, which is never static, but always in a process of transformation of its objects and of minds.

4. The concept of substance

What are these substances that chemistry creates and studies? Here the evolution of chemistry for Bachelard has, once again, a very important lesson to impart on philosophy. For him, the philosophical concept of substance that has dominated philosophy must be profoundly revised on the basis on what chemistry has shown us in its history. Historically, Bachelard argued, chemists have been instinctively naïve, and realist about the substances that they studied. Early chemists simply followed common knowledge, and as a consequence they were substantialist. For Bachelard, the idea of substance comes quite naturally to human beings, indeed, it has a 'seductive charm', and it is linked to the equally natural – and equally

18 Gaston Bachelard, *La philosophie du non. Essai d'une philosophie du nouvel esprit scientifique* (Paris: Presses Universitaires de France, 1988 [1940]), p. 58.
19 Bachelard, *Le pluralisme cohérent de la chimie moderne*, p. 22.
20 Bachelard, *Le rationalisme appliqué*, p. 8.
21 Bachelard, *La philosophie du non*, p. 53.
22 Bachelard, *Le matérialisme rationnel*, p. 33.
23 Ibid., p. 121.

un-scientific – realistic instinct; indeed, Bachelard calls realism 'the only innate philosophy there is'.[24] However, he regarded substantialism and realism as epistemological obstacles: natural instincts to be overcome in order to produce scientific knowledge. He believed that in fact substantialism still crept into chemistry textbooks, so contradicting the real teachings of chemistry. He quoted a handbook saying that 'menthol, menthone and menthyl acetate all smell of mint'. For Bachelard, many educated readers would find this statement obvious, as if the endings of the names of the three substances just added supplementary functions to an essential quality, belonging to the substance 'mint'. But in fact, Bachelard explained, 'derivatives of the same chemical body can have different properties'.[25] Modern chemistry has developed a conception of substance that says 'no', to use a Bachelardian expression, not only to its own past, but also to metaphysics' conception of substance. For him, metaphysics could only have a certain notion of substance because the 'elementary conception of physical phenomena was content to study a geometrical solid characterized by general properties'.[26] Metaphysics was constructed on elementary physics. He proposed to substitute metaphysics with 'metachemistry', in which the substances of reference are chemical substances.[27]

For Bachelard one of the most important revolutionary steps that chemistry has taken over metaphysics is that, in modern chemistry, substance cannot be thought of as something immutable and existing in its own right. The objects that chemistry studies for him are not all 'real in the same way, substance does not have, at all levels, the same coherence; *existence is not a one-toned function*; it cannot affirm itself, everywhere and in the same tone, all the time'.[28] It should be born in mind that the 'coherence' that substance may or may not have crucially depends on a number of factors, including the scientists' methods, the techniques that enable them to create new objects and new substances (in the chemical sense), and their place in the history of their science. Moreover, chemical substances for Bachelard are processes, rather than something fixed. As shown in the example of common salt, chemical elements change dramatically when they react with each other, and create something new. They cannot be thought of as 'substances' in the philosophical sense. Chemistry is the science of transformations; this means that its objects are historical not only in the sense explained in the previous section, which is a sense that they share with all scientific objects. All scientific objects are created in history, but the objects of chemistry have time inscribed in them, as they are processes. Chemistry studies processes, so time is part of what it investigates.

24 Gaston Bachelard, *The Formation of the Scientific Mind*, trans. Mary McAllester Jones, intr. Cristina Chimisso (Manchester: Clinamen Press, 2002 [1938]), p. 136.

25 Ibid., pp.119-20.

26 Bachelard, *The Philosophy of No*, p. 45.

27 Bachelard, *La philosophie du non*, p. 53; on metachemistry, see A. Nordmann, "From metaphysics to metachemistry" in *Philosophy of chemistry. Synthesis for a new discipline*, ed. D. Baird and E. Scerri, (Dordrecht: Kluwer, 2010).

28 Bachelard, *The Philosophy of No*, p. 46.

Bachelard thought that the temporality of chemistry is particularly clear when we consider chemical processes in which there is exchange of energy. Energy is for Bachelard an integral part of a substance, and the latter is not more or less real than the former. Energy exchanges determine material modifications, and material modifications produce energetic exchanges. These processes are temporal, and for Bachelard, time marks the substance via energy. He concluded that we can no longer think of substance as something outside time.[29] His conclusion is not restricted to chemistry, indeed it is meant to be a lesson that chemistry gives philosophy, which, as mentioned, for Bachelard should let go of metaphysics, and embrace metachemistry.

5. The role of chemistry in Bachelard's philosophy

It is without doubt that chemistry played a crucial role in Bachelard's philosophy of science, and indeed in his philosophy. However, can we claim that chemistry is directly responsible for his ideas? Were his conceptions of scientific object, substance, and analysis and synthesis only dictated by chemistry and its history? An objection to the unique, or at least decisive, influence of chemistry is undoubtedly the importance that modern physics had in his thought. As I have discussed, Bachelard wanted to substitute metaphysics with metachemistry, because metaphysics was based on obsolete notions derived from elementary physics. However, he believed that modern physics, and especially the theory of relativity and quantum mechanics, should also be the basis of our philosophical thinking. In the metaphysics/metachemistry dichotomy there is a historical asymmetry, rather than one based on different sciences. In several respects, it seems that Bachelard drew similar lessons from modern chemistry and modern physics. Can we still claim an irreplaceable role for chemistry in his philosophy? I believe so, but at the same time I do not think that it is possible to make general considerations, as the relative weight of chemistry in his philosophy considerably varies depending on the particular aspect or idea.

As far as his conception of substance is concerned, Bachelard discussed it with reference both to chemistry and physics, sometimes inevitably together, as in his 'Lumière et substance' (1934). There he discussed the philosophical revolution of photochemistry; which Bachelard regarded as liberation from the previous substantialist approach. Physics, and mathematical physics in particular, however, is part of his discussion. He argued for the 'irrationality' of the concept of substance, and criticised the Aristotelian distinction between substance and attributes when

29 Bachelard, *La philosophie du non*, pp. 66-7. Bachelard discussed the concept of substance extensively. In addition to the works cited in this section, see also Bachelard, *Le pluralisme cohérent de la chimie moderne*; Bachelard, 'Lumière et substance' [1934] in Gaston Bachelard, *Etudes* (Paris: Vrin, 1970); Bachelard, *L'activité rationaliste de la physique contemporaine*.

discussing microphysics: he remarked that an electron cannot be seen as a small body (substance) that is negatively charged (the charge would be its attribute).[30] However, chemistry appears to have been crucial for his criticism of the philosophical concept of substance and to have directly inspired his view of the 'historicity' of substance, which suggests that instead of a substance with a fixed identity we should rather think of processes.

It would not be even-handed to argue that Bachelard owed his conception of the scientific object as the product of rationality and of technical processes only to chemistry, to the exclusion of microphysics. Indeed, his reflections on this issue are made about both sciences, even in his early works. On the one hand, *Les intuitions atomistiques* (1933) brings out the implications of chemical practice; on the other hand, in 'Noumène et microphysique' (1931) he had already made clear that in microphysics the object is both rational and technically produced. It is in 'Noumène et microphysique' that he employed the famous expression 'cogitatur ergo est', which highlights the rational character of the scientific object, and it is here that he introduced for the first time the even more famous concept of 'phenomenotechnique', which highlights that the new phenomena are not simply 'found' but rather 'invented' and 'constructed'.[31] He would elaborate on the concept of phenomenotechnique in several of his subsequent works, variously discussing physics, chemistry and modern science in general.[32] It is without doubt that chemistry, which understands the world by way of creation of substances, greatly impacted on Bachelard's view of scientific object; however, microphysics here I think played at least an equally important role.

Finally, I believe that the role that he assigned to analysis and synthesis in the production of knowledge was especially inspired by chemistry, a science that so directly 'synthesises' new substances. The concept of synthesis stayed at the core of Bachelard's philosophy, and he applied it both to the works of physics (wave mechanics in particular) and crucially to the historical advancement of science.[33] There are also other 'lessons' that chemistry for Bachelard gives particularly well, and that I have no space to discuss here. Just to mention one, it seems to me that for

30 Bachelard, *L'activité rationaliste de la physique contemporaine*, Chapter 2.

31 Bachelard, 'Noumène et microphysique' [1931-2] in Bachelard, *Etudes.*

32 See Gaston Bachelard, *Le nouvel esprit scientifique* (Paris: Presses universitaires de France, 1991 [1934]), p. 17; Bachelard, *La formation de l'esprit scientifique*, p. 61; Bachelard, *Le rationalisme appliqué*, pp. 168-9; Bachelard, *L'activité rationaliste de la physique contemporaine*, p. 92. On 'phenomenotechnique', see Teresa Castelão-Lawless, "Phenomenotechnique in Historical Perspective: its Origins and Implications for Philosophy of Science", *Philosophy of Science* 62 (1995); Hans-Jörg Rheinberger, "Gaston Bachelard and the Notion of 'Phenomenotechnique'", *Perspectives on Science* 13, no. 3 (2005); C. Chimisso, "From Phenomenology to Phenomenotechnique: the role of early twentieth-century physics in Gaston Bachelard's philosophy", *Studies in History and Philosophy of Science* 39, no. 3 (2008).

33 See for instance Bachelard, *L'activité rationaliste de la physique contemporaine*, pp. 21ff.

him chemistry is particularly apt to overcome the epistemological obstacle of general knowledge.[34] Chemistry deals with specific substances and with specific reactions in a way that physics arguably does not. Indeed, Bachelard argued that the study of what is general has been far more obscure in chemistry than in physics, and certainly more difficult to carry out.[35] The 'coherent pluralism' of chemistry appears to afford this science, in Bachelard's eyes, a philosophically productive perspective on the world. Overall, I think that the role of chemistry is central to Bachelard's philosophy, and indeed this science gave him some specific philosophical 'lessons'. However, it would be wrong to downplay the role of physics. Arguably, it is precisely his double interest that partly shaped his philosophy, and his interest in chemistry influenced the way he looked at physics and vice versa.

6. Conclusion: The Construction of the Philosophical Object

In order for a discussion about the impact of a science on philosophy to be precise there is a fundamental question that should be addressed. What do we really mean when we say that a certain science directly influenced or even prompted a certain philosophy, or philosophical idea? If we take a science as a 'given' in a philosophical discussion, we are making a great simplification. The claim that chemistry can directly suggest a philosophy rather than another is equivalent to saying that a certain natural object would suggest a certain scientific theory. In my analogy, I referred to the natural object rather than the scientific object with intent. Just as scientists construct their own object, so philosophers construct theirs. If we believe that 'chemistry' and 'history of chemistry' are the same objects for a variety of different philosophers, we regard them as given, rather than constructed. Bachelard's history of science was deliberately and openly an epistemological history; indeed he theorized the 'pedagogical' necessity of a 'judged' history in which error is distinguished from truth, what is sterile from what is fecund. This is a history from the point of view of the present; as a consequence it can change depending on the latest scientific achievements. For Bachelard, this epistemological history is different from the history of the historians of science, who should not judge their object and who in effect produce a 'palaeontology' of a scientific mind which no longer exists.[36] Bachelard was particularly explicit about the construction of the objects of knowledge and the reconstruction of the history of science. However, in different ways, this clearly applies to all philosophers. Many French philosophers focused on chemistry, and unsurprisingly did not come up with the same philosophical ideas. The object of their study was not in fact the same. There is no space here to discuss this in detail, but a couple of examples may give an

34 Bachelard, *La formation de l'esprit scientifique*, Chapter 3.

35 Bachelard, *Le pluralisme cohérent de la chimie moderne*, p. 20.

36 Bachelard, *L'activité rationaliste de la physique contemporaine*, Chapter 1.

idea. Bachelard regarded modern chemistry as eminently removed from common sense, instinctive realism and substantialism. Emile Meyerson, to mention one of the philosophers more closely connected with chemistry, and a past professional chemist, regarded chemistry as a development of common sense. Bachelard regarded history of chemistry as exhibiting epistemological ruptures, and as 'saying no' to its own past. Meyerson regarded history of chemistry as continuous. For Bachelard chemical syntheses of new substances exemplify the modern and productive rationality of chemistry, which overcomes the old conceptions of substance as well as of analysis as the method of knowledge. And yet, in Meyerson's eyes, chemical syntheses are not, from the point of view of reason, formally different from common sense, or indeed from what Lucien Lévy-Bruhl – a firm reference for both Bachelard and Meyerson, and their contemporaries – called 'primitive mentality'.[37] One of the most-often quoted passages by Meyerson is the one in which he argues that chemical formulas (he used the example of Na + Cl = NaCl) express the same type of partial identification that Lévy-Bruhl described regarding Aboriginal Australians, when they 'identified' with parrots. Lévy-Bruhl interpreted the Australians' move as indifference to contradiction, and as therefore displaying a different mentality from modern 'logical' thought. Meyerson on the other hand saw a strong continuity between the two (although he believed that native people were wrong about the content of their partial identification, while scientists were right).[38] Chemistry for him is something rather different than it is for Bachelard. The latter argued that, based on our present knowledge, we tend to interpret faint glimmers of understanding as fully developed scientific ideas; for him this is what Meyerson did when regarded the reasoning of pre-Lavoisien chemists and of modern chemists as similar, which in turn made him think that the mind works in the same way in all times.[39] For Bachelard, Meyerson's view of history of science led him to his 'wrong' conception of the mind.

37 Lucien Lévy-Bruhl, *Les fonctions mentales dans les sociétés inférieures* (Paris: Alcan, 1910); Lucien Lévy-Bruhl, *La mentalité primitive* (Paris: Alcan, 1922); Lucien Lévy-Bruhl, *L'âme primitive* (Paris: Presses Universitaires de France, 1996 [1927]).

38 Emile Meyerson, *Du cheminement de la pensée* (Paris: Alcan, 1931), tome I, pp 81-88; see also Lucien Lévy-Bruhl and al., "L'âme primitive. Séance du 1er juin 1929", *Bulletin de la Société française de Philosophie* 29 (1929) (Meyerson sent a letter to the meeting); Joseph LaLumia, *The Ways of Reason: A Critical Study of the Ideas of Emile Meyerson* (London: Allen & Unwind, 1966), pp. 54ff; Bensaude-Vincent, "Chemistry in the French Tradition of Philosophy of Science", p. 641; Chimisso, *Gaston Bachelard*, pp. 170ff; Bernadette Bensaude-Vincent, "Emile Meyerson, chimiste philosophe", in *L'histoire et la philosophie des sciences à la lumière de l'œuvre d'Émile Meyerson (1859-1933)*, ed. Eva Telkès-Klein and Elhanan Yakira (Paris: Honoré Champion, 2010).

39 Bachelard, *Le rationalisme appliqué*, p. 9. See also Bachelard, *La formation de l'esprit scientifique*, p. 7. Comparisons between Bachelard and Meyerson abound, and are found in a range of publications, from 'classic' treatments of Bachelard's philosophy (Dominique Lecourt, *Marxism and Epistemology. Bachelard, Canguilhem and*

Chemistry suggested different views of the mind and of knowledge to Bachelard and Meyerson; it was in fact a rather different enterprise for them. Bachelard learned his philosophical trade at the Sorbonne, under Léon Brunschvicg, who analyzed history of science in order to study how the mind works. He learned to observe chemistry, and indeed physics – sciences that he had studied and taught – by bearing in mind philosophical questions and assumptions that his teacher had already developed. The idea of breaks in the history of thought and indeed in the history of the mind was certainly not new, although of course Bachelard elaborated it in a novel way, for instance by using psychoanalysis. Brunschvicg had already argued that the Kantian categories are in fact historical, and that human beings reason differently in different times; indeed he argued that immutable *a prioris* of knowledge were 'a superstition' that threatened to reduce Kantian philosophy to impotence.[40] His sciences of reference were mathematics and physics and he detected great changes in how the mind works by studying their history. Just to mention the more recent ones, non-Euclidean geometries and the theory of relativity had revolutionized the philosophical conceptions of space and time, once thought to be the immutable conditions of our experience.[41] And indeed Bachelard invoked Brunschvicg's authority in order to reject Meyerson's 'static conception' of the human mind.[42] On the other hand, Brunschvicg in 1921 put it to Meyerson that salt ($NaCl$) is different from chorine and sodium and their combination shows 'originality' (rather than partial identification, as Meyerson believed); it is not difficult to see where Bachelard's views developed from.[43] When Bachelard started analysing the history and practice of chemistry in order to draw philosophical lessons, he obviously was not a mere receptacle for the lessons of chemistry. He did so not only with specific philosophical questions about the history of the mind and the discontinuity of history, but also with a philosophical training that had suggested to him that the history of science narrates the history of the mind and that discontinuities – or revolutions – characterise this history. This philosophical perspective shaped his objects of study: the history of chemistry and physics. These 'rectified' objects in turn changed the way Bachelard thought about the mind and knowledge. Chemistry as I have argued in the previous section had a decisive impact on Bachelard's philosophy. However, we must not forget that his 'chemistry' and 'history of chemistry' were inevitably 'rectified' philosophical objects.

Foucault (London: NLB, 1975)) to recent studies of Meyerson's: Frédéric Fruteau de Laclos, *L'épistémologie d'Emile Meyerson. Une anthropologie de la connaissance* (Paris: Vrin, 2009).

40 Léon Brunschvicg, "History and Philosophy", in *Philosophy and History. Essays presented to Ernst Cassirer*, ed. Raymond Klibansky (Oxford: Clarendon, 1936), p. 33.

41 Léon Brunschvicg, *L'expérience humaine et la causalité physique* (Paris: Alcan, 1922), pp. 458, 550 and passim.

42 Bachelard, *Le rationalisme appliqué*, p. 9.

43 Léon Brunschvicg and al., "L'intelligence est-elle capable de comprendre? Séance du 24 février 1921", *Bulletin de la Société française de Philosophie* 21, no. 3-4 (1921).

The objects of his philosophy, on the one hand, and his philosophical ideas on the other shaped each other, indeed dialectically produced each other. I think that in this case, we can in fact apply Bachelard's view of scientific practice to his own philosophical practice.

Department of Philosophy
Arts Faculty
The Open University
Walton Hall
Milton Keynes MK7 6AA
United Kingdom
c.chimisso@open.ac.uk

Thomas Uebel

Carnap's *Aufbau* and Physicalism: What Does the "Mutual Reducibility" of Psychological and Physical Objects Amount to?

The present paper is part of a larger project of investigating how far puzzling questions about Carnap's philosophical deflationism – as expressed most prominently in "Empricism, Semantics and Ontology"[1] – can be answered by reference to his own preferred position in areas upon which this meta-philosophical position can be expected to have a bearing. For that project the explorations below provide a starting point; on the present occasion they will, I hope, be found to be of independent interest. At issue is the question of what makes Carnap's *The Logical Structure of the World* (the *Aufbau*) a *pre*-physicalist work.[2]

1 Two versions of the intertranslatability thesis

In the *Aufbau* Carnap sought to develop "the conceptual system of unified science" by determining the logical place of all scientific concepts in a "constructional" or "constitutional system" that was built up from a minimal number of elementary concepts and gained its significance solely due to its formal structure (§§ 1-2).[3] Construction theory equated an object or concept *x* being *reducible* to an object or concept *y* with both *x* being *constituted* by *y* and *x* being *eliminable* by *y*, and Carnap went on to characterise the relation of reducibility or constitutability or eliminability of *x* by *y* as *y* giving *necessary and sufficient conditions* for *x*, with the evidential relation of *y* being an *indicator* for *x* as special case (§§ 35, 47, 49).[4] In just this sense of providing constructional definitions, Carnap declared

1 Rudolf Carnap, "Empiricism, Semantics and Ontology", *Revue Internationale de Philosophie* 4 (1950) 20-30, rev. and repr. in Carnap, *Meaning and Necessity*, 2nd Edition, Chicago: University of Chicago Press, 1956, pp. 205-221.

2 Rudolf Carnap, *Der logische Aufbau der Welt*, Berlin: Weltkreis, 1928, 2nd ed. Hamburg: Meiner, 1961, repr. 1998. Trans. *The Logical Structure of the World / Pseudoproblems in Philosophy*, Berkeley: University of California Press, 1967, repr. Chicago: Open Court, 2003. (All page or paragraph references below not further qualified are to the latter.)

3 Note that Carnap's "Konstitutionstheorie" is variably translated as "construction theory" or "constitution theory" and "Konstitutionssystem" as "constructional" or "constitutional system": here the term of each of these pairs are treated as interchangeable. Construction theory is the theory of constructional systems.

4 Carnap noted that "in construction theory we sometimes speak of constructed objects,

M.C. Galavotti et al. (eds.), *European Philosophy of Science – Philosophy of Science in Europe and the Viennese Heritage*, Vienna Circle Institute Yearbook 17,
DOI 10.1007/978-3-319-01899-7_3,

physical objects constitutable out of psychological ones and psychological objects constitutable out of physical ones (§57) and spoke of the "mutual reducibility" of psychological and physical objects (p. 93). This two-way relation of constitutability between physical and psychological objects made it "logically possible" to develop two constructional systems of concepts, one where psychological objects were constructed from physical ones and one where physical objects were constructed from psychological ones (ibid.).

Given that these two constructional systems employ, at their fundamental levels, of physical and psychological objects respectively, it is tempting to render Carnap's assertion of the mutual reducibility of physical and psychological objects as an "intertranslatability thesis". Here we must be careful, however, for it turns out that this thesis comes in two versions, a weak and a strong one. The two versions interpret Carnap's claim of mutual reducibility differently.

The weak version simply says that (i) there exist two constitutional systems, one taking elements and relations of the physical domain as basic and one taking elements and relations of the psychological domain as basic, such that (ii) in the system with a physical basis the reduction of psychological to physical objects and in the system with a psychological basis the reduction of physical to psychological objects are effected. Moreover, (iii) there exists a recursive procedure for each constitution system to furnish statements that are extensionally equivalent to statements of the other (they have the same truth value). In consequence of (i)-(iii), neither of the two constitution systems can be held to possess overall primacy.[5] The strong version of the intertranslatability thesis derives from Carnap's assertion of the mutual reducibility of physical and psychological objects also the further claim that (iv) the basic language of the system with a physical basis and the basic language of the system with a psychological basis are mutually translatable. It follows that (v) for all statements formulatable in one system an extensionally equivalent one can be formulated in the other. (The strong and the weak versions are not distinguished by whether statements of the two systems are extensionally equivalent or not.)

One further factor has to be taken into account. Carnap distinguished between two types of constructional systems with a psychological basis: one with "autopsychological objects" as its fundamental domain (first-person experiences) and

---

sometimes of constructed concepts, without differentiating" and added that "construction theory employs a neutral language and maintains that objects are neither 'created' nor 'apprehended' but 'constructed'" (p. 10). In the system of the *Aufbau* (others are possible) the elementary concepts pertain to the contents of experiences and their connections, but the aim was to secure "an intersubjective, objective world … which is identical for all observers" (p. 7).

5 In this sense the intertranslatability thesis was invoked for expository purposes in my *Empiricism at the Crossroads*, Chicago: Open Court, 2007, p. 38. The distinction between the weak and the strong versions of the intertranslatability thesis was not drawn there, however.

one with "hetero-psychological objects" as its fundamental domain (other minds) (§ 58). So to be true to the *Aufbau*, strong intertranslatability asserts the mutual translatability of the language speaking of physical objects with the languages speaking of either type of psychological object (and the extensional equivalence of all of their statements).

Did Carnap in the *Aufbau* endorse the intertranslatability thesis only in its weak or also in its strong version? The answer bears crucially on the relation of his views to his later physicalism. Given that Carnap already in the *Aufbau* affirmed the possibility of a constitution system with a physical base alongside one with a phenomenalist base, the question arises what distinguishes his position there from his later physicalism. Strong intertranslatability would place him in fairly close proximity.

As always, interpreters of the *Aufbau* must decide how to weigh what many from Quine onwards regard as Carnap's failure to provide eliminative definitions, first of the relation "is at" holding between colour qualities and so-called "world points" in objective space and time (§§ 125-127) and then of any objects and relations built on this basis.[6] I propose to discount this perceived failure for present purposes for my concern here is not with the success or otherwise of Carnap's project in the *Aufbau* as a whole – nor with the thorny question of how Carnap's *Aufbau* project is to be interpreted – but with certain internal aspects of the construction of the world there attempted. So in pursuit of my question I shall speak, with Carnap, of reducibility relations even when, in fact, none have been established. I begin by describing somewhat further Carnap's derivation of the two constructional systems before pressing the question of the intertranslatability of the languages fundamental to them.

2 The two constitutional systems

Consider how Carnap argued for the logical possibility of two constitution systems. On the one hand, physical statements were translatable into psychological ones. "If a physical object were irreducible to sensory qualities and thus to psychological objects, this would mean that there are no perceptible indicators for it. Statements about it would be suspended in the void; in science, at least, there would be no room for it. Thus all physical objects are reducible to psychological ones." (p. 92) On the other hand, psychological statements were translatable into physical ones. For this Carnap offered two arguments. One turned on the assumption of psycho-physical parallelism: "There is a univocal correspondence between each property of the psychological process and some (even though entirely dif-

6 See W. V. O. Quine, "Two Dogmas of Logical Empiricism", *Philosophical Review* 60 (1951) 20-43, rev. and repr. in Quine, From *a Logical Point of View*, Cambridge, Mass.: Harvard University Press, 2nd ed 1960, pp. 20-46, at p. 40.

ferent) property of the brain process. Thus every statement about a psychological object is translatable into a statement about physical objects." But Carnap noted that this translation was only "in principle possible" since the "correlation problem of the psycho-physical relation" had "not yet been solved" sufficiently to "indicate a general rule of translation" (ibid.).[7] Carnap therefore relied on a second argument which turned on the existence of the "expression relation". "Every hetero-psychological process is in principle recognizable, that is, it can either be inferred from expressive motions or else questions can be asked about it. (It can be reported.) Thus, every statement about a psychological object can be transformed into a statement about those indicators. Thus it follows that all psychological objects can be reduced to expressive motions (in the wider sense), i.e. to physical objects." (p. 93)[8]

Carnap suggested that it would be advantageous to explore both types of constructional systems which he said were logically possible because they served different purposes. "A materialistic constructional system has the advantage that it uses as its basic domain the only domain (namely the physical) which is characterised by a clear regularity of its process." It follows that "from the standpoint of empirical science the constructional system with [a] physical basis constitutes a more appropriate arrangement of concepts than any other." (p. 95) However, it would be false to claim – as behaviourism did, according to Carnap – that "this ordering of objects is also a correct reflection of the epistemic relations" (p. 96). Taking physical objects and concepts to be basic failed to respect the epistemic priority of the auto-psychological. This was decisive for Carnap's choice. "From an epistemological viewpoint (in contradistinction to the viewpoint of empirical science), we are led to another arrangement of concepts, namely to a constructional system with [an] auto-psychological basis." (p. 95) Carnap concluded that "science as a whole … needs both an experiential and a materialistic derivation of concepts." (p. 96) Due to his earlier decision to reconstruct the epistemic relations between the objects of cognition, however, the development of a constitution system with a physical basis was not pursued. Even though Carnap later on noted three possible physical bases (§ 62) he stated: "We cannot, at this time, give an explicit characterization of this system and its importance for science." (p. 95)

As indicated, Carnap's choice of system form followed from his prior commitment to develop a system of objects constructed "with respect to epistemic primacy" (p. 94; cf. § 54). Note that in that system the auto-psychological and the hetero-psychological objects played different roles in relation to the physical objects. The hetero-psychological objects were reducible to physical objects while the latter in turn were reducible to auto-psychological objects. So it was not the same type of psychological object that was both reducible to and constitutive of

7 For a discussion of Carnap's acceptance of psycho-physical parallelism, see sect. 4 below.

8 Note that Carnap's argument for logical reducibility in either direction was underwritten by a verificationist principle that, in the *Aufbau*, was not specified.

physical objects. This asymmetry in the reduction relations between the physical objects and the two types of psychological objects is significant. It obtained not due to the noted unsatisfactory state of the correlation problem, such that translations of auto-psychological states were less readily forthcoming than those of hetero-psychological states, but it reflected a particular assumption of the order of "epistemic priority", namely the assumption that knowledge of other minds was gained only through others' physical behaviour and that knowledge of physical states of affairs was gained only through one's own experience ("methodological solipsism") (§§ 54, 58-60, 64).

3 Strong intertranslatability challenged

Strong intertranslatability takes Carnap's remark that physical and psychological objects are "mutually reducible" to suggest that languages basic to the constitution systems of the *Aufbau* are intertranslatable across the board. This suggestion trades on treating the psychological as interchangeable under the different guises of first-person and third-person mental attributions, as indeed we do in ordinary parlance. Does the suggestion hold up?

Carnap did not offer an explicit reduction of auto-psychological to physical objects anywhere in the *Aufbau*. On one way of looking at it, therefore, the strong intertranslatability thesis disappears under analysis as a misleading and ultimately false conceptualisation of what the *Aufbau* attempted, for the psychological pole of the strong intertranslatibility thesis is not univocal. After all, it is not the language which speaks of auto-psychological objects that is reducible to the language speaking of physical objects and it is not the language that speaks of hetero-psychological objects that the latter reduces to.

On another way of looking at it, however, strong intertranslatability appears retained. Since suitably chosen pairs of first- and third person statements will be co-extensional, advocates of strong intertranslatability might want to overrule the asymmetry of explicit reduction relations on such grounds. Carnap's constitution systems aimed only for the extensional equivalence of the concepts constituted, for their "logical" content or value, not for their "sense" or "epistemic value" (§ 50); they aimed for "logical translations", not for "translations of sense" (§ 51). Because of this, the substitutivity of first-person pronouns and third-person pronouns or names suffices to underwrite strong intertranslatability. Auto-psychological claims are a subset of hetero-psychological claims.

Against this, opponents of strong intertranslatability would argue that construction theory is interested in conceptual reductions according to one or another principle from a set of basic concepts; supplementary considerations in support of such reductions are irrelevant for this purpose. To be sure, Carnap argued that it is possible to constitute hetero-psychological objects from physical ones, namely on the basis of manifestations of the designation and expression relations and these manifestations typically will include a relevant first-person report. But this only shows that the possibility of some kind of same-saying relation between a person's first-person statements about psychological states of affairs and mental state attributions to that person is presupposed in some evidential capacity in Carnap's scheme for reconstructing other minds. It says nothing about the reducibility of auto-psychological to physical objects.

Against this in turn, advocates of strong intertranslatability might respond that their notion of intertranslatability was not intended as an intra-systemic notion – as the previous counter would seem to suggest – but as one that relates different constitution systems. They would argue that our ordinary practice in attributing mental states to each other supports the claim that talk of auto- and hetero-psychological objects can be interchangeable and that therefore the basic language of the constitution system with auto-psychological objects at its base is after all intertranslatable with the basic language of the constitution system with physical objects at its base. In this fashion, advocates of strong intertranslatability would claim, a complement to the explicit reduction of physical to auto-psychological objects is provided and the worrying asymmetry of the absence of an explicitly noted converse reduction relation is trumped.

It is, I think, highly unlikely that Carnap would have agreed to this reasoning. To begin with, reducibility across constitution systems remained an unspecified notion. Logical relations between concepts were only defined within one constitution system. Note that it was with reference to intra-systemic relations that Carnap stated that a certain type of conceptualisation may be basic to our intuitive knowledge of a given fact, but that this knowledge may be demonstrated to depend logically on other facts from which it can be constituted (§ 75). Just that was the case for our knowledge of other minds and it explained why hetero-psychological objects were dispensable as supra-physical objects (even though our intuitive conception of them did not depict them as physical things). But no equivalent demonstration of logical dependence of the facts it spoke about on physical facts had been given for the language speaking of auto-psychological objects. Moreover, we have to distinguish between auto-psychological objects as states of a subject as they are constructed at higher levels of the constitution system (after the construction of "my body") and the subject-less auto-psychological objects at the lower levels. In an epistemologically oriented system with an auto-psychological basis, the logical dependence of the facts it spoke about on physical facts could not even arise in that system. Auto-psychological objects of the lower levels of this constitution system (the ones at issue in our query about intertranslatability) remained irreducible to physical objects.

It might be wondered, of course, whether constitution systems could be built to make possible reductions that were not executed in the *Aufbau*. (I will consider a relevant proposal below.)[9] As regards the *Aufbau* as it stands, however, it seems that the failure to provide a reduction of auto-psychological objects to physical ones cannot be gainsaid. So the strong intertranslatability thesis cannot be sustained because (iv) is false: the basic language of the constitution system with an auto-psychological base was not shown to be reducible to the basic language of the constitution system with a physical base.

4 Auto-psychological exceptionalism probed

Does it matter that strong intertranslatability is not sustained in the *Aufbau*? It does – for the simple reason that were one to believe it to hold one would misunderstand Carnap's view of mental-physical relations in the *Aufbau*. Note that the weak intertranslatability thesis, according to which two radically different constitution systems may nevertheless formulate some pairs of extensionally equivalent statements, still holds. The important point is that in the *Aufbau* Carnap went no further for (v) is also false: it is not the case that for all statements formulatable in the constitution system with an auto-psychological base an extensionally equivalent one be formulated in the constitution system with a physical base. If strong intertranslatability were true then Carnap would have been, already in the *Aufbau*, a kind of proto-physicalist. What distinguishes this position from *bona fide* physicalism is this. Bona fide physicalism is, first, a meta-linguistic thesis that, second, insists on observing the distinction between the correct formal mode and the misleading material mode of speech and, third, it accords primacy to the physical language (declaring it to be the only universal language such that all other languages can be translated into it). By contrast, the proto-physicalist position, while already a meta-linguistic thesis, does not recognise the distinction between modes of speech or accord primacy to either of the languages speaking of physical or auto-psychological objects but still grant their full intertranslatability.[10] The asymmetry in the reduction relations between the basic languages of the two constitution systems in the *Aufbau* means that Carnap was not yet even a proto-physicalist.

---

9 Carnap considered one in which physical objects were reduced to hetero-psychological ones. He concluded that while that may be logically possible, doing so would not allow one to respect the assumed order of epistemic priority (§§ 58, 63-64). But note that the reducibility of auto-psychological to physical objects would also not have been established. That could only be effected in a constitution system with a physical basis,

10 For the *bona fide* type of physicalism, see, e.g., Rudolf Carnap, "Die physikalische Sprache als Universalsprache der Wissenschaft", *Erkenntnis* 2 (1932) 432-465, trans. *The Unity of Science*, London: Kegan Paul, Trench Trubner & Co, 1934, repr. Bristol: Thoemmies, 1994. What is here called "proto-physicalism" finds expression only partially in the first, unpublished and later heavily revised draft of Carnap's "Die physikalische Sprache als Universalsprache der Wissenschaft" (RC 110-03-22, ca. May 1930, Archive of Scientific Philosophy, Hillman Library, University of Pittsburgh); for discussion see my *Empiricism at the Crossroads*, op. cit, Ch. 6.

It might be thought that the conclusion that Carnap was neither a *bona fide* nor a proto-physicalist in the *Aufbau* should not really surprise us: as we saw, Carnap endorsed psycho-physical parallelism and parallelism is not identity. However, since Carnap's physicalism was a meta-linguistic and not an ontological thesis – specifying relations between (types of) languages and not (types of) objects – our conclusion does not follow simply from his endorsement of parallelism. So let us distinguish also quasi- and proto-physicalism: unlike the latter, quasi-physicalism is not yet a meta-linguistic thesis but one that speaks of reduction relations between objects and affirms the mutual reducibility of physical and auto-psychological objects.[11] That psycho-physical parallelism and quasi-physicalism do not have significant features in common is not at all clear. (But was the Aufbau even quasi-physicalist?)

It is important to understand Carnap's characterisation of the mind-body relation as psycho-physical parallelism correctly. Carnap did not affirm any philosophical theory thereof. He explicitly rejected what he considered the then live philosophical options: interactionst dualism, non-interactionist parallelism and identity theory in the sense of double-aspect theory (§ 22). When he spoke, as we saw, without qualification of "a univocal correspondence between each property of the psychological process and some (even though entirely different) property of the brain process" and held that "all types of psychological processes have physical parallels (in the central nervous system)", Carnap endorsed psycho-physical parallelism purely as an empirical hypothesis. For him, psycho-physical parallelism simply asserted a correlation which to determine the relata of was the job of brain physiology, psychology and psychopathology (§ 21).[12] To go further than to ask "between what objects the relation obtains" (and presumably under what circumstances the relation does so) and to speculate philosophically about "what it is between the correlated objects by virtue of which they are connected" Carnap counted as switching, illegitimately, to an "essence question", in short, as switching from empirical science to "metaphysics" (p. 35).

11 What is here called "quasi-physicalism" – for now the system with a physical basis was held to be equally indispensible (not only also possible as in the *Aufbau*) – would have found expression in Rudolf Carnap, "Die alte und die neue Logik", *Erkenntnis* 1 (1930) 12-26, trans. "The Old and the New Logic" in A. J. Ayer (ed.), *Logical Positivism*, New York: Free Press, 1959, 113-146, at pp. 143-144, if at that time Carnap also held auto-psychological objects to be reducible to physical objects. Doubts about this are raised below.

12 This was by no means an eccentric view at the time: for a discussion of the wide acceptance of psycho-physical parallelism as an empirical hypothesis in 19th and early 20th century Germany and Austria – and of the different philosophical interpretations it was often given all of which Carnap rejected – see Michael Heidelberger, "The Mind-Body Problem in the Origin of Logical Empiricism: Herbert Feigl and Psychophysical Parallelism", in Paolo Parrini, Wesley Salmon, Merrilee Salmon (eds.), *Logical Empiricism. Historical and Contemporary Perspectives*, Pittsburgh: University of Pittsburgh Press, 2003, pp. 233-262.

Construction theory was a formal science replacement of such "by virtue of which" questions. "To the extent to which it is necessary and possible for science, the psycho-physical problem can be clarified in the indicated way on the basis of construction theory. … a question which goes further cannot even be formulated within science (i.e. formulated with scientific, that is to say, constructable, concepts)." (p. 271) Since the *Aufbau* lacks the framework within which to raise reductionist questions about auto-psychological objects, Carnap may appear to have been within his rights to call the issue closed. But note that his affirmations of psycho-physical parallelism do nothing to dispel our residual puzzlement. If all types of psychological processes have physical parallels, then psychological processes involving first-person awareness must also have physical parallels. The construction-theoretical replacement for metaphysical embellishments of psycho-physical parallelism was the mutual reducibility of psychological and physical objects. So why should auto-psychological objects resist reduction to the physical objects? Why was Carnap not even a quasi-physicalist (as defined above)? How are we to make sense of the fact that, as matters are set up in the *Aufbau*, the question of the reducibility of auto-psychological to physical objects cannot even be raised?

Here is a simple explanation of why no reduction of auto-psychological to physical objects was even attempted in the *Aufbau*. That is that in the *Aufbau* Carnap did not, as he stated himself, develop a constitution system with physical objects at its base. (The reduction of hetero-psychological objects was part of the reduction to an auto-psychological base.) Without a physicalist or materialist constitution system in operation, of course, it made no sense even to raise the issue. So there is no mystery: there is nothing more to the inability of the *Aufbau* to pronounce on the reducibility of the auto-psychological to the physical than that, as a mere matter of contingent fact, its author did not choose or felt unable to develop a physicalistic constitution system in which it could have been pursued.

The nagging question here, of course, is why the author should have decided this or felt that way. When he outlined the three possible bases he indicated no reason why such constructions should run into difficulties. Carnap's candidates for basic elements and relations of the system with a physical basis were these: (i) electrons (including protons) and their spatio-temporal relations; (ii) space-time points of the four-dimensional continuum and their relative locations in the continuum and the many-one relations between real numbers and space-time points corresponding to the individual components of certain functions; and (iii) world points as elements of world lines of physical points (following Minkowski) and relations of coincidence and local time-order (§ 62). About (i) he remarked how in it first the properties of the electromagnetic field can be "defined through implication statements about the acceleration of electrons", then atoms of all chemical elements can be constructed as "certain constellations of electrons" and then gravitation "through implication statements about the acceleration of atoms". He concluded: "The derivation of the remaining constants and other concepts of physical science then no longer offers difficulties of principle since, in physics, they all can be reduced to magnetic fields, electrons and gravitation. The perceptible physical things and properties can then easily be constructed from the things and properties of physical science, since they are uniquely determined by them." (p. 99, trans. amended) Interestingly, Carnap also noted that the basic elements of (i) were said to be constructable out of the elements of (ii) and the basic elements of (ii) out of the elements of (iii). All of this amounts to a pretty impressive sketch of a constructional system on a physical basis.

So what was the problem here that held Carnap back? The outlines of two concerns are barely discernible. First, concerning the possible bases, Carnap may have been reluctant to propose a constructional system that employed non-eliminative definitions, so close to – or even at – the very fundamental level. Now I said that I would disregard that his constructional system with an auto-psychological base tolerated non-eliminative definitions higher-up (as in the step from constructed qualities to their ascriptions to points in space and time), but Carnap may well have preferred more solid reductions at the base of constitution systems in general. Second, very much higher up in the system, Carnap may have been worried by our ignorance about details of the psycho-physical correlation which he assumed all along. Unless one were to believe that mental states simply consisted in dispositions to behave in certain ways, the construction of mental states in a system with a physical basis ought not advert essentially to expression and similar relations and instead rely directly on the psycho-physical correlation – but the details of that were still largely unknown.

There is much plausibility to this explanation of why the Aufbau features no reduction of auto-psychological to physical objects. To be sure, it faces the difficulty that the constitution system with an auto-psychological physical basis also ran into trouble at the fundamental level. As Carnap noted, its basic relation of remembered similarity was not uniquely determined purely in logical or structural terms – as all terms should be in a constructional system of the sort he sought to build – and required what he himself noted was a highly problematic expedient to overcome the problem (§§ 153-155). While it might be wondered whether the problems Carnap saw arising for a system with a physical basis were really so very much more severe, it must be noted that a complement of this basic relations problem is also likely to effect a constitution system with a physical basis, so the two problems just adduced are at least additional ones that Carnap may have felt he could do without for the time being. (After all, he often enough stressed that the *Aufbau* was but a sketch.)

So could Carnap have been a some kind of physicalist if he had wanted to? It might seem that if the explanation just offered were accepted, then he could have been one in principle though in practice he faced very considerable problems. But we have to be careful once more. As noted, *bona fide* and proto-physicalism are meta-linguistic theses and – like the intertranslatability thesis – introduce talk of relations between languages that is, in fact, alien to the *Aufbau* which speaks not physical or psychological languages but objects. So the view that Carnap could have reduced auto-psychological to physical objects had he wanted to would only translate into quasi-physicalism, but would not allow for strong intertranslatability. Moreover, the difficulties Carnap saw facing a constitution system with a physical basis must not be underestimated. To be sure, he could have developed one if he had known how to, but the point is that he did *not* know how to. (All he knew – or thought he knew – was that it was logically possible.) So it is not the case that, on any ordinary understanding of these words, Carnap could have been even only a quasi-physicalist if he had wanted to.

Indeed, I hesitate to take it as established that Carnap would have been happy to reduce auto-psychological objects to physical objects, even though he was happy to so reduce hetero-psychological objects. As far as I can see, the possibility cannot be ruled that such a reduction was precisely one of the problematic issues that Carnap saw arising for a constructional system with a physicalist basis. Since Carnap never discussed the matter in the *Aufbau*, any interpretation of his position here is reduced to arguing from circumstantial evidence. It strikes me as one such piece of significant evidence that still in 1930 and 1931 Carnap felt that he had had independent reasons to exempt the auto-psychological from reduction to the physical. (By then he had started to talk in terms of the relation between different languages so talk of intertranslatability or failure thereof comes natural.) Both in unpublished drafts and in discussions of the Circle he held that the auto-psychological language resisted translation into the physical language because some of its non-negligible content was lost thereby.[13] If those worries that Carnap had in 1930 and 1931 were not new ones – as I suspect they were not – then we can conclude that the constructional lacuna of the *Aufbau* concerning the reduction of auto-psychological to physical objects was not unpremeditated.

Of course, Carnap's own objections still in early 1931 were soon overcome by his introduction of the distinction between the formal and the material modes of speech. As he once put it: "Only due to this sharp distinction and the rejection of the material mode has the elimination of the dualism of the two languages become possible."[14] But that, as they say, is a story for another day (one that, as it happens, begins with Carnap's *Scheinprobleme*, a booklet for a wider audience dating from later in 1928).[15]

13 For documentation and discussion of these worries and of how they were overcome, see my *Empiricism at the Crossroads*, op. cit., Chs. 6-7.

14 "Erst auf Grund dieser Trennung und der Verwerfung der inhaltlichen Redeweise ist aber die Überwindung des Dualismus der beiden Sprachen möglich." Carnap to Neurath, 2 March 1932, RC 029-12-60/61, p. 2. (Quoted by permission from the Archive of Scientific Philosophy, Hillman Library, University of Pittsburgh.) The previous sentence makes clear that the dualism in question is that of the physical language and the auto-psychological protocol language.

15 Rudolf Carnap, *Scheinprobleme in der Philosophie*, Berlin: Weltkreis, 1928, repr. (ed. by G. Patzig) Frankfurt a. M., 1966 and (ed. by T. Mormann) Hamburg: Meiner, 2004. Trans. in *The Logical Structure of the World/Pseudoproblems in Philosophy*, Berke-

5 Conclusion

We saw that the mutual reducibility of physical and psychological objects that Carnap talked about in the *Aufbau* equivocates on the psychological side between auto-psychological and hetero-psychological objects: strictly speaking, there obtained no mutual reducibility between physical objects and either of them on their own in the constitution system actually developed there. Most importantly, no reduction of auto-psychological to physical objects was effected. If this is correct, then the fact that Carnap nowhere denied the reducibility of the auto-psychological language in the *Aufbau* does not at all betoken positive belief in its reducibility – it rather turns on the fact that the issue of the relation between the auto-psychological and the physical language could not even be raised.

One aspect of concern about the reduction of the auto-psychological to the physical has hardly been touched upon. It might be wondered to what extent the conception of mind-body relations adopted by Carnap was responsible for his evident qualms about that reduction. As the longer story not told here would show, however, the pressure to change the position adopted in the *Aufbau* with regard to the auto-psychological does not appear to have arisen in response to these types of considerations. Rather, external pressure sprang from epistemological concerns expressed by colleagues (Neurath) about the sustainability of methodological solipsism, while Carnap's own physicalist turn (in several stages) was owed to a combination of new insights about how to philosophise about logic and language and scientific knowledge – ultimately coalescing in his later philosophical deflationism. Nevertheless, focussing upon how mind-body relations were conceptualised in the *Aufbau* has thrown light, I hope, on the details of its constitution system that helped to prompt the so-called protocol sentence debate at the end of which Carnap had become what he had not even conceived of when he wrote the *Aufbau*: not just a quasi- or proto- or even a *bona fide* physicalist, but a physicalist who discarded, at least for the practical purpose of reconstructing the languages of current science, the use of auto-psychological protocol languages altogether.[16]

Philosophy, School of Social Sciences
University of Manchester
Manchester M13 9PL
UK
thomas.uebel@manchester.ac.uk

ley: University of California Press, 1967, repr. Chicago: Open Court, 2003.

16 For helpful discussions and comments on an earlier draft I wish to thank Richard Creath and Sean Crawford. I also wish to thank an anonymous referee for constructive suggestions. Crawford's "The Myth of Logical Behaviourism and the Origins of the Identity Theory" (in Michael Beaney (ed.), *The Oxford Handbook of the History of Analytic Philosophy*, Oxford: Oxford University Press, forthcoming) prompted the present return to the pre-physicalistic Carnap.

Claude Debru

On the Relationship between Neuroscience and Philosophy: the Case of Sleep and Dreaming

In this paper, I will try to examine from both philosophical and epistemological points of view the major developments of neurophysiological research on sleep and dreaming during the last sixty years. In the nineteen fifties and sixties, neurophysiologists and psychophysiologists had great hopes regarding the contribution of sleep and dreaming studies to the understanding of consciousness mechanisms. These hopes were not entirely fulfilled. Progress in the field of sleep research was achieved thanks to developments in disciplines like neurophysiology properly said, neurochemistry, clinical medicine, etc. Today, more modern EEG Imaging techniques bring new and interesting results on cerebral correlates of conscious experience. The purpose of the following analysis is to show how some problems could be solved while others remain open for future investigation, be it experimental or (hopefully) philosophical. I will start with a sketch of the history of the field since the nineteen fifties.Then I will present some epistemological remarks on experimental approaches in neurophysiology, and I will go on with discussions on both psychophysiological parallelism and physiological functions.

I Historical Sketch

Major developments in the field of sleep physiology took place in the nineteen fifties. At that time, sleep physiology was well established as a physiological specialty. The major representative of the field in America was Nathaniel Kleitman, the author of an influential monograph, *Sleep and Wakefulness*[1], and the founder of a school in sleep psychophysiology and clinical medicine at the University of Chicago.[2] One of his collaborators, Eugene Aserinsky, studied brain activity in the newborns and discovered sleep phases which were characterized by rapid eye movements (REM). The so-called "rapid eye movement sleep" (REM sleep) was much different for the already recognized slow wave sleep, characterized by slow eye movements as well as by slow electroencephalographic waves. This observation made in 1953 was really the starting point of a new research domain, with a new distinction within sleep between slow wave sleep and rapid eye move-

1 Nathaniel Kleitman, Sleep and Wakefulness, Chicago: University of Chicago Press 2nd edition, 1963.

2 For a more detailed analysis of the material presented in this paper, see Claude Debru, Neurophilosophie du rêve, Paris : Hermann, 2nd edition, 2006.

M.C. Galavotti et al. (eds.), *European Philosophy of Science – Philosophy of Science in Europe and the Viennese Heritage*, Vienna Circle Institute Yearbook 17,
DOI 10.1007/978-3-319-01899-7_4,

ment (REM) sleep. This new structure of sleep was further substantiated thanks to another observation made by William Dement in 1955. He discovered that the rem-phase was also characterized by a rapid electroencephalographic pattern. This new discovery helped to confirm the rem-sleep concept, and moreover opened the opportunity of studying in an objective way the content of psychical activity during this well-defined sleep phase. Dement woke up sleepers during this phase and asked them questions about the content of their mental activity. He proposed that rem-sleep could be the physiological correlate of dreaming and confirmed this hypothesis while analysing the dream reports. This was the starting point of a long story, full of uncertainties and misinterpretations regarding the difference in mental activity between slow wave sleep and rem-sleep, which is apparently still not solved.

This new phenomenon, rem-sleep, took place within the cycle of sleep and wakefulness – or according to Kleitman's terminology the "basic rest-activity cycle" (BRAC). Since rem-sleep supposedly corresponded to dreaming, the idea was to interpret rem-sleep as a kind of intermediate state between slow-wave sleep and wakefulness. Rem-sleep was understood as light sleep, rather than as deep sleep. Kleitman's BRAC theory was certainly responsible for this interpretation, which was later criticized. This interpretation is an example of the influence of a preexisting theoretical context in the interpretation of a new phenomenon.

The same phenomenon was rediscovered in 1958 in a very different context, the context of physiological research on the basic mechanisms of brain plasticity and conditioning in the Pavlovian spirit. This kind of research was performed by Michel Jouvet, a neurosurgeon working at the Faculty of Medicine of the University of Lyon. Jouvet looked for the neurophysiological mechanisms of conditioning in deeper brain structures like the famous Moruzzi-Magoun ascending activating reticular system of the brain stem, which was held responsible for wakefulness. He performed these experiments on cats, whose cortex was totally or partially removed in order to study the electrical activity of deeper structures. In these structures (especially in the reticular system), he discovered an electrical activity of a particular kind, characterized by spikes. He discovered also that the muscle tone was completely abolished during these episodes. When the experimenter keeps parts of the cortex, the cortex shows a rapid electrical activity. The contrast between a fast activity in the brain and the disappearance of peripheral activity led Jouvet to coin the phrase "sommeil paradoxal" (paradoxical sleep). In the following developments, I will use both terminologies of rem-sleep (for American works) and paradoxical sleep (for French works), due to the differences in basic philosophical orientations and typical problems between both research schools.

Indeed the newly coined term "paradoxical sleep" involved a particular interpretation regarding the nature of the phenomenon. Jouvet continued his exploration of the brain and made a systematic study of the electrical activity of other structures in sleep. He came to the conclusion that this sleep phase is characterized by an intense internal brain activity and the active suppression of any peripheral motor or sensory activity, in the sense that a strong noise does not awake the sleeping animal. Thus the so-called "paradoxical sleep" should be considered as a deeper sleep than slow wave sleep, not as a light sleep – contrarily to the Dement-Kleitman concept. Consequently, Jouvet did not classify the paradoxical phase as an intermediate state between sleep and wakefulness – but he emphasized the strong peculiarities of this state and he classified it as a third state of alertness of the central nervous system, together with slow wave sleep and wakefulness. Jouvet's original observations were made independently from the work of Kleitman's school because of lack of communication in the fifties. After that, there was a kind of division of labor between the more clinical and psychophysiological work made in America and the more physiological and biological work made in France on the same physiological phenomenon. The physiological problem of sleep became more and more subdivided into different specialties, which made an integrative view of the functions more and more difficult to elaborate and justify. However, psychophysiology and neurophysiology were never completely separated. The reason is the well known role of major neurotransmitters like serotonin in the regulation of psychic phenomena like mood. The cerebral (including electrophysiological and neurochemical) correlates of conscious experience became common parlance among neurophysiologists during the developments I am going to describe, so that the epistemological and scientific dimensions and the philosophical dimensions of sleep research cannot be entirely separated..

II Epistemology

The way neurophysiological and neurochemical work developed in the sixties and seventies is extremely revealing for scientific methodology in general, and particularly for epistemology regarding the search for causes in systems whose complexity degree was unknown. Indeed, very few neurotransmitters were known in the nineteen sixties: acetylcholine, monoamines like serotonin, dopamin, epinephrin and norepinephrin, and the P-substance. Now there are more than one hundred substances of various sorts, including neuropeptides, acting as neurotransmitters or neurohormones in the brain. The extreme intrication of structures and molecular agents in the brain is a matter of fact, which was not really recognized or even imagined in the sixties and early seventies. This is a reason why it is really difficult to judge fairly the attempts which were made during this period.

An important step was made in the sixties when researchers realized that the nuclei which were involved in the genesis of paradoxical sleep were the richest ones in monoamines in the brain. This was done thanks to the use of the histoflurescence technique by Annica Dahlstrom and Kjell Fuxe in 1965. The idea came, that serotonin could play a role in the sleep states, together with other monoamines, which could possibly play a different kind of role. An enormous amount of pharmacological work was performed during this period in order to give arguments about the possible involvement of serotonin in the preparation and triggering of the paradoxical phase. Here we meet an important argument : no final and decisive proof of the involvement of serotonin as a necessary and sufficient condition of paradoxical sleep could be provided, in spite of at first sight most convincing evidence. The belief that a proof of a necessary and sufficient condition could obtain turned out to be a philosophical prejudice.

The basic arguments regarding causality were pharmacological – which means that unsuspected problems of specificity could occur. As a matter of fact, the results of the experiments which we are going to expose are presently not entirely understood. The first step in the experimental reasoning was the inhibition of serotonin (5-HT) biosynthesis from its immediate precursor 5-hydroxytryptophan (5-HTP). This synthesis is inhibited by parachlorophenylalanine (pCPA). When pCPA is given to the experimental animal in a single dose (acute administration), paradoxical sleep disappears. When 5-HTP is given, paradoxical sleep reappears. Now the question is, whether suppression and restoration of a physiological phenomenon by suppression and restoration of a biochemical factor are sufficient proofs of the role of this factor as a necessary and sufficient condition of this phenomenon. If we follow the traditional view of experimental medicine as exposed for instance by Claude Bernard, this would be certainly the case. But this conception did not allow to make real progress in the understanding of the brain's workings. Furthermore, due to continuous attempts at reaching better and better proofs, contrary evidence began to accumulate, and the overall picture became extremely complex and increasingly impossible to understand. A most revealing experiment was chronic administration of pCPA which induces a total depletion of 5-HT in the brain. Under these conditions, sleep reappears. Thus, serotonin cannot be considered as a necessary condition of sleep. In the early nineteen eighties, the serotonergic theory of sleep was in a deep epistemological crisis. In such a situation, Michel Jouvet decided to set up a crucial experiment, in the sense outlined by Duhem and Popper. Indeed, he wanted to get some kind of certainty in a situation characterized by an increasing uncertainty due to very diverging evidence.

The basic principle of the experiment he performed is well-known in physiology. It is a fluid transfer from the body of an animal to another animal. The first animal was rendered insomniac by an instrumental technique which would prevent it to sleep and thus would allow hypnogenic factors to accumulate in its brain. The second animal was rendered insomniac by pCPA, so that its brain did not contain any 5-HT. A certain amount of cerebrospinal fluid taken from the first animal was given to the second animal, which consequently recovered its sleep. The biochemical analysis of the transferred fluid showed that it contained no trace of serotonin nor of any serotonin precursor. The inescapable conclusion of the whole evidence is that serotonin is not a necessary condition of sleep but only a sufficient condition. This whole body of evidence is still poorly understood, thirty years after the most crucial experiment (a negative experiment, in the Duhem fashion), and the exact role of the serotonin molecule is still much in doubt.

The search for other unknown biochemical sleep factors (other than the already known factors endowed with more particular roles in sleep mechanisms, like acetylcholine) began then – a very difficult one indeed. Much later, a big success was reached in a cooperation between a genetical and a biochemical approach in the study of the genetical disease narcolepsy-cataplexy in the famous breed of narcoleptic dogs raised by William Dement. This study made by Emmanuel Mignot at Stanford revealed the role of a recently discovered hypothalamic peptide, hypocretin, in sleep mechanisms. The lack of hypocretin receptors in man produces narcolepsy-cataplexy.[3] However, hypocretin is only one factor among many other factors which play a role in the control of the sleep cycle. It is clear from this extremely sketchy account that the traditional epistemology of experimental physiology and medicine regarding causal explanation is no more suitable to capture the level of complexity of such a broad and intricate physiological system as the one of sleep and wakefulness. The traditional view of a major cause in the form of a particular biochemical factor triggering a particular phase was rather misleading.

The modeling of the complex network of various excitatory and inhibitory actions needs certainly a preliminary identification of the various factors involved and of their particular role in all physiological aspects of such a complex state as the paradoxical phase. But the next question is: how to understand the behavior of such complex, multifactorial systems, whose dynamics is governed by a particular combination of excitatory and inhibitory actions and is thus very precisely controlled? To describe and understand these systems, the old causal language is no more suitable. This point was already made by Ernst Mach in his book *The Analysis of Sensation* (1886). Mach made a very strong statement, when he wrote in a paragraph about causality that causality belongs to a primitive and even pharmaceutical view of the world which should be replaced by a mathematical analysis in terms of functions of several variables. This is exactly what happened in biology in recent years, when mathematicians collaborating with biologists introduced non linear differential systems in order to describe all sorts of interactions, including interactions between different brain structures involved in sleep mechanisms.

III Philosophical remarks on psychophysiological parallelism and cerebral correlates of conscious experience

The complex picture of sleep biochemical mechanisms is only one part of the overall picture of sleep. It gives only a broad account of the physiological basis of mental activity during sleep, in terms of various particular biochemical factors,

3 See Claude Debru, op. cit., p. 376-377.

anatomical structures and physiological functions, from the brain stem to the cortex. Psychophysiology on the other hand tries to fill the gap between brain states and mental states, by hypothesizing correlations at the various levels of brain and mind activity. The dynamics of specialized neural networks is quite relevant in that respect, since it provides a link between physiology and psychology at an appropriate level. This state of affairs leads us to make comments about questions of a more philosophical sort.

When Dement proposed that rem-sleep could be the physiological basis of dreaming, he created big hopes regarding the relevance of the classical philosophical framework of psychophysiological parallelism in order to establish strong correlations between mental and physiological states. Indeed, the discovery of rem-sleep appeared as a particularly favorable occasion for investigating these correlations in both directions: for making assumptions about the psychological correlates of physiological phenomena, and vice versa for making assumptions about the physiological correlates of psychological, conscious or even unconscious processes. However, in the case of sleep and dreaming, the research programme based on the assumption of psychophysiological parallelism did not allow to reach firm conclusions.

In the early sixties, William Dement started from the assumption of the rem-sleep/dreaming equivalence (in the sense that the two words designate the same phenomenon) and used rem-sleep with its clearcut temporal definition to investigate the possible relationships between dreaming, psychiatry and psychoanalysis. He asked such questions as: Is there a difference in the structure of rem-sleep and in the content of dreaming between normal and psychotic people? Is Freud's hypothesis about the function of dreaming as the guardian of sleep really sound? Can sleeplessness produce psychotic symptoms during wakefulness? Dement used all kinds of experimental situations, including long-term sleeplessness, in order to investigate these questions. He found something he was not looking for: when the participants in the sleep deprivation experiment are allowed to go back to sleep, their amount of rem-sleep is much bigger than expected. This so-called "rebound effect" is a clear sign that rem-sleep should have an important physiological function. Another consequence of Dement's studies was that Freud's hypothesis about dreaming as the guardian of sleep was not confirmed, since the rem-sleep phase ends up in an awakening.

Along the same line of strongly parallelistic thinking, some psychophysiologists and psychiatrists like David Foulkes made use of the important distinction emphasized by the Italian neurophysiologist Giuseppe Moruzzi between phasic and tonic processes. Phasic processes are strongly discontinuous, like the rapid eye movements or the electrical spikes recorded in the deep brain structures. Tonic processes are maintained in a continuous fashion, like muscular atonia. It was assumed that these two different kinds of processes serve different psychophysiological functions. David Foulkes went further, and proposed to interpret the difference between phasic and tonic processes in terms of Freud's metapsychology. The phasic processes would correspond to the primary visual experience, and the intervals of ocular rest would correspond to a secondary cognitive elaboration. Phasic and tonic processes would thus correspond to primary and secondary processes in Freud's sense. Some researchers went even further in this attempt at applying metapsychological concepts in order to interpret physiological phenomena. On the other hand, physiologists were rather skeptical about these attempts, which could become easily very naïve. It is true that at that time, in the nineteen sixties, rem-sleep was perhaps the most appropriate field of neurophysiological research for producing new facts and arguments in the old philosophical discussion of the mind-brain problem. How far could researchers go in demonstrating a strong and detailed correlation between physiological phenomena (like rapid eye movements) and aspects of the reported subjective experience of the dreamer – for instance by trying to correlate the direction of the rapid eye mouvements and the direction of the gaze corresponding to the reported experience (the so-called "scanning hypothesis" – which could neither be entirely established not entirely rejected).

When we look at these attempts after fourty years, we may conclude that only very loose relationships could be established. Even the relationship between rem-sleep and dreaming became more and more problematic. From the beginning, psychophysiologists asked the question whether there would be a mental activity going on in slow wave sleep, of a different kind from the dream, which was supposed to occur only in rem-sleep. A better definition of dreaming was needed. The analysis of dream reports showed that they were mostly characterized by features like a vivid visual experience, an emotional content, a narrative etc., while slow wave sleep mental activity was pictured as more abstract, conceptual etc.

However, the issue of mental activity during sleep became more and more controversial. An increasing number of psychophysiologists claimed that genuine dreaming may occur during slow wave sleep as well as during rem-sleep. Perhaps the electrophysiological signs which were used at the very beginnng of the field were not subtle or specific enough in order to substantiate psychophysiological correlations in this case. However, this field of study was never entirely abandoned, and new results about the comparison between all states of alterness including waking, slow wave sleep, rem-sleep and its variety lucid dreaming could be gained more recently. The analysis of dream reports compared with wake-state thinking made by Allan Hobson showed interesting features about cognition in dreams (cognition within the dream scenario rather than about the dream). Psychophysiogival correlations are substantiated in a much more precise way thanks to techniques like single-unit recordings, multi-channel EEG and brain imaging. Allan Hobson made an EEG study of lucid dreaming (a variety of dreaming in which the dreamer is aware of the fact that he is dreaming). The frequency analysis shows that the reflective awareness peculiar to lucid dreaming does correspond to significant electrophysiological differences from typical rem-sleep activity, and to other differences from waking.[4] William Dement said once that psychophysiological parallelism was more useful than ever. Progress in that direction is a matter of fact. More objective and precise definitions are needed and more thorough studies of the subtle structure of the states of alertness are also needed to go further in that direction.

IV Functional hypothesis

Nevertheless, there are other ways of looking at the whole physiology of sleep and to ask different kinds of questions. Developmental physiology, comparative physiology, evolutionary physiology, behavioral physiology may produce significant results in order to propose answers to the question of the biological function or functions of paradoxical sleep. In this respect, Michel Jouvet made an important discovery in the sixties. During the course of his lesion experiments, he destroyed the nucleus *locus coeruleus*, whose function is to inhibit muscular activity during paradoxical sleep and thus to create the muscular atonia which is typical of this state and which played an important role in its definition. When the experimenter destroys the nucleus *locus coeruleus*, the experimental animal (a cat) starts performing during all paradoxical phases various behavioral activities which are typical of its species, like following an imaginary mouse or catching an imaginary prey, or showing all signs of fear, anger, etc. This remarkable discovery lead to a new definition of the paradoxical phase as inhbited behavior, and to a new question about the function of this state as internal expression of externally inhibited behaviors, which reappears regularly during sleep. Is there a relationship between behavior, sleep, and dreaming? Which theoretical ideas could be introduced here? In 1967, a computer scientists working for the US Air Force, Edmund Dewan, proposed that rem-sleep could be a reprogramming process of genetical properties in the brain. Indeed, it was tempting to interpret rem-sleep or paradoxical sleep as a recurring expression of genetical behavioral programmes, due to the behavioral content of paradoxical sleep and to its recurring character. In 1972, Michel Jouvet proposed to interpret paradoxical sleep as "genotypic arousal".[5] This remarkable phrase synthesizes all known facts and hypotheses about paradoxical sleep at that time. Jouvet's discovery of the behavioral aspect of paradoxical sleep and Dewan's hypothesis of a reprogramming process could merge in the new hypothesis of paradoxical sleep as a reprogramming of specific genetical behavioral proper-

4 Ursula Voss, R. Holzmann, I. Tuin, John Allan Hobson, "Lucid dreaming: a state of consciousness with features of both waking and non-lucid dreaming", *Sleep* 32, 9, 2009, pp. 191-200.

5 Michel Jouvet, "The role of monoamines and acetylcholine containing neurons in the regulation of the sleep-waking cycle", *Ergebnisse der Physiologie* 64, 1972, pp. 270-272.

ties. This theory was typical of the growing interdisciplinarity between genetics, computer science, and neuroscience in the late sixties.

The ensuing developments of this theory were sometimes disappointing and sometimes inspiring. Disappointing results: the suppression of paradoxical sleep has no influence on such a fundamental behavior as the maternal behavior in rats. Exciting results: in rats which belong to different genetical breeds endowed with different learning abilities, paradoxical sleep suppression is followed by a decrease of the difference. Consequently, paradoxical sleep could help maintaining more subtle individual genetical differences in the brain. The discussions about the relationships between paradoxical sleep and memory are going on.

There were other kinds of arguments in favor of the reprogramming hypothesis. A sleep pattern similar to paradoxical sleep is predominant before birth during brain maturation, when neuronal networks are created. Many scientists from different backgrounds, such as Francis Crick or Gerald Edelman, became interested in paradoxical sleep and proposed their own visions. The whole field is structured by the question of the relationships between genetics and epigenetics. Paradoxical sleep coud play the role of a kind of regulating device acting between subtle brain properties and day-life experience which may affect them: a kind of regulator of the individuation process in the brain.

In spite of many uncertainties which remain in this highly speculative field of biological function or functions (polyfunctionality being a major attribute in biology), the whole sleep research has developed in many directions, including very practical, medical and pharmacological ones – and here the new field of neuroethics comes in. If we would become able to manipulate in a pharmacological way the reprogramming functions, this could have very adverse behavioral effects – and this is certainly something to look after.

Chair of Philosophy of Science
Ecole normale supérieure
45 rue d'Ulm,
F-75005 Paris
France
claude.debru@ens.fr

RICHARD CREATH

(ANTI-)METAPHYSICS IN THE THIRTIES: AND WHY SHOULD ANYONE CARE NOW?

We live in a metaphysical age. And I do not mean just that too many people still believe *The Prophecies of Nostradamus* and/or the horoscopes found in most local newspapers. It is a metaphysical age among philosophers – even among those who shun horoscopes and are frankly embarrassed to find Nostradamus so prominently displayed in the metaphysics section of their campus bookstore. Nowadays, distinguished philosophers in prestigious departments proudly call themselves metaphysicians. They all know, of course, that Carnap and his Viennese friends campaigned in the late twenties and early thirties against metaphysics. These distinguished current philosophers and most of our professional contemporaries find this complete rejection of metaphysics baffling and misguided at best. They cannot imagine what Carnap was doing, and they doubt that it is worth the finding out. When they reflect that Carnap said that metaphysics was utterly without cognitive content, that it actually made no claims at all, they find this so outrageous as to be not to be worth considering.

I have no intention of trying to convince anyone that Carnap was correct in any of this. I am not sure that 'correct' is the word we should be looking for, and I have my own qualms about what Carnap was saying in this period. But I think we can, first, come to understand better than we generally do what Carnap was up to in his campaign against metaphysics in the early thirties. We can begin to see what his sources and motivations were, what he took the issues to be, and what arguments he offered. Again, this is not to convince anyone that he was correct, but instead that there is a genuine philosophic issue that Carnap was addressing and that his response to this issue was not nearly as dismissible as it is usually taken to be. All this will take up the bulk of my remarks here, but in the last part of this paper I want to suggest that the worries that Carnap was addressing ought to worry us still and that Carnap's response to these issues has much to teach us even now.

PRECEDENTS

However much Viennese philosophy between the wars is new, there is nothing new about attacks on metaphysics. Carnap inherits a tradition stretching back over centuries. Many if not most of the figures of the scientific revolution inveighed against scholastic metaphysics. Hume famously urged us at the end of the *Enquiry* to commit volumes of "school metaphysics" to the flames. Such volumes

M.C. Galavotti et al. (eds.), *European Philosophy of Science – Philosophy of Science in Europe and the Viennese Heritage*, Vienna Circle Institute Yearbook 17,
DOI 10.1007/978-3-319-01899-7_5,

could contain nothing but sophistry and illusion. Kant warned us against saying anything substantive about things-in-themselves. Why? Because they were beyond the reach of any possible science. Of course Kant did not oppose what *he* would call metaphysics. But Kant's argument against making substantive claims about things-in-themselves is, in broad strokes, the same argument that Carnap and his Viennese friends used against what *they* called metaphysics. Hume and Kant were enlightenment figures who opposed mere speculation in part because it could make no real progress and in part because such speculations aided and abetted superstitions that were dangerous both to the individual and to the community. As we shall see, Carnap has both of these concerns.

In the 19th century, and well into the 20th, many of what we now regard as empirical sciences broke off from philosophy. This changed the dynamic of discourse in various ways. Scientists (The term was newly coined in the 19th century.) were more willing to express their impatience with the tradition they left behind. Often they spoke of metaphysics in a pejorative way. It is speculative they said, and each speculation can be overturned by the next. There is no way to choose among them so the controversies are endless and wearisome.

Philosophers also reacted to the departure of the empirical sciences – and in various ways. Some emphasized that *philosophy* has pride of place, that philosophy provides the indispensible and substantive foundation for science. In short, science is not science till it has been certified by the philosophers. Often it is said that philosophy in general and metaphysics in particular has its own methods for learning what the world is (really) like and that it uses these methods to reveal truths that are deeper and more important than anything to which empirical science could aspire.[1] One still hears similar claims made on behalf of various branches of the humanities. These ambitious claims for the superiority of philosophy seem to me generally to involve particularly narrow conceptions of science and its methods.

I find this talk of the inherent superiority of philosophy silly, but no doubt it can be and was matched by an equal and opposite silliness that asserts that philosophy is worthless, that it has no role to play in our knowledge of the world.

Throughout the 19th century and into the 20th there was another tradition within philosophy that did *not* try to distance itself from science. This included at least the 19th century positivists, some neo-Kantians, the American pragmatists, and Bertrand Russell – quite conspicuously in some moods. Friedman (2012) and Richardson (1997) have separately traced these developments; I need not do so here. But before we move on I note only that some of these scientific philosophers considered themselves (and were) metaphysicians. This should be enough to remind us that there are many different flavors of metaphysics, and what Carnap attacked under that title might not be what someone else is defending. In fact there is a very considerable distance between Carnap's primary targets and the best rep-

1 For a later example see (Adler, 1941).

resentatives of contemporary analytic metaphysics. Carnap might still have had some reservations about these contemporaries, but they would not be the targets he had primarily in mind.

So far we have seen that there was a very long tradition of attacks on metaphysics. But what often seems to set Carnap apart, especially from 1928 to 1934, is that he not only claimed that metaphysics (as he understood it) was misguided and doomed to failure, he also insisted that the so-called statements of metaphysics said nothing at all. This general idea, the idea that some apparent statements do not really say anything at all, is not entirely new either. Russell's solution to the paradox that bears his name is basically to say that statements that seem to say that a certain class contains itself are not grammatically well formed and so do not say anything. Wittgenstein was to use this general idea extensively in the *Tractatus* where there was a very long list of things that could not be said, quite possibly including most of the *Tractatus* itself. And Wittgenstein is comparatively clear in taking this attitude toward traditional metaphysics as well. (1921/1922, 6.53) Certainly Carnap read him so. Now any claim as to exactly what Wittgenstein meant is bound to be risky, a risk that we do not need to take here. My point in all this, rather, is that Carnap had ample precedents both for rejecting metaphysics and for the device of labeling something as unintelligible, as without cognitive content of any kind, even when that was something that many or most people thought made perfectly good sense.

Motives

In these precedents there are two related motives for wanting to avoid metaphysics and for wanting others to avoid it too. Carnap shared both. One of these motives appears in an almost sociological difference between the scientific and philosophical communities. Like many commentators Carnap saw science as international, cooperative, and progressive. Sometimes theories are overturned but for the most part this change is in favor of clearly better theories. In metaphysical philosophy, Carnap said, each scholar builds up his or her own system. The philosophers work at cross purposes. Their controversies never end or get anywhere. Fashions change, of course, but that is not the same as progress. Avoiding endless controversies is perhaps motive enough. The sociological difference between the communities arises, or so Carnap thinks, from the fact that the theses of metaphysics cannot be rationally justified, i.e., cannot be shown to be more likely to be true than apparently conflicting theses. (Carnap, 1928a/1967, xvii) As a result, there are no rational grounds for preferring one theory to another, grounds that might ultimately result in progress.

The lack of rational justification, often hidden by the obscurity of the language, lies behind another of Carnap's motivations. It is, in its way, political. In the last paragraph of the preface of the *Aufbau* Carnap asserts that there is an inner kinship between his anti-metaphysical orientation and

> movements which strive for meaningful forms of personal and collective life, of education, and of external organization in general … It is an orientation which acknowledges the bonds that tie men together, but at the same time strives for free development of the individual. (Carnap, 1928a/1967, xviii)

The manifesto, *Wissenschaftliche Weltauffassung* (Hahn et al., 1929) makes the political character of Carnap's anti-metaphysical stance even clearer.[2] The obscurity of metaphysical claims hides their lack of justification. And such claims and such obscurities are used in turn to protect and defend religious, economic, and social superstitions. Baldly and polemically put, we are enslaved by ideas that we use without understanding, and clearheaded science could set us free. The tie between clarity and science on one hand and Carnap's longing for "meaningful forms of personal and collective life" on the other could not be closer.

The *Pseudoproblems* Moment

Let us turn then to Carnap's own arguments of the period as a way of getting clearer about just what he was rejecting and why. I will take Carnap's *Pseudoproblems in Philosophy* (1928b/1967) as my primary text, though I am talking about the general view of some other essays on into the early thirties as well, especially *Wissenschaftliche Weltauffassung*, *Pseudoproblems* came out in 1928, the same year as the *Aufbau*. But in his autobiography (1963) Carnap correctly grouped *Pseudoproblems* with the physicalist papers of the early thirties.

The nominal topic of *Pseudoproblems* is the longstanding debate between Realism and Idealism as Carnap understands these two positions. And the issue between these two camps, as Friedman nicely summarizes it, is one of "which entities and processes – physical or psychological, objective or subjective – are ultimately real". (Friedman, 2007, p. 133) Carnap takes both sides seriously. While he ultimately distances himself from both sides, there are no smart-aleck remarks. And unlike his "Overcoming of Metaphysics Through the Logical Analysis of Language" (1932/1959),[3] there is no suggestion here that either Realists or Idealists are merely musicians without musical ability. Note that Carnap's topic is not Nominalism *vs* Platonism, an issue that would be taken up seriously in the 40s. Nor was the topic realism about atoms and molecules, that is, Scientific Realism *vs* Instrumentalism, an issue that Carnap would take up in the 50s.

There is a long list of things that Carnap's argument will take for granted. He takes them for granted not because he thinks that they are obvious but because

2 See especially (Uebel, 2012).

3 This departs slightly from the English title used in the Pap translation.

he thinks that the various parties to the dispute already agree on them. We need to identify these to-be-taken-for-granted ideas so we can bracket them in order to focus instead on what Carnap thought he was adding to the discussion. Of course, *we* might disagree with some of the ideas that Carnap takes as not under dispute between the Realists and Idealists. Fine, but considering those issues would deflect us from seeing what Carnap was trying to do and assessing his line of argument.

So what does Carnap take for granted? First, he assumes the broad structure and viability of the *Aufbau*. That book had conceived of our knowledge (and concepts) as falling within four broad levels: the autopsychological (concerning my inner states), the physical, the heteropsychological (roughly other minds, or better, minds considered in the third person), and the cultural. Each level, save the first, was to be constituted out of the previous ones. *Pseudoproblems* takes this enterprise for granted but narrows its focus to the middle construction, that is, to the construction of the heteropsychological out of the physical. The construction represents a certain logical/epistemological relation between the levels.

Second, Carnap takes all parties to agree that all the evidence we have to go on about other minds is physical. This is about evidence not ontology. It is neither philosophical behaviorism nor physicalism in the modern sense. It is even more liberal than methodological behaviorism. But presumably Realists and Idealists agree that when grading their students they have nothing to go on but the noises, inscriptions, and other behavior of these students. Of course, our actual evidence at any given time is always limited, and there are likely to be alternative hypotheses that are equally good so far. But all these niceties, so bothersome to the philosopher of science, are beside the point in the present context.

Third, Carnap takes the Realists and Idealists to have agreed on scientific methods, on the patterns of scientific inference. These can be as sophisticated, holistic, coherentist, or hypothetico-deductive as you like, but they are not under dispute in the debate between Realism and Idealism. Of course, there *can be* all kinds of methodological differences among scientists of whatever stripe. Carnap is taking for granted that Realists and Idealists will start with the same observations (however differently they interpret them), use the same instruments, get the same outputs from the instruments, and reach the same conclusions at the observational level (again, however differently they may interpret them). Within scientific philosophy one can of course raise the question of what the patterns of rational inference ought to be. But this is logic in a broad sense, and this is not now at issue. The issue at hand does not depend on the details of what those inference patterns are.

Fourth, it is taken for granted that on the basis of this agreed on evidence and these agreed on principles of inference we do in fact know a lot about what others are thinking. Sometimes we know what our students are thinking – and sometimes they should be ashamed of themselves. The point is not that we know everything about other minds, but rather that we do know something via the agreed on scientific means.

Furthermore, both metaphysical parties, the Realists and the Idealists, agree that out beyond this domain of scientific agreement there is a further content, a proposition if you will, about which Realists and Idealists disagree, namely about what is ultimately real. By hypothesis, this is out beyond whatever science has access to. Moreover, the two parties agree that, as metaphysicians, they do have access to this content – though they come down on different sides. It is precisely this claim to go beyond any possible science that Carnap calls metaphysics and precisely this to which he objects. Going beyond science was the whole point of doing metaphysics, the whole basis of its claim to be independent of and superior to the sciences. There is, of course, still the option of scientific philosophy that Carnap would welcome.

The list of assumptions so far does, I think, help to clarify what Carnap takes to be the issue. But it does not settle it. To this we now turn.

The issue on which Carnap focuses concerns the status of the claims of the metaphysician that are agreed to go beyond (or behind or deeper than) what one could get to by legitimate inference from legitimate evidence. The issue is not whether they stand outside the broad domain of science but what to make of them given that they do. Carnap's conclusion in the 1930s was that these supposed claims in fact say nothing at all. Now of course metaphysicians found this offensive, and they still do.

Along the way to that conclusion Carnap tells a revealing fable. Suppose two geographers, Sue a Realist and Sam an Idealist, were to travel to darkest Arizona. (The details of Carnap's story have been changed so that I know what I am talking about.) The object of their quest is to determine whether there are any mountains there higher than 3800 meters above sea level. They find one, Humphreys Peak, use their previously calibrated equipment, and determine that this peak is a little more than 3850 meters above sea level. Both geographers are well trained and know how to use their measuring equipment. They write reports that read identically. So far they are doing science; so far they agree. But when they resume their work as philosophers they add to their respective reports. The Realist adds to her report the sentence 'The mountain in Arizona that we measured and found to be more than 3800 meters tall is also real.' And the Idealist adds to his report the sentence 'The mountain that we measured is not real, for only perceptions and conscious processes are real.' What shall we make of this dispute? The scientific facts are clear enough. But having exhausted all of the scientific facts, there is nothing left that would be relevant, there is nothing left that even could be relevant for addressing the metaphysical issue – if there really is an issue. It is not hard to see why the empirically minded folk who separated their work from philosophy in the 19th century would find such metaphysical wrangling so exasperating. There is no way even to begin to settle the issue. Such disputes are just wearisome.

Carnap's conclusion, of course, is that for precisely this reason there is no issue; each of the added sentences is without any meaning at all. This is a straightforward use of what is nowadays called the verifiability theory of meaningfulness.

The formulation of verifiability that Carnap gives here is fairly liberal requiring only that the evidence support the claim rather than completely verify it. Verifiability was to be given a large number of different formulations over the years, and the successive counterexamples and reformulations became a fairly large cottage industry.

I have no intention of arguing for the correctness of Carnap's then current version of the verifiability principle. Even Carnap did not try to prove in *Pseudoproblems* or anywhere else that the principle is correct. What he did do was to challenge his interlocutor to come up with a better criterion of meaningfulness and not just the pious hope that one would be forthcoming. Everyone admits that some noises do not make meaningful claims, and Carnap, like many logicians, was committed to the grammatical gambit of Russell's type theory. Carnap was not unbreakably wedded to the formulation in *Pseudoproblems*, for he was to refine it considerably over the years.

There are, however, two elements of that principle, that are worth defending. Naturally, that is more than I can do in a few minutes. In any case, the first of these defense worthy elements is its broad empiricism. This says that observational evidence is all the evidence there is. When it comes to an independent world, making any other unsupported claims would just be guessing.

Another element of verifiability that is also worth defending is the idea of linking meaning (including meaningfulness) to an epistemological structure. I think it is wise (I do not say correct) to take the meaning of an expression to be its role with respect to an epistemic structure. If two expressions play the same role, they are synonymous. If an expression plays no epistemic role, it has no meaning. If its role is wholly unconnected to empirical evidence, then it has no empirical meaning. Now all of this taken together is only a strategy and not a detailed account. But it is appealing as a first approximation. The general idea in one form or another has been central to pragmatism from Peirce to Sellars and Quine, however much all three may have considered themselves metaphysical realists. Certainly, Quine was perfectly happy to call himself a verificationist throughout his career.[4] The idea of linking meaning and epistemology survives as well in contemporary inferential role semantics. And some form of it might be essential to any responsible talk of meaning or meaningfulness at all, that is to say, to any attempt to answer Quine's challenge to any serious theory of meaning whatever.

Having said that elements of the verifiability strategy are worth defending, I am happy to concede that apart from the engaging story of the two geographers and the challenge for others to come up with a better criterion there is a certain baldness to Carnap's empiricism and to the verifiability principle. They have struck more than one reader as, well, dogmatic. This makes them somewhat unstable in the sense that what is dogmatically proposed can be rejected equally dogmatically. This is a point that Carnap himself quickly appreciated. By late 1932 Carnap

4 See for example, (Quine, 1986, pp. 154-5).

adopted a Principle of Tolerance. This removes the appearance of dogmatism and stabilizes both the empiricism and the approach to meaning, though I cannot go into detail here.

Even without appealing to the Principle of Tolerance, one can state Dr. Carnap's cure for metaphysics in a single word: 'clarity'.[5] One must demand clarity of ourselves and encourage it in others. What this clarity consists in is explicitness about what is to count as evidence and what sentences are to follow from what sets of other sentences. This would give us the structure of an epistemology. The hope is that when all this is explicit, obscurity has no place left to hide.

Now

All of this discussion of the late twenties and early thirties may now seem archaic to the point of irrelevance. Carnap and the figures with whom he contended are all long since dead. Now few if any people talk about Idealism (in the philosophic sense). Now we all have iPhones, and we are science friendly. Right? Why worry about metaphysics now?

Well, at least Carnap's political motivation is still with us. The world is still full of superstitions and superstitious people. We still have a superfluity of horoscope readers, religious and ethnic bigots, evolution and climate change deniers, radio talk show hosts, and, in the U.S., candidates for high public office. A certain amount of belligerent ignorance is to be expected, but, alas, it does not seem to be diminishing.

I noted earlier that there are now many prestigious analytic metaphysicians, at least in the best American departments. But I also noted that these people were not at all like the main targets of Carnap's campaign. They try to be clear, rigorous, and scientific. The best of them are impressively smart and accomplished logicians. This can also be said of distinguished representatives of continental traditions. It is interesting in this connection that in the field of meta-ontology there is renewed interest in Carnap's ideas and even in the idea that some apparent ontological disagreements might be disagreements at all. No wonder then that contemporary analytic metaphysics does not see itself in Carnap's portrait. To a certain extent this defense of contemporary analytic metaphysics succeeds. Carnap is rejecting metaphysics in one sense, and our best contemporaries exemplify it in a different sense.

But in shedding the verifiability principle – and it has largely been dropped by our contemporaries – many philosophers shed as well the sense that they needed to articulate a detailed epistemology and to place their own views within it. But as

5 It could be argued that his cure for metaphysics would require a healthy empiricism as well. Perhaps, but I suspect that he would call clarity the primary requirement and that he expected that to bring empiricism along with it.

welcome as the articulation of an epistemology would be, it is not by itself enough. Beyond that there is another problem that is still more pervasive. When Carnap moved to his mature philosophy in the mid-thirties, he realized that the general problem he had been working was not exclusively between the good-guy scientists and the bad-guy metaphysicians. It also infects disputes among *us*, that is, among all philosophers, however scientific, and among different varieties of empiricists – where those disputes are taken to be disagreements over matters of independent fact. Even when analytic philosophers put forward a logic or an ontology or even a philosophy of science as the uniquely correct one, Carnap says that we begin to engage in a wearisome controversy from which there is no way out. We have, in short, fallen into an unwise metaphysics. Here – and now – the cure is still the clarity provided by explication and the tolerance to let others speak differently than we do.

In this Carnap is offering more than a negative commentary on the old-style metaphysics of the thirties or on the metaphysics of our own time. He shows us a positive conception of what philosophy can be, of how to push the philosophic enterprise forward in a way that is both scientifically useful and conducive to "meaningful forms of personal and collective life". A full exploration of this, however, is a story for another day.

Bibliography

Adler, Mortimer J., 1941, "God and the Professors", Science, Philosophy and Religion: A Symposium, (New York: Science, Philosophy and Religion in Their Relation to the Democratic Way of Life, Inc.), 120-138.

Carnap, Rudolf, 1928a/1967, Der logische Aufbau der Welt, (Berlin-Schlachtensee: Weltkreis-Verlag); translated as The Logical Structure of the World by Rolf A. George, (Berkeley: University of California Press).

——, 1928b/1967, Scheinprobleme in der Philosophie: Das Fremdpsyche und der Realismusstreit, (Berlin-Schlachtensee: Weltkreis-Verlag); translated as Pseudoproblems in Philosophy by Rolf A. George, (Berkeley: University of California Press).

——, 1932/1959, "Überwindung der Metaphysik durch logische Analyse der Sprache", Erkenntnis, 2: 219-241; translated as "The Elimination of Metaphysics Through Logical Analysis of Language" by Arthur Pap, in Logical Positivism, A. J. Ayer, ed., (New York: Free Press), 60-81.

——. 1963, "Intellectual Autobiography", in The Philosophy of Rudolf Carnap, Paul A. Schilpp, ed., (La Salle, IL: Open Court Publishing), 1-84.

Friedman, Michael, 2007, "The Aufbau and the Rejection of Metaphysics", in The Cambridge Companion to Carnap, Michael Friedman and Richard Creath, eds., (Cambridge: University of Cambridge Press), 129-152.

——. 2012, “Scientific Philosophy from Helmholtz to Carnap and Quine”, in Rudolf Carnap and the Legacy of Logical Empiricism, Richard Creath, ed., (Vienna: Springer), 9-19.

Hahn, Hans, Rudolf Carnap, and Otto Neurath, 1929, Wissenschaftliche Weltauffassung, (Vienna: Artur Wolf Verlag).

Quine, W.V.O., 1986, “Reply to Roger F. Gibson, Jr.”, in The Philosophy of W.V. Quine, (La Salle, IL: Open Court Publishing), 155-57.

Richardson, Alan, 1997, “Toward a History of Scientific Philosophy”, in Perspectives on Science, 418 – 451.

Uebel, Thomas, 2012, “Carnap, Philosophy, and ‘Politics in its Broadest Sense’”, in Rudolf Carnap and the Legacy of Logical Empiricism, Richard Creath, ed., (Vienna: Springer), 141-146.

Wittgenstein, Ludwig, 1921/1922, “Logisch-Philosophische Abhandlung”, Annalen der Naturphilosophie, 14: 185-262; translated as Tractatus Logico-Philosophicus by C. K. Ogden, (London: Routledge & Kegan Paul).

School of Life Sciences 874501
Arizona State University
Tempe, AZ 85287-4501
USA
creath@asu.edu

Maria Carla Galavotti

Probabilistic Epistemology: A European Tradition

Abstract

Probabilistic epistemology holds that probability is an essential ingredient of science and human knowledge at large, and that induction is a necessary constituent of the scientific method. Developed in some detail by a number of authors including Patrick Suppes, Richard Jeffrey and Brian Skyrms, this view has been embraced by so many, as to gradually become predominant. While probabilistic epistemology has been growing, awareness of its origins was somehow left behind. Probabilistic epistemology is usually seen as a product of the encounter of logical empiricism with American pragmatism. Without denying the impact of such an encounter, it is argued that the probabilistic approach was already part of the European scenario before the dissolution of logical empiricism in the late 1930s. Notably, such an approach is not linked to a particular interpretation of probability, having been endorsed by upholders of different probability notions, including Janina Hosiasson, Frank Plumpton Ramsey, Bruno de Finetti, Harold Jeffreys, and Hans Reichenbach, to mention only those authors whose contribution is addressed here. The list could extend to the French probabilists Émile Borel and Maurice Fréchet, and others not considered in this survey. The work of so many authors operating at the four corners of Europe in the first decades of the 19th century reflects a European tradition in probabilistic epistemology, somewhat overshadowed in the literature by the attention paid to logical empiricism.

1. About probabilistic epistemology

In broad terms, probabilistic epistemology can be taken as the view that *probability* is an essential ingredient of science and human knowledge at large, and that *induction* is the building block of the scientific method. In Patrick Suppes' words: "it is probabilistic rather than merely logical concepts that provide a rich enough framework to justify both our ordinary ways of thinking about the world and our scientific methods of investigation".[1] This attitude is rooted in the conviction that the ideals of truth and certainty are "chimeras" that should be replaced by the recognition that knowledge is imbued with *uncertainty*. Borrowing another passage from Suppes: "certainty of knowledge – either in the sense of psychological immediacy, in the sense of logical truth, or in the sense of complete precision of

1 Patrick Suppes, *Probabilistic Metaphysics*. Oxford: Blackwell 1984, p. 2.

M.C. Galavotti et al. (eds.), *European Philosophy of Science – Philosophy of Science in Europe and the Viennese Heritage*, Vienna Circle Institute Yearbook 17,
DOI 10.1007/978-3-319-01899-7_6,

measurement – is unachievable. It is the responsibility of a thoroughly-worked-out empiricism to include an appropriate concept of uncertainty at the most fundamental level of theoretical and methodological analysis. Probabilistic methods provide a natural way of doing so".[2]

The conviction that probability enters all stages of a comprehensive analysis of scientific knowledge goes hand in hand with a *constructivist* approach, according to which such an analysis should start from experimental methodology and move upwards to the formation of hypotheses, models and theories. This implies that the *context of discovery* be regarded as an object of investigation no less important than the *context of justification*, which was considered by logical empiricists the sole concern of philosophy of science. Also in this connection Suppes takes a clear-cut position, claiming that measurement and statistical techniques "for constructing and abstracting empirical structures in all domains of science"[3] are essential ingredients of the representation of scientific theories. A similar attitude involves a shift in emphasis from laws to models and notions like invariance, symmetry and stability. Moreover, the notion of *context* becomes crucial within this approach.

Probabilistic epistemology emphasizes *prediction* as the main task of science, ascribing to *human action* a crucial role within the characterization of important categories of knowledge. Moreover, *success* is taken as the canon for the justification of the inductive method. Another ingredient of probabilistic epistemology is a *dynamic view of knowledge*, which often goes hand in hand with a *pluralistic notion of rationality*. As Richard Jeffrey put it: "standards of rationality are cultural artifacts, bodies of practice resulting from the concurrence of congeries of factors".[4]

In the first decades of the 20th century a probabilistic approach to epistemology was shared by a number of philosophers and scientists operating in distant parts of Europe. Their work points to the existence of a philosophical trend that might be labelled "European pragmatism" in view of its pragmatist flavour. The following sections contain a sketchy recollection of the ideas put forward by some of the main upholders of a probabilistic approach to epistemology.

2. Janina Hosiasson (1899–1942)

The Polish philosopher Janina Hosiasson, now almost forgotten, was deemed by Richard Jeffrey "one of probabilism's clearest early voices".[5] Hosiasson graduated

2 *Ibid.*, p. 99.

3 Patrick Suppes, "Empirical Structures", in: Eberhard Scheibe (Ed.), *The Role of Experience in Science*. Berlin-New York: Walter de Gruyter 1988, p. 23.

4 Richard Jeffrey, *Probability and the Art of Judgment*. Cambridge: Cambridge University Press 1992, p. 11.

5 Richard Jeffrey, *Depois do Empirismo Logico. After Logical Empiricism*. Lisbon: Ed-

in Warsaw in 1926 with a dissertation on the "Justification of Inductive Reasoning", and was very active in the international arena in the years before 1942, when she was executed for being a Jew during the Nazi occupation of Poland. Hosiasson was a representative of the Lvov-Warsaw School, which took shape at the end of the nineteenth century around the person of Kazimierz Twardowski. Among the representatives of the School who exercised the strongest influence on Hosiasson were her tutor Tadeusz Kotarbiński and Stefan Mazurkiewicz, who published an axiomatization of probability in 1932, a year before Kolmogorov's seminal work.[6] The main focus of Hosiasson's writings are the foundational problems related to probability, induction and confirmation. As portrayed by Ilkka Niiniluoto, she was "the first and only member of the Lvov-Warsaw School who was a professional confirmation theorist by international standards".[7]

Following Mazurkiewicz, and also John Maynard Keynes' *Treatise on Probability*, Hosiasson embraces an epistemic notion of probability, but her viewpoint is closer to subjectivism than to logicism. She defines probability as *justified belief*, and claims that knowledge can at best aim at justified belief, while absolute truth and completeness of knowledge are unreachable. Moreover, she takes probability as a basis for decision and action, and claims that mathematical expectation is its major assumption. In her article "Why do we Prefer Probabilities Relative to Many Data?" Hosiasson addresses the question: "How do we account of probabilities in particular cases?"; her answer being that "in a considerable number of cases in ordinary life we take account of them by considering the amount of something which could be said to be a mathematical expectation".[8] Under Mazurkiewicz's influence Hosiasson takes an axiomatic approach, and assumes mathematical expectation as an axiom, rather than a principle of psychology, as Ramsey did. By contrast, she shares Ramsey's pragmatist stand in claiming that "taking gains or mathematical expectations into consideration, could be considered as an epistemological answer only from a pragmatist point of view".[9]

Hosiasson also addresses the problem of weight, already raised by Keynes. In an attempt to answer the question whether given two probabilities of equal degree one should prefer the course of action supported by a greater body of knowledge, she maintains that the acquisition of new evidence pays in expectation. As argued

---

ições Colibri 2002, p. 2.

6 See Stefan Mazurkiewicz, "Zur Axiomatik der Wahrscheinlichkeitsrechnung", in: *Comptes rendues des séances de la Société des Sciences et des Lettres de Varsovie, Classe III*, 25, 1932, pp. 1-4.

7 Ilkka Niiniluoto, "Induction and Probability in the Lvov-Warsaw School", in: Katarzyna Kijania-Placek and Jan Woleński (Eds.), *The Lvov-Warsaw School and Contemporary Philosophy*. Dordrecht–Boston: Kluwer 1998, p. 330.

8 Janina Hosiasson Lindenbaum, "Why do we Prefer Probabilities Relative to Many Data?", in: *Mind* 40, 1931, p. 30.

9 *Ibid.*, p. 36.

in the next section, this result was actually proved by Ramsey, anticipating subsequent literature by several decades.

Notably, in a footnote contained in the same article, Hosiasson writes: "I am greatly indebted [...] to an unpublished paper by Mr. F. P. Ramsey on "Truth and Probability" which the kindness of Mr. Braithwaite has enabled me to read. I had, however, previously thought independently on similar lines".[10] This is a remarkable claim and evidence that Hosiasson was one of the first to think of probability in subjectivist terms. As reported by Daniela Gromska in a biographical sketch,[11] Hosiasson spent some time in Cambridge on a fellowship around 1929–1930. There, most certainly she interacted with a number of renowned scholars including George Edward Moore, John Maynard Keynes, William Ernest Johnson, Charlie Dunbar Broad and Harold Jeffreys. Most likely, Ramsey untimely death in January 1930 prevented the two from meeting.

Hosiasson's article "On Confirmation" outlines a probabilistic solution to the raven paradox based on the possibility of discriminating between instances of the paradoxical and non-paradoxical kind, the idea being that a non-paradoxical instance of a certain hypothesis increases its prior probability to a greater degree than a paradoxical instance.[12] In *Studies in the Logic of Confirmation* Hempel shows appreciation of Hosiasson's discussion, albeit his aim is to define confirmation in qualitative, non-probabilistic, terms.[13] By contrast, in 1966 Suppes proposed a solution to the paradox along Bayesian lines claiming to have borrowed the leading idea from Hosiasson.[14]

Hosiasson's perspective includes a view of knowledge as "an aggregate of opinions justified, actual and connected – in an adequate degree".[15] When reading this claim, one is immediately reminded of Ramsey's view of knowledge as belief formed in a "reliable way", and indeed also in this connection Hosiasson's perspective bears some resemblance to Ramsey's. Though she explicitly does not address the collective character of knowledge, Hosiasson seems to share Ramsey's

10 *Ibid.*, p. 30.

11 See Hosiasson's biographical sketch in Daniela Gromska, "Philosophes Polonais morts entre 1938 et 1945", in: *Studia Philosophica* 3, 1948, pp. 31-97 (especially pages 57-59).

12 See Janina Hosiasson Lindenbaum, "On Confirmation", in: *Journal of Symbolic Logic* 5, 1940, pp. 133-148.

13 See Carl Gustav Hempel, "Studies in the Logic of Confirmation", in: *Mind* 54, 1945, pp. 1-26 and 97-121. Reprinted with changes in *Aspects of Scientific Explanation and Other Essays in the Philosophy of Science*. New York: The Free Press 1965, pp. 3-51.

14 See Patrick Suppes, "A Bayesian Approach to the Paradoxes of Confirmation", in: Jaakko Hintikka and Patrick Suppes (Eds.), *Aspects of Inductive Logic*. Amsterdam: Noth-Holland 1966, pp. 198-208.

15 Janina Hosiasson Lindenbaum, "Theoretical Aspects of the Advancement of Knowledge", in: *Synthèse* 7, 1948, p. 253.

conviction that knowledge is belief of a special sort, namely belief that is not just entertained by individuals, but is apt to be shared by the community of men.[16]

3. Frank Plumpton Ramsey (1903–1930)

Unanimously considered the founder of modern subjectivism, Ramsey had a genuinely probabilistic and pragmatist view of knowledge rooted in the conviction that people hardly ever entertain a belief with certainty, and usually act under uncertainty. Probability is regarded by Ramsey as degree of belief and defined with reference to personal preferences assumed to obey the principle of mathematical expectation, which is taken as a "law of psychology".[17] After giving probability a psychological foundation, Ramsey specifies a set of axioms fixing a criterion of *coherence*,[18] adding that coherent degrees of belief satisfy the laws of probability. Coherence is thereby the only condition of acceptability that needs to be imposed on degrees of belief, and the overall criterion of rationality of belief and decision. The crucial link between probability and degree of belief provided by coherence is the cornerstone of subjective probability. In fact, consistency guarantees the applicability of the notion of degree of belief, which can therefore qualify as an admissible interpretation of probability. By showing that from the assumption of coherence one can derive the laws of probability Ramsey paves the way to a fully-fledged subjectivism.

Almost in the same years in which Ramsey developed his view of probability, a further step was made by Bruno de Finetti, who supplied the "static" definition of subjective probability in terms of coherent degrees of belief with a "dynamic"[19] dimension, obtained by joining subjective probability with exchangeability within the framework of the Bayesian method. Though this crucial step was actually made by de Finetti, there is evidence that Ramsey's probabilistic epistemology included a dynamic view of knowledge. This is testified by the already mentioned result contained in his note "Weight or the Value of Knowledge", where Ramsey is able to prove that collecting evidence pays in expectation, provided that acquiring the new information is free, and shows how much the increase in weight is. As pointed

16 For more on Hosiasson's contribution see Maria Carla Galavotti, "A Tribute to Janina Hosiasson Lindenbaum: a Philosopher Victim of the Holocaust", in: Roberto Scazzieri and Raffaella Simili (Eds.), *Migration of Ideas*. Sagamore Beach: Watson 2008, pp.179-194.

17 Frank Plumpton Ramsey, "Truth and Probability", in: *Philosophical Papers*, Hugh Mellor (Ed.). Cambridge: Cambridge University Press 1990, p. 70.

18 Ramsey uses the term "consistency".

19 This terminology is borrowed from Sandy Zabell, "Ramsey, Truth and Probability", in: *Theoria* 57, 1991, pp. 210-38. For a comparison between Ramsey and de Finetti on subjective probability, see Maria Carla Galavotti, "The Notion of Subjective Probability in the Work of Ramsey and de Finetti", in: *Theoria* 57, 1991, pp. 239-59.

out by Nils-Eric Sahlin and Brian Skyrms, Ramsey's note on weight anticipates subsequent work by Savage, Good, and others.[20] We also know from Ramsey's note "Rule of Succession"[21] that he was intrigued by the notion of "equiprobability of all permutations", namely the property of exchangeability, about which he learned from William Ernest Johnson, the Cambridge logician who should be accredited for having grasped the significance of this important notion before de Finetti's systematic work.[22]

In his note "Knowledge" Ramsey embraces a pragmatist conception of knowledge as belief "obtained by a reliable process".[23] The process by which knowledge is obtained rests on induction, and the criterion for the reliability of inductive inferences is identified with their success. As Ramsey states: "this is a kind of pragmatism: we judge mental habits by whether they work".[24] Induction is regarded as "a useful habit" which it is reasonable to adopt because mankind would be helpless without it.

In the same vein, Ramsey accounts for the truth of theories in pragmatist terms. In this connection he holds the view, whose paternity is usually attributed to Charles Sanders Peirce, that theories which gain "universal assent" in the long run are accepted by the scientific community and taken as true. Along similar lines he characterizes a "true scientific system" with reference to a system to which everyone's opinion, grounded on experimental evidence, will eventually converge. According to this pragmatically oriented view, chance attributions, like all general propositions belonging to theories – including causal laws – are not to be taken as propositions, but rather as "variable hypotheticals", or "rules for judging",[25] apt to provide the user with a tool to meet the future.

---

20 See the commentary in the "Preamble" by Nils-Eric Sahlin to Frank Plumpton Ramsey, "Weight of the Value of Knowledge", in: *The British Journal for the Philosophy of Science* 41, 1990, pp. 1-3. For a discussion of Ramsey's result see also Brian Skyrms, "Discovering 'Weight, or the Value of Knowledge'", in: Maria Carla Galavotti (Ed.), *Cambridge and Vienna. Frank P. Ramsey and the Vienna Circle*. Dordrecht: Springer 2006, pp. 55-65.

21 In Frank Plumpton Ramsey, *Notes on Philosophy, Probability and Mathematics*, Maria Carla Galavotti (Ed.), Naples: Bibliopolis, 1991, pp. 279-281.

22 For a discussion of Johnson's "Permutation postulate" and its relevance for Bayesian statistics see Sandy Zabell, "W.E. Johnson's 'Sufficientness' Postulate", in: *The Annals of Statistics* 10, 1982, pp. 1091-1099.

23 In Frank Plumpton Ramsey, *Philosophical Papers*, *cit.*, p. 110.

24 *Ibid.*, p. 93.

25 See especially Ramsey's note "General Propositions and Causality", in: *Philosophical Papers, cit.*, pp. 145-163. For more on Ramsey's probabilistic epistemology see Maria Carla Galavotti, *Philosophical Introduction to Probability*. Stanford: CSLI 2005, which also contains a detailed exposition of the views of the authors whose work is discussed in the following sections.

4. Bruno de Finetti (1906–1985)

Starting from the late 1920s, Bruno de Finetti developed a radical form of probabilism that can be described as a blend of pragmatism and the kind of empiricism that is today called anti-realism. Authors like Ernst Mach, Henri Poincaré, and the Italian pragmatist Giovanni Vailati exercised a strong influence on him. After rejecting the notions of truth, determinism and "immutable and necessary" laws, de Finetti reaffirms a conception of science as a product of human activity ruled by probability. The fundamental conviction inspiring de Finetti's perspective is well-described by the following passage, taken from his paper *Probabilismo*, which he considered his philosophical manifesto:

> no science will permit us to say: this fact will come about, it will be thus and so because it follows from a certain law, and that law is an absolute truth. Still less will it lead us to conclude skeptically: the absolute truth does not exist, and so this fact might or might not come about, it may go like this or in a totally different way [...] What we can say is this: I foresee that such a fact will come about, and that it will happen in such and such a way, because past experience and its scientific elaboration by human thought make this forecast seem reasonable to me.[26]

As suggested by the title of his well-known article "La prévision, ses lois logiques, ses sources subjectives",[27] for de Finetti prediction is the main purpose of both science and everyday life. Given that prediction is made possible by probability, and that it is always referred to a subject, being the product of his experience and convictions, the conceptual tool we need is the subjective theory of probability. Accordingly, probability is taken as degree of belief subject to the only constraint of coherence. Jeffrey deemed de Finetti's subjectivism "anti-rationalist" in the sense that "probability does not correspond to a self- proclaimed 'rational' belief, but to the effective personal belief of anyone".[28] Insofar as coherence is satisfied, disagreement among different evaluations of probability based on the same amount of information is admitted.

In connection with the operative definition of probability, de Finetti takes a pluralistic attitude which admits that it can be produced in various ways. As an

26 Bruno de Finetti, "Probabilism", in: *Erkenntnis* 31, 1989, p. 170 (English edition of "Probabilismo", in: *Logos*, 1931, pp. 163-219).

27 Bruno de Finetti, "La prévision: ses lois logiques, ses sources subjectives" in: *Annales de l'Institut Henri Poincaré* 7, 1937, pp. 1-68. English edition "Foresight: its Logical Laws, its Subjective Sources", in: Henry Kyburg, jr. and Howard Smokler (Eds.), *Studies in Subjective Probability*. New York–London–Sydney: Wiley 1964, pp. 95-158; second modified edition Huntington (N.Y.): Krieger 1980, pp. 53-118.

28 Bruno de Finetti, "Recent Suggestions for the Reconciliation of Theories of Probability", in: Jerzy Neyman (Ed.), *Proceedings of the Second Berkeley Symposium on Mathematical Statistics and Probability*. Berkeley: University of California Press 1951, p. 218.

alternative to the betting scheme he developed a definition in terms of the qualitative relation "at least as probable as", while in the last period of his production he preferred to adopt scoring rules based on penalties.[29]

De Finetti's most important achievement is the result known as the "representation theorem" (although he did not like this terminology). In a nutshell, this shows that the adoption of Bayes's method, taken in conjunction with the property of exchangeability, leads to a convergence between degrees of belief and frequencies. This makes subjective probability applicable to statistical inference, which according to de Finetti can be entirely based on it – a conviction shared by the neo-Bayesian statisticians.

For the subjectivist de Finetti probability is always definite and known; there are no unknown, or true, "objective" probabilities. While discarding objective probability as a metaphysical concept, de Finetti takes seriously the problem of the objectivity of probability evaluations. In this connection, he insists that one should not conflate the *definition* of probability with its *evaluation*. De Finetti defines probability as subjective *degree of belief*, whereas the evaluation of probability depends on *two ingredients*: "the objective component, consisting of the evidence of known data and facts; and (2) the subjective component, consisting of the opinion concerning unknown facts based on known evidence".[30] The subjective component can include myriad elements like expertise, personal abilities, past experience, and so on. Obviously, in some contexts the objective component will be preponderant and the subjective elements will play little or no role at all, while in other contexts subjective elements will have a greater influence on probability evaluations, especially when only scanty information is available.

5. Harold Jeffreys (1891–1989)

Reputedly one of the last century's most prominent geophysicists and a pioneer of the study of the Earth, Jeffreys also left a mark in other fields, like seismology and meteorology, and, last but not least, probability. He also developed a deeply probabilistic epistemology, inspired by the conviction that "probability is the most fundamental and general guiding principle of the whole of science".[31] For Jeffreys science is "a branch of the subject-matter of probability",[32] in fact probability has

29 See Bruno de Finetti, *Teoria delle probabilità*. Torino: Einaudi 1970. English edition *Theory of Probability*. New York: Wiley 1975.

30 Bruno de Finetti, "The Value of Studying Subjective Evaluations of Probability", in: Carl-Axel Staël von Holstein (Ed.), *The Concept of Probability in Psychological Experiments*. Dordrecht-Boston: Reidel 1974, p. 7.

31 Harold Jeffreys, *Scientific Inference*. Cambridge: Cambridge University Press 1931, p. 7. Reprinted with "Addenda" 1937. (2nd edition 1957, 3rd edition 1973).

32 *Ibid.*

a larger scope than science because the process of acquiring knowledge is permeated with probability since its first steps.

Jeffreys regards induction as the cornerstone of the scientific method. In this spirit he rejects all attempts to reduce induction to deduction, including the attitude of orthodox statisticians who "appear to regard observation as a basis for possibly rejecting hypotheses, but in no case for supporting them".[33] Jeffreys's inductivism revolves around Bayes's theorem, of which he claims that it "is to the theory of probability what Pythagoras' theorem is to geometry".[34]

As to the definition of probability Jeffreys is a logicist, although in certain respects his position is closer to subjectivism. Unlike subjectivists Jeffreys believes that "on a given set of data p we say that a proposition q has in relation to these data one and only one probability. If any person assigns a different probability, he is simply wrong".[35] By contrast, like de Finetti he refuses the notion of "unknown probability" and criticizes John Maynard Keynes for holding that not all probabilities are known.

A most interesting aspect of Jeffreys's probabilism is a constructive attitude according to which the notions of *objectivity*, *reality* and *causality* should be defined on probabilistic grounds. This process, which originates from our sensations, is inductive and requires the use of probability. Empirical laws are established in a similar way; a general proposition can become a law only after the rules of induction "have compared it with experience and attached a high probability to it as a result of that comparison". In this procedure lies "the only scientifically useful meaning of 'objectivity'".[36] In other words, the objectivity of laws is made to depend on scientific methodology, and the same holds for the related notion of "chance", namely the kind of probability that occurs in those situations in which "given certain parameters, the probability of an event is the same at every trial, no matter what may have happened at previous trials".[37]

Similarly, the notions of reality and causality are built on experience through statistical method. A useful notion of reality obtains when some scientific hypotheses receive from the data a probability which is so high, that on their basis one can draw inferences whose probabilities are practically the same as if the hypotheses in question were certain. Hypotheses of this kind are taken as certain in the sense that all their parameters "acquire a permanent status". In such cases, we can assert the associations expressed by the hypotheses in question "as an approximate rule".[38] Finally, causal analysis is based on statistical testing. It starts by consider-

---

33 Harold Jeffreys, *Theory of Probability*. Oxford, Clarendon Press 1939, p. 5. (2nd edition 1948, 1961, 1983).

34 Harold Jeffreys, *Scientific Inference*, *cit*., p. 7.

35 *Ibid*, p. 10.

36 Harold Jeffreys, *Theory of Probability*, *cit.*, p. 336.

37 Harold Jeffreys, *Scientific Inference*, 2nd edition, *cit*., p. 46.

38 Harold Jeffreys, "Scientific Method, Causality, and Reality", in: *Proceedings of the Aristotelian Society*, New Series, 37, 1937, p. 69.

ing all variation to be random and proceeds to detect correlations which are the more precise, the better their agreement with observations. This procedure leads to asserting laws, which are eventually accepted because "the agreement (with observations) is too good to be accidental".[39] Granted that the fundamental problem of scientific method is finding a good agreement between our hypotheses and experience, Jeffreys's judgment is that in solving this problem "causality does not help; but it turns out that the theory of probability does".[40]

His constructivist and pragmatist attitude, together with his conviction that science is fallible and that there is a continuity between science and everyday life, place Jeffreys among the representatives of probabilistic epistemology. Besides, Jeffreys admission that empirical information can be "vague and half-forgotten", a fact that "has possibly led to more trouble than has received explicit mention",[41] anticipates recent literature on probability kinematics, which contemplates the application of Bayesian inference to uncertain information.[42]

6. Hans Reichenbach (1891–1953)

A member of the "Berlin Society for empirical philosophy", Reichenbach was one of the first to criticize the verifiability theory of meaning embraced by logical empiricists, in the conviction that it is probability, not truth, that allows for a reconstruction of scientific knowledge in tune with scientific practice. In Reichenbach's words: "the ideal of an absolute truth is an unrealizable phantom".[43]

Reichenbach identifies prediction as the task of scientific knowledge and claims that "the theory of knowledge is a theory of prediction".[44] Since statements about the future are uncertain, probability is an essential ingredient of knowledge and prediction, and so is the inductive method, which is inextricably entwined with probability. In fact, the theory of probability is regarded as a "theory of propositions about the future".[45]

39 *Ibid.*, p. 62.

40 *Ibid.*, p. 78. For more on the epistemology of Jeffreys see Maria Carla Galavotti, "Harold Jeffreys' Epistemology between Logicism and Subjectivism", in: *The British Journal for the Philosophy of Science* 54, 2003, pp. 43-57.

41 Harold Jeffreys, *Scientific Inference*, 3rd edition 1973, *cit.*, p. 406.

42 The literature on this topic includes contributions by Jeffrey, Skyrms, van Fraassen, Lewis, and many others. See, for instance, Richard Jeffrey, *Probability and the Art of Judgment*, *cit.*, and Brian Skyrms, *The Dynamics of Rational Deliberation*. Cambridge, Mass.: Harvard University Press 1990.

43 Hans Reichenbach, "La philosophie scientifique: une esquisse de ses traits principaux", in: *Travaux du IX Congrès International de Philosophie*, Paris: Hermann 1937, p. 90.

44 *Ibid.*, p. 89.

45 Hans Reichenbach, "Logistic Empiricism in Germany and the Present State of its Problems." *The Journal of Philosophy* 6, 1936, p. 159.

Reichenbach's approach to probability revolves around the frequency theory, grounded on the principle that degrees of probability can never be ascertained *a priori*, but only *a posteriori*. The method by which degrees of probability are attained is "induction by enumeration". This "is based on counting the relative frequency [of a certain attribute] in an initial section of the sequence, and consists in the inference that the relative frequency observed will persist approximately for the rest of the sequence; or, in other words, that the observed value represents, within certain limits of exactness, the value of the limit for the whole sequence".[46] Such a procedure is reflected by the "Rule of induction", namely the frequentist canon for assessing probabilities as limiting values of observed frequencies. Induction by enumeration is the cornerstone of the *method of concatenated inductions*, whose fundamental feature is that of being *self-corrective*. This method is the core of Reichenbach's probabilistic epistemology, characterized by a dynamic view of knowledge as a continuous interplay between experience and prediction, as echoed by the title of one of his major works.[47]

An important feature of Reichenbach's perspective is his Bayesianism, which qualifies as strictly objective because it requires that priors be determined on the basis of frequencies alone.

Reichenbach described his own perspective "as a further development of ideas which originated in pragmatism",[48] and in fact his epistemology represents a unique blend of empiricism and pragmatism, originating from an unshakable trust in empiricism combined with the deeply held conviction that our knowledge is uncertain and relies on induction. In addition to the emphasis placed on prediction, a key feature of Reichenbach's pragmatism is the importance ascribed to action. In the first place, action plays an essential role within Reichenbach's theory of meaning, which revolves around the idea that "there is as much meaning in a proposition as can be utilized for action".[49] Moreover, action enters the justification of induction, which is based on the success of the inductive method. Since probability statements serve the purpose of guiding decisions leading to actions, they will be justified if it can be shown that they are the best possible guide to action. It turns out that the frequency interpretation is amenable to this kind of pragmatic justification because it satisfies what Reichenbach calls the "principle of the greatest number of successes", namely it leads us to act in the most successful possible

46 Hans Reichenbach, "The Logical Foundations of the Concept of Probability", in: Herbert Feigl and Wilfrid Sellars (Eds.), *Readings in Philosophical Analysis*. New York: Appleton-Century-Crofts, 1949, p. 351 (English modified edition of "Die logischen Grundlagen des Wahrscheinlichkeitsbegriffs", *Erkenntnis* 3, 1933, pp. 401-425).

47 Hans Reichenbach, *Experience and Prediction*. Chicago- London: The University of Chicago Press 1938. For more on Reichenbach's probabilistic epistemology see Maria Carla Galavotti, "On Hans Reichenbach's Inductivism", in: *Synthèse* 181, 2011, pp. 95-111.

48 Hans Reichenbach, *Experience and Prediction, cit.*, p. 69.

49 *Ibid.*, p. 80.

way. Induction is therefore justified on pragmatic grounds, in view of the attainment of what is taken to be its end, namely the formulation of good predictions.

7. Conclusion

The authors whose work has been outlined share a cluster of ideas characterized by a pragmatist stance. These ideas reflect the authors' commitment to a probabilistic view of epistemology which heralds recent literature by several decades. It is noteworthy that their probabilistic approach is not linked to a particular interpretation of probability as the authors considered here embrace different notions of probability. Moreover, their commitment to a probabilistic view of epistemology overrides the objectivism/subjectivism and realism/anti-realism disputes.[50]

Department of Philosophy
University of Bologna
Via Zamboni 38
40126 Bologna
Italy
mariacarla.galavotti@unibo.it

50 Some further considerations on the influence of pragmatism on the debate on the foundations of probability are to be found in Maria Carla Galavotti, "Probability and Pragmatism", in: Dennis Dieks, Wenceslao Gonzalez, Stephan Hartmann, Thomas Uebel and Marcel Weber (Eds.), *Explanation, Prediction, and Confirmation*. Dordrecht: Springer 2011, pp. 499-510.

Michael Esfeld

Reductionism today

Abstract

The paper outlines the main argument for ontological reductionism in today's discussion, claims that ontological and epistemological reductionism (theory reduction) stand or fall together and finally sketches out how today's most widespread form of reduction, namely functional reduction, can be developed into a fully-fledged theory reduction, thus taking up the programme of the Vienna circle in today's philosophy.

1. Introduction

The Vienna circle is associated with a strong reductionist programme in the philosophy of science, seeking to reduce all scientific theories to theories formulated in a physical vocabulary and to reconstruct the statements formulated in that latter vocabulary on the basis of protocol sentences and logics and mathematics. In particular, Carnap's *Logischer Aufbau der Welt* (1928) is evidence of this strong reductionist programme. This reductionist programme was considerably weakened when the heritage of the Vienna circle survived in the logical empiricism in the United States in the 1950s and the 1960s. Seminal works such as the Oppenheim and Putnam (1958) and Nagel (1961, in particular ch. 11) still pursue reductionism in the sense of the reduction of the theories of the special sciences to fundamental physical theories, but there no longer is the intention to reconstruct the vocabulary of physics on the basis of protocol sentences and logics and mathematics.

With the demise of logical empiricism, reductionism became unfashionable. However, during the last two decades or so, reductionism has come back on the agenda, but the focus is on ontology rather than on epistemology (theory reduction), since there always have been and still are strong arguments in favour of ontological reductionism. The thesis of this paper is that one cannot have a coherent ontological reductionism without epistemological reductionism, that is, establishing the principled possibility of reducing the theories of the special sciences to basic physical theories. In that sense, the spirit of the Vienna circle is still with us.

In the next section, I outline the main contemporary argument for ontological reductionism. I then explain why this argument is incomplete unless ontological reductionism is joint with theory reduction and sketch out how theory reduction can be defended in view of the objections that have been raised against it since the

M.C. Galavotti et al. (eds.), *European Philosophy of Science – Philosophy of Science in Europe and the Viennese Heritage*, Vienna Circle Institute Yearbook 17,
DOI 10.1007/978-3-319-01899-7_7,

demise of logical empiricism (section 3). The material for this paper is based on the book Esfeld and Sachse (2011), which elaborates on the case for a comprehensive, but *conservative reductionism*. In this paper, I take for granted that reductionism is conservative instead of eliminativist: its aim is to show how the entities to which the special sciences are committed can exist in a physical world and how the theories of the special sciences can be true and can incorporate laws and explanations in such a world instead of having merely a heuristic or pragmatic value.

2. Ontological Reductionism

Suppose that it is possible to define a basic physical domain of the world unambiguously: let it consist in all and only those physical properties that occur at space-time points. In order to obtain a complete microphysical description of the world, one would thus have to quantify over all space-time points and specify which physical properties occur at these points. Suppose now that the whole microphysical domain of the world is duplicated (see Figure 1). In other words, an operation takes place that duplicates the whole space-time including all and only those physical properties that are instantiated at space-time points. The world w^* thus created hence is microphysically identical with the real world w. Does w^* contain all that there is in w, that is to say, all the organisms, all the biological, psychological, social, etc. properties that there are in w, including the Vienna circle, the conference in Vienna in December 2011 and the book in which this paper is published? In other words, is w^* a duplicate *simpliciter* of w?

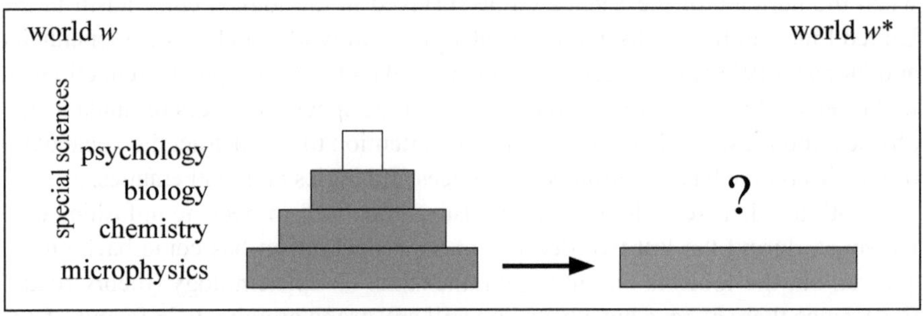

Figure 1: the world w* *on the right is an exact and complete microphysical duplicate of the world* w *on the left. Does* w* *contain everything that there is in* w?

There are good reasons to answer this question in the affirmative. We know that all objects that exist in the real world have developed from microphysical objects and are composed exclusively of microphysical objects. There can hence be no objects that exist in w, but that are absent in w^*. However, do complex macroscopic objects possess in w^* all the qualitative properties that they possess in w?

In other words, is the operation that consists in projecting the whole domain of microphysical properties from *w* to *w** sufficient to guarantee that all the biological, psychological, social, etc. properties that there are in *w* exist also in *w**? Note that there is no question of a deterministic dynamics here: we stipulate that all the microphysical properties in the *whole* space-time of *w* be copied to *w**. The issue of what the development of the world *w* in time is like, whether it is deterministic or not, has therefore no bearing on this question.

If a certain biological, psychological, social, etc. property existed in *w*, but missed in *w**, we would search for a reason for this difference. That search would take us beyond the domains of biological, psychological, social, etc. properties: according to all what we know about the world, it is not possible that a duplicate of the world lacks only one phenotypic property – say, the yellow colour of the petals of a certain individual plant, these petals being red instead of yellow in *w** – without there being also a genetical or an environmental difference between the duplicate of the world (*w**) and the world (*w*). However, if there is such a difference, there also is some molecular and consequently some microphysical difference between *w** and *w* as well. Hence, in this case, *w** would not be an exact microphysical duplicate of *w*, but would differ from *w* in some microphysical detail.

By the same token, according to all what we know about the world, it is not possible that a duplicate of the world lacks only one psychological property – say, the thought of Barack Obama on 13 February 2012 that winning the presidential election in November 2012 will not be an easy matter. If this psychological property were absent in *w**, there would be further psychological differences between Barack Obama in *w** and Barack Obama in *w*, since any thought is linked up with further thoughts, as well as with emotions and finally with actions. Consequently, there would be some neurobiological difference or other between *w** and *w* in the brain state of Barack Obama at the indicated time as well as some behavioural difference and thus some molecular and finally some microphysical difference. Again, according to all what we know about the world, it is not possible that a duplicate of the world lacks only one economic property – say, that the Dow Jones Index slightly rises on 13 February 2012. If this economic property were absent in *w** – in other words, if the Dow Jones Index developed in a different manner in *w** on that day –, there would be some difference in the intentional attitudes and actions of persons between *w** and *w*, and thus some neurobiological and behavioural difference and hence finally some molecular including some microphysical difference. Consequently, in this case, *w** would again not be an exact microphysical duplicate of *w*.

We can sum up this reasoning in the following manner: there is no biological, psychological, social or economic difference without there also being a microphysical difference. In other words, everything that there is in the world globally supervenes on the microphysical domain. All biological, psychological, social or economic properties that are instantiated in the world have some effects (make a difference to the world), and they can have their effects only by also having physical effects down to microphysical effects.

However, in making this statement, we face a serious problem: for any physical change, there is a complete physical cause (insofar as there is a cause at all). Any physical change comes under physical laws, and these laws contain only physical variables. Even if these laws are not deterministic, but only probabilistic, they indicate the complete probabilities for the physical change in question. No biological, psychological, social or economic variables can influence the probabilities for the occurrence of certain physical changes without these probabilities being at the same time completely fixed by physical variables. This causal completeness of physics is known since the advent of Newtonian mechanics and employed since Leibniz' *Monadology* (1714, § 80) to refute dualist interactionism. We can illustrate this problem in the following figure (see Kim 1998, ch. 2, and 2005, ch. 2, for an elaborate argument):

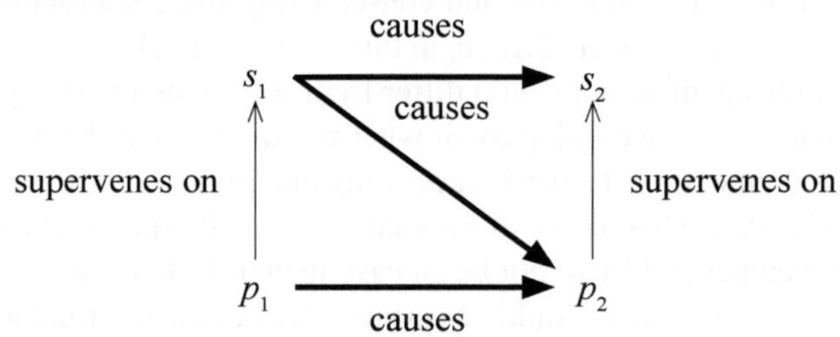

Figure 2: The domain of the properties with which the special sciences deal – the domain of biological, psychological, social, economic properties, etc. – globally supervenes on the domain of physical properties. A property token s_1 *of a special science causes another property token* s_2 *of a special science and thereby causes also a physical property token* p_2 *(the supervenience base for* s_2*). However,* p_2 *also has a complete physical cause* p_1.

There are exactly two possibilities to solve this problem if one admits global supervenience. The first possibility consists in maintaining that the properties (in the sense of property tokens) with which the special sciences deal are not identical with physical properties (that is, $s_1 \neq p_1$ and $s_2 \neq p_2$). Consequently, the properties of the special sciences *overdetermine* all their effects. For all physical effects of properties of the special sciences there is a complete physical cause. Furthermore, the supervenience of the domain of the properties of the special sciences on the domain of physical properties implies that there are sufficient physical conditions for the occurrence of any property token of a special science in the world. In other words, if s_2 supervenes on p_2, then p_2 is a sufficient condition for the existence of s_2: given p_2, s_2 cannot fail to exist.

However, it is not the case that physical properties also overdetermine the effects in the domain of the special sciences. The situation is not symmetrical. There are purely physical effects for which there are no causes in the domain of the special sciences. But there is nothing in the domain of the special sciences for which there are not completely physical sufficient conditions. Consequently, it is only the supposed causal efficacy of the properties of the special sciences that results in the mentioned overdetermination. The conclusion therefore is the following one: a situation (or a possible world) in which the properties of the special sciences do not cause anything and thus are epiphenomenal would be indiscernible from the situation depicted in figure 2.

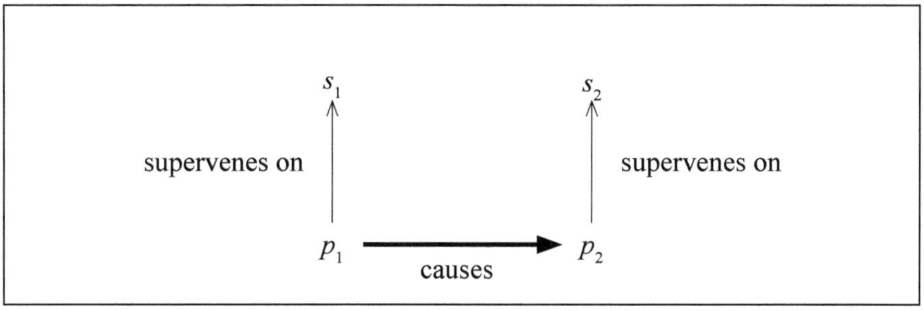

Figure 3: The properties of the special sciences being epiphenomenal.

Overdetermination is hence not a solution of the above mentioned problem that is in the position to establish the causal efficacy of the properties with which the special sciences deal. This result is independent of the theory of causation that one favours (for a contrary view see Ladyman 2008). For all physical property tokens, there are complete physical causes (insofar as there are causes at all). The laws in which types of physical properties figure are strict laws. The laws of the special sciences, by contrast, are never strict. Even if one endorses a theory of causation according to which causal relations are completely captured by counterfactual propositions, there is the mentioned failure of symmetry, since the laws of nature figure prominently among the truth conditions of the counterfactual propositions in question. That is to say, the counterfactual propositions that are about causal relations among physical property tokens have a privileged status, since they are backed up by strict laws. There are of course also true counterfactual propositions linking supervenient property tokens with subsequent physical property tokens. However, there is no argument visible why these propositions should express a causal relationship (see Esfeld 2010 for an elaboration and Harbecke 2011 for a counter-argument).

The other possibility to solve the above mentioned problem consists in maintaining that the properties in the sense of the property tokens with which the special sciences deal are *identical with* physical properties. More precisely, each property token in the domain of a special science is identical with a configuration of physical property tokens. Supervenience is compatible with identity: the relationship of supervenience does not exclude that all the properties in the domain of supervening properties are identical with properties in the domain of the supervenience base. Instead of the three arrows of causation drawn in figure 2 above, there hence is only one causal relation; but there are different descriptions of the property tokens standing in the causal relation in question:

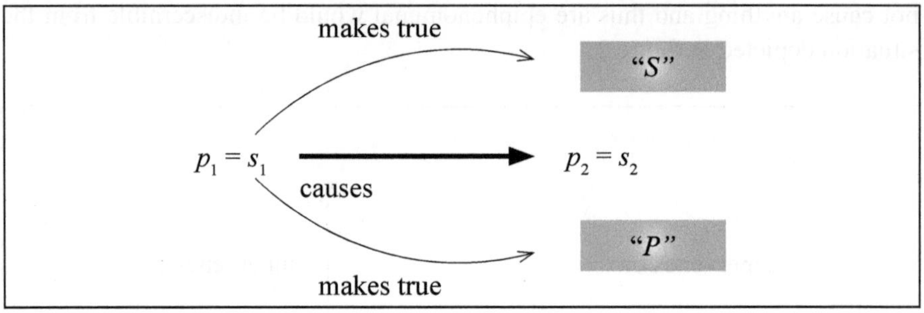

Figure 4: The property tokens of the special sciences are identical with physical property tokens. One and the same property tokens make true descriptions of different types.

However, it is often objected that if one takes the properties with which the special sciences deal to be identical with physical properties, then instead of vindicating the causal efficacy of the former, one *de facto* eliminates them, retaining *only* physical properties in one's ontology. But eliminativism is distinct from the identity theory: identity is a logical relation that is symmetric. If all the properties with which the special sciences deal are identical with physical properties, then some physical properties *are* properties of the special sciences. In general, if all *A*s are identical with *B*s, then both *A*s and *B*s exist, and some *B*s are identical with *A*s. Furthermore, it does not make sense to ask whether a given object in the domain of a special science brings about certain effects in virtue of its physical properties or in virtue of its special sciences' properties, since both are the same on the identity theory. In general, if the property of being *A* is the same as the property of being *B*, then all the effects that an object brings about qua being *A* are the effects that it brings about qua being *B*, and *vice versa*.

The identity claim as such does hence not provoke any eliminativism worry. The point at issue behind the eliminativism objection raised against the identity theory is the question how the properties with which the special sciences deal can be identical with physical properties. That question is well-taken. Even if the identity of biological, psychological, social or economic properties with physical properties in the sense of token identity is a fact, it is not a brute fact, but we need a theory that explains how it is possible for that identity to obtain. Functionalism is such a theory. If the properties with which the special sciences deal are functional properties, their essence is a causal role, that is, consists in bringing about certain effects given normal conditions. If the physical properties are also causal properties (that is, if their essence is or includes the disposition to bring about certain effects), then the causal role of certain complex configurations of physical properties can, given normal conditions, be identical with the causal role that is the essence of certain properties of the special sciences (see Esfeld and Sachse 2011, ch. 2, for an elaborate argument for this view).

In sum, if one accepts the premises of global supervenience and causal completeness of the physical domain, then one can build up a strong argument for *ontological reductionism*, that is, for the view that all the properties that are instantiated in the world are either physical property tokens or identical with (complex) configurations of physical property tokens.

3. Theory reduction

Ontological reductionism as set out in the previous section can also be accepted by some people who see themselves as physicalists, but who reject reductionism, thus endorsing the position that is known as non-reductive physicalism. Even Fodor (1974), in his famous argument against the unity of science as proposed by Oppenheim and Putnam (1958), can be read as accepting token identity, that is, the claim that every token of a special sciences' property is identical with some (complex) physical token. What Fodor (1974) – and non-reductive physicalists in general – reject, is type identity: there is no identity between types of the special sciences – such as biological or psychological types – and physical types. The reason is multiple realization: tokens coming under one and the same type of a special science can come under widely different physical types. However, this position provokes the following objection: insofar as there is only a reduction of tokens, but not of types, the problem remains whether the properties on which the special sciences focus cause anything qua biological, or psychological properties, etc.

Let us briefly consider the background of that objection: Donald Davidson (1970) claims in his famous paper "Mental events" that mental events are identical with physical events. More precisely, all events admit a physical description, and some events admit also a mental description. It is not possible to reduce the mental to a physical description. This position is known as anomalous monism – monism, because all events are physical, anomalous, because there is no nomological connection between the mental and the physical description of events, which would enable a reduction of the former to the latter. This position is widely recognized to fail due to the following objection: it cannot show that events cause anything insofar as they are mental events. Fred Dretske highlights this problem by conceiving the following example: the voice of a soprano singer causes a thin glass to shatter.

This effect occurs in virtue of the amplitude and frequency of the sounds. The meaning of the sounds is irrelevant to this effect (Dretske 1989, pp. 1-2). The same applies to events insofar as they are mental in Davidson's anomalous monism according to a widely recognized objection (see the papers in Heil and Mele 1993).

The ontological reductionism sketched out in the preceding section differs from Davidson's anomalous monism in that what is identical are not events, but property tokens. Nonetheless, Paul Noordhof (1998) objects to David Robb (1997) that in the same way as it is reasonable to ask whether a Davidsonian mental event causes anything qua mental, it is reasonable to ask whether a mental property token causes anything qua being a token of a mental type. Robb (2001) retorts that if identity is applied to those entities in virtue of which an object or event causes something, namely property tokens, it makes no sense to raise the qua-question for these entities, since they are already the most fine-grained ones.

Even though that reply is correct, there remains a problem. Let us assume for the sake of the argument that all that exists in the world are particulars (objects having property tokens); types then are concepts that seize salient similarities among the property tokens that objects have, whereby such salient similarities can amount to natural kinds. Multiple realization then is the epistemological fact that tokens coming under one single concept of the special sciences often come under different physical concepts. The concepts of the special sciences and the corresponding physical concepts differ not only in meaning, but they are also not coextensive. Consequently, it is not possible to reduce the concepts of the special sciences to physical concepts.

However, in that case, the problem that haunts Davidson and that Noordhof raises against Robb reappears: it has to be possible to relate the different descriptions in a systematic, reductive manner, if they are descriptions that are made true by one and the same token in the world and if each of them is to have a scientific quality (that is, to provide for law-like generalizations that are projectible, support counterfactuals, yield causal explanations, etc.). Otherwise, it could not be vindicated that these descriptions are about the *same* entities in the fine-grained sense of the same tokens instead of being about *different* properties that objects have. Consequently, the ontological reductionism set out in the preceding section cannot stand on its own. *Reductionism cannot be had in a piecemeal way*. Ontological and epistemological reductionism (theory reduction) stand together, or fall together.

But how can a theory reduction be possible given multiple realization? Nagelian reduction (Nagel 1961, ch. 11) has been superseded by functional reduction (see notably Lewis 1994 and Kim 1998, ch. 4, 2005, ch. 4 & 5). Let T_1 be a theory of a special science and T_2 be a physical theory that covers the domain of objects with which the special science in question deals. Functional reduction then proceeds in three steps:

1) One defines the property types in the domain of T_1 in a functional manner by indicating in terms of T_1 notably the characteristic effects of the tokens that

come under these types – to put it differently, the causal roles that tokens of these property types exercise.

2) One looks for realizers of these causal roles in the domain of the properties of T_2. The realizers of the functionally defined property types may differ in physical composition.

3) One explains in each case – that is, for each token – why there is a functional property falling in the domain of T_1 instantiated by describing how a configuration of properties in the domain of T_2 present in the situation under consideration brings about the effects that are characteristic of the functional property type in the domain of T_1 in question.

Functional reduction hence offers in each case, multiple realization notwithstanding, a causal explanation of why there is a property token present falling in the domain of T_1 by telling us how the effects that are characteristic of the property type in question are brought about (cf. what Chalmers 1996, pp. 42-51, calls a reductive explanation). Functional reduction thereby explains why there are properties falling in the domain of T_1 instantiated in the world and thus shows how T_1 is about salient properties. Functional reduction supports one-way conditionals of the sort that everything that comes under a physical type P_1 also comes under a special science type S ($\forall x\ (P_1x \rightarrow Sx)$); but it does not support biconditionals linking the types P_1 and S, since some tokens that come under S come under the physical type P_2 instead of the physical type P_1.

Though one-way conditionals are sufficient for the discovery of realizer types and reductive explanations, they are not sufficient for reducing T_1 to T_2, even if the domain of objects of T_1 is a proper part of the domain of objects of T_2 (T_2 may be a fundamental and universal physical theory). The reason is that one cannot deduce the laws of T_1 from the laws of T_2: there are no concepts available in T_2 that are coextensive with the concepts proper to T_1 and in which the laws of T_2 can be formulated, insofar as they are pertinent for that part of the domain of objects of T_2 that is identical with the domain of objects of T_1. Consequently, one cannot deduce the laws of T_1 from laws of T_2: the concepts figuring in fundamental and universal laws of nature (such as e.g. the laws of gravity or electromagnetism) are too general in order to deduce the laws of T_1 from these laws, and the concepts proper to T_2 that seize particular realizer types of property types of T_1 and laws or law-like generalizations formulated in terms of these concepts are too specific to capture the property types on which T_1 focuses: if there is multiple realization, several concepts proper to T_2 are needed to cover the extension of a single concept proper to T_1.

Nonetheless, there is a possibility to turn the mentioned one-way conditionals into biconditionals. Consider configurations of physical property tokens that all come under a single special science type S, but that make true different physical descriptions (P_1, P_2, P_3, etc.). Each of the physical descriptions P_1, P_2, P_3, etc. expresses a minimal sufficient condition to bring about the effects that characterize S under normal conditions. However, configurations coming under P_1 differ from configurations coming under P_2 and from configurations coming under P_3 in their physical composition. Differences in physical composition imply differences in the way in which the effects characterizing *S* are brought about. For each such difference, environmental conditions are conceivable in which that difference is relevant even on the level of abstraction at which *S* is situated.

Consider classical genetics. Natural selection explains why there is multiple realization in the domain of classical genetics (see Papineau 1993, p. 47): depending on the environmental conditions, only some of the causal powers of a given molecular configuration, realizing a property of the type *S* of classical genetics, are pertinent for selection. Against this background, it is reasonable that the proper concepts of classical genetics abstract from molecular differences. There are for instance molecular differences among DNA sequences possible that, under certain cellular conditions, do not amount to phenotypic (functional) differences. Nonetheless, these molecular differences are different ways to bring about the effects that define *S*. But this implies that there is at least one difference in the production of side effects that are systematically linked with the main effects in question – such as different causal interactions with the molecular environment within the cell during the causal process from a gene to the production of its characteristic phenotypic effects. For any such difference in side effects, there is a molecular environment possible in which that difference leads to a detectable functional difference within the scope of classical genetics and the evolutionary context because any such difference may become pertinent to selection in certain environments (see Rosenberg 1994, p. 32). Consequently, that difference can in principle also be considered in terms of the concepts that are proper to classical genetics. Hence, for any type *S* of T_1 (that is multiply realized by P_1, P_2, P_3, etc.), it is possible to conceive functional sub-types S_1, S_2, S_3, etc., taking those side effects into account.

These sub-types are no longer multiply realizable, since any molecular difference that is relevant to distinguish between different types of realizers leads to specific functional differences. The functionally defined sub-types of a special science hence correspond to one physical type each. They are nomologically coextensive with physical types and thus make it possible to reduce theories of a special science to physical theories in a functional manner.

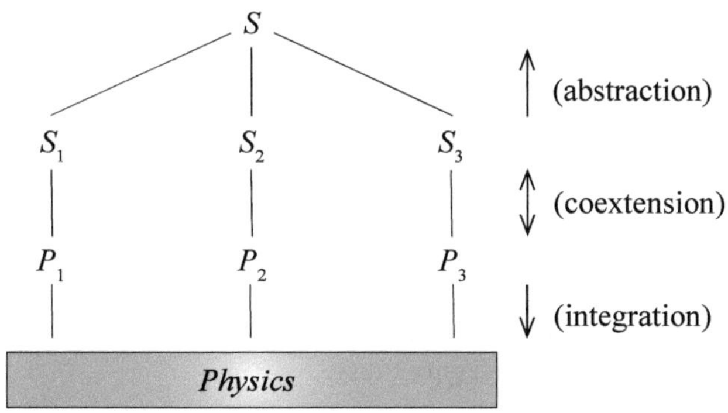

Figure 5: The reduction of a theory of a special science to a physical theory via functional sub-types.

More precisely and more generally speaking, (1) within a physical theory T_2, one builds the concepts P_1, P_2, P_3, etc. capturing the differences in composition among the physical configurations that are all described by the same concept *S* in T_1. (2) One makes *S* more precise by building functional sub-concepts (sub-types) S_1, S_2, S_3, etc. of *S*, seizing the systematic side effects linked to the different ways of producing the effects that define *S*. Provided that one such functionally defined sub-concept can be construed for each type of realizer of *S* in such a way that the former grasps the functional differences to which the latter give rise under certain circumstances, it follows that these sub-concepts S_1, S_2, S_3, etc. are nomologically coextensive with the concepts P_1, P_2, P_3, etc. (3) One can reduce any concept *S* of T_1 to T_2 via S_1, S_2, S_3, etc. and P_1, P_2, P_3, etc. Starting from T_2, one builds P_1, P_2, P_3, etc. and then deduces S_1, S_2, S_3, etc. from P_1, P_2, P_3, etc. given the nomological co-extension. One gains then *S* by abstracting from the conceptualization of the functional side effects contained in S_1, S_2, S_3, etc., retaining only the main functional specification they have in common, which is nothing but the functional definition of *S*. This abstraction step depends on what the world is like – that is, what salient normal environmental conditions there are – rather than on our heuristic and practical aims. It enables thereby to highlight genuine causal similarities in the world that are brought out by the special sciences. As regards the laws, one can formulate the laws of T_1 in terms of S_1, S_2, S_3, etc. by adding more functional details. Given the nomological coextension, one can deduce these sub-type laws from the laws of T_2, couched in terms of P_1, P_2, P_3, etc. and then gain the laws of T_1 formulated in terms of *S* by a theory-immanent abstraction from functional details (that are not relevant in many environmental contexts) (see Esfeld and Sachse 2011, chapters 4 and 5, for a detailed account).

In conclusion, functional reduction, thus conceived, can achieve the same as Nagelian reduction, namely a fully-fledged theory reduction. As mentioned above, ontological and epistemological reduction stand or fall together. Nonetheless, one can still have both, as envisaged in the Vienna circle, although today's argumentative focus is different.

References

Carnap, Rudolf (1928): *Der logische Aufbau der Welt*. Leipzig: Meiner.

Chalmers, David J. (1996): *The conscious mind. In search of a fundamental theory*. New York: Oxford University Press.

Davidson, Donald (1970): "Mental events". In: L. Foster & J. W. Swanson (eds.): *Experience and theory*. Amherst: University of Massachusetts Press. Pp. 79-101. Reprinted in D. Davidson (1980): *Essays on actions and events*. Oxford: Oxford University Press. Pp. 207-225.

Dretske, Fred I. (1989): "Reasons and causes". In: J. E. Tomberlin (ed.): *Philosophical Perspectives 3: Philosophy of mind and action theory*. Oxford: Blackwell. Pp. 1-15.

Esfeld, Michael (2010): "Causal overdetermination for Humeans?" *Metaphysica* 11, pp. 99-104.

Esfeld, Michael and Sachse, Christian (2011): *Conservative reductionism*. New York: Routledge.

Fodor, Jerry A. (1974): "Special sciences (or: The disunity of science as a working hypothesis)". *Synthese* 28, pp. 97-115.

Harbecke, Jens (2011): "Mind in a Humean world". Forthcoming in *Metaphysica*, DOI 10.1007/s12133-011-0086-2.

Heil, John and Mele, Alfred (eds.) (1993): *Mental causation*. Oxford: Oxford University Press.

Kim, Jaegwon (1998): *Mind in a physical world. An essay on the mind-body problem and mental causation*. Cambridge (Massachusetts): MIT Press.

Kim, Jaegwon (2005): *Physicalism, or something near enough*. Princeton: Princeton University Press.

Ladyman, James (2008): "Structural realism and the relationship between the special sciences and physics". *Philosophy of Science* 75, pp. 744-755.

Lewis, David (1994): "Lewis, David: Reduction of mind". In: S. H. Guttenplan (ed.): *A companion to the philosophy of mind*. Oxford: Blackwell. Pp. 412-431. Reprinted in D. Lewis (1999): *Papers in metaphysics and epistemology*. Cambridge: Cambridge University Press. Pp. 291-324.

Nagel, Ernest (1961): *The structure of science. Problems in the logic of scientific explanation*. London: Routledge.

Noordhof, Paul (1998): "Do tropes resolve the problem of mental causation?" *Philosophical Quarterly* 48, pp. 221-226.

Oppenheim, Paul and Putnam, Hilary (1958): "Unity of science as a working hypothesis". In: H. Feigl, M. Scriven and G. Maxwell (eds.): *Concepts, theories, and the mind-body problem. Minnesota Studies in the philosophy of science. Volume 2*. Minneapolis: University of Minnesota Press. Pp. 3-36.

Papineau, David (1993): *Philosophical naturalism*. Oxford: Blackwell.

Robb, David (1997): "The properties of mental causation". *Philosophical Quarterly* 47, pp. 178-194.

Robb, David (2001): "Reply to Noordhof on mental causation". *Philosophical Quarterly* 51, pp. 90-94.

Rosenberg, Alexander (1994): *Instrumental biology or the disunity of science*. Chicago: University of Chicago Press.

Department of Philosophy
University of Lausanne
Dorigny
CH-1015 Lausanne
Switzerland
Michael-Andreas.Esfeld@unil.ch

Wlodek Rabinowicz and Lina Eriksson

Betting Interpretation and the Problem of Interference[1]

It has long been common to identify an agent's degrees of belief with her betting rates. Here is this *betting interpretation* (BI) in a nutshell: A bet on a proposition *A* with price *C* and a non-zero stake *S* (= the monetary prize to be won by the bettor if *A* is true) is said to be *fair* for an agent iff the latter is willing to take each side of the bet, to buy the bet as to sell it. Assuming that such a bet on *A* exists and that the *C*/*S* ratio is constant for different fair bets on *A*, this ratio is the agent's *betting rate* for *A*. The rate in question is taken to be the agent's degree of belief in *A* (his credence in *A*, or – to use yet another terminology – his subjective probability for *A*).

BI is an instance of operationalism. It makes degrees of belief observable and measurable. It also opens up for pragmatic arguments for various rationality constraints on beliefs such as standard probability axioms, conditionalization, reflection, etc. An agent whose degrees of belief violate these constraints will have betting rates that make her vulnerable to exploitation.

There is an actualist and a dispositionalist version of BI. On the former, degrees of belief are fixed by the agent's actual betting behavior, while on the latter, the more plausible one, they are fixed by the agent's dispositions to bet. It is the dispositionalist version that we shall focus on. Note that the identification of *degrees* of belief with betting rates does not force the dispositionalist to identify beliefs themselves with dispositions to bet. He can instead assume that an agent's betting dispositions supervene on her beliefs.

Although operationalism has fallen from grace in most other fields, it has shown amazing resilience regarding degrees of belief. The objections to operationalism in this area mostly concern cases in which agents are judged to have beliefs that they aren't disposed to manifest in behavior, or cases in which agents have reasons to bet that are unconnected with what they believe. Troublesome as these objections are, it is arguable that attitudes to bets still *could* represent the agents' degrees of belief; the objections only show that in this or that situation, they don't.

We propose to consider a more fundamental problem for BI: There is a sense in which degrees of belief *cannot* be interpreted as betting rates. The bets we

1 This is a short version of our joint paper "The Interference Problem for the Betting Interpretation of Degrees of Belief", which is forthcoming in *Synthese*. We wish to thank the participants of the conference on *Philosophy of Science in Europe – European Philosophy of Science*, in Vienna 2011, for stimulating comments. We are grateful to an anonymous referee for helpful comments.

M.C. Galavotti et al. (eds.), *European Philosophy of Science – Philosophy of Science in Europe and the Viennese Heritage*, Vienna Circle Institute Yearbook 17,
DOI 10.1007/978-3-319-01899-7_8,

are disposed to accept do not reflect our current unconditional degrees of belief in various propositions. Whether a bet on *A* would be accepted or not does not depend on the agent's degree of belief in *A* but rather on the degree of belief she would have if she were confronted with this bet proposal, or – more generally – the degree of belief she would have if she were in a position to bet on *A*. Assuming the conditionalization model for belief change, this means that whether the agent would be willing to bet depends on her current *conditional* beliefs concerning *A* on the supposition that she has an opportunity to make this bet. Furthermore, her disposition to bet also depends on the expected effects the act of betting would have on the truth of the proposition to be betted on. Both these factors imply that finding oneself in a betting situation might alter one's expectations in important ways. Consequently, the identification of (unconditional) degrees of belief with betting rates is a mistake. The reason is, to put it shortly, that we need to take into consideration potential interferences that bet opportunities and betting itself might create with regard to the proposition to be betted on. It is because of this *interference problem* that the agent's degree of belief in *A* cannot be interpreted as her betting rate for *A*. This suggestion will be developed in what follows.

The problem as such isn't new. Ramsey, for example, was right on to this idea when he wrote:

> … the proposal of a bet may inevitably alter [one's] state of opinion; just as we could not always measure electric intensity by actually introducing a charge and seeing what force it was subject to, because the introduction of the charge would change the distribution to be measured. (Ramsey 1931 [1926], p. 172)

We will show that this problem comes up in several different versions.

Causal relations between bets and the propositions betted on

Consider this case: I am fairly confident that I will fall asleep tonight. But were I to bet on this, I might lie awake all night, trying to get to sleep and failing. So, I am not confident that I would fall asleep tonight if I were to bet that I will. (See Hájek 2009.)

The problem can be of two sorts. First, trying to win the bet might be counter-productive. Trying to fall asleep makes it harder to do so. There are benefits that can be gained as side-products, but not when they are deliberately targeted. Second, it might be that it is thinking of the bet you have made rather than trying to make true the proposition betted on that is counter-productive. If you bet you will be happy next week, this can make you less happy in that period if you are going to worry that you might lose your bet or if you are going to regret for some reason that that you have made this bet. Of course, it doesn't have to work out in this way. Feeling pretty certain that you will be happy, the expectation of the gain from your bet might add to your happiness. Taking the bet might thus lead your beliefs in either direction, making it either more or less likely that the proposition you have betted on is true. However, the crucial point is that no matter in which direction you go, the bet can influence your probabilities for what you are betting on.

Falling asleep is something we can't do on purpose (if we don't have sleeping pills). *Not* falling asleep, however, is something we *can* control, at least for a time. If you place a bet that you won't fall asleep tonight, you can do something to make it happen. While in the absence of the bet you are confident you will fall asleep tonight, your probability decreases if you bet that you won't. This point generalizes: our probabilities for actions under our control can differ from our probabilities for these actions if we bet on them. This spells trouble for the betting interpretation.

Some proponents of BI have argued that we should bite the bullet and accept that we cannot have subjective probabilities for the choices we are about to make. Practical deliberation crowds out self-prediction (cf. Levi 1997, pp. IX and 81). Wolfgang Spohn presented this idea as follows:

> The agent's readiness to accept a bet on an act does not depend on the betting odds but only on his gain. If the gain is high enough to put this act on the top of his preference order of acts, he will accept it, and if not, not. (Spohn 1977, p. 115)

Explanation: On BI, the agent's subjective probability for a proposition A equals the price-stake ratio C/S in a fair bet on A. But now suppose that A describes an action that stands at the agent's disposal. If the agent bets on A, then, since A is something she can make true if she so chooses, she can count on a sure net gain $G = S-C$. We can represent the expected utility of the bet on A as $\mathrm{eu}(A)+G$, where $\mathrm{eu}(A)$ stands for the expected utility of A in the absence of the bet. Now, assume for simplicity that there is just one action, B, that is an alternative to A. Suppose that $\mathrm{eu(B)} > \mathrm{eu(A)}$. However, as long as $\mathrm{eu}(B) < \mathrm{eu}(A) + G$, the rational agent will accept the bet on A. But, and this is crucial, the size of $G = S-C$ is not fixed by the ratio C/S. When we vary C and S while keeping the ratio C/S constant, G can decrease so much that the bet on A loses its attraction. As Spohn puts it: "If the gain is high enough to put this act on the top of his preference order of acts, [the agent] will accept it, and if not, not". This sensitivity to the size of G conflicts with the assumption that C/S is constant for different fair bets on a given proposition. But then, in the absence of such a constant ratio, there is no well-defined betting rate for A, which means – on the betting interpretation – that the agent lacks a subjective probability for that proposition.

The argument above implicitly presupposes that the agent is *certain* she will perform the action if she bets on it. If there is some doubt on this account, the agent might still abstain from taking the bet even if the net gain G makes the sum $\mathrm{eu}(A) + G$ larger than $\mathrm{eu}(B)$. For in the presence of such uncertainty she would not be assured of winning the bet in the first place.

However, a slightly modified argument goes through even in the absence of certainty (cf. Rabinowicz, 2002, p. 101). The net gain *G* could swing the balance so much to *A*'s advantage as to make it profitable for the agent to accept the bet, as long as she considers it sufficiently *likely* that she would perform *A* as a result. There is no need of certainty, if *G* is sufficiently large.

This argument, if correct, proves more than Spohn has intended, however (cf. Rabinowicz, 2002, pp. 101ff). At least in principle, it shows that for *all* of one's future actions, and not just for those that are subject to current deliberation, their subjective probabilities (*and* the subjective probabilities of events that these actions could influence) cannot be interpreted as betting rates. Betting on them might make them sufficiently likely for the bet to be profitable, if G is large enough. While friends of BI might be prepared to deny that the agent has subjective probabilities for the actions she currently deliberates upon, extending this claim to all future actions of the agent would go too far. Surely, we do have degrees of belief concerning things we will do in the future. To be able to allow for this, while still denying that we have degrees of belief for the actions about which we currently deliberate, friends of BI would need to make use of some pretty heavy idealizations, such as, say, positing that – insofar as future actions are concerned – the agent expects to forget her bets after having made them, before she is going to start to her future deliberation on how to act. Forgotten bets on future actions do not create interferences, since they are not going to influence the actions themselves. Clearly, however, the more counterfactual and outlandish idealization is required, the more problematic becomes the claim that BI gives an account of what degrees of belief *are*.

We'll come back to the issue of idealization in the penultimate section.

Evidential relations between bets and the propositions they concern

Another class of putative counterexamples to BI depends on the fact that betting, or finding oneself in a position to bet, might sometimes carry information bearing on the proposition on which the bet is made. This can happen in at least four different ways: The information might be carried by someone offering you a bet, by your acceptance of that bet, by your own bet offer, or by your bet offer being accepted.

(i) Examples of the first kind are easy to find. In general, an offer of a bet will change one's probability for the proposition to be betted on if one has grounds to suspect that the person who has made the offer has access to superior information.

Bradley and Leitgeb (2006) argue that the examples in which the offer of a bet changes your credence aren't yet worrying for BI, as they do not concern credences that are relevant for behavior. The relevant credences are those you have *ex post*, after having been offered the bet. And your credence ex post does equal your betting rate: You are willing to bet at the rate that corresponds to your changed degree of belief. It is for this reason they move to a more elaborate case, "Forgery": Suppose a fair coin will be flipped. Whether it lands heads or tails, you will be offered a bet on Heads. But you know that, if the coin will land heads, the bet will use fake money (both your banknotes and the bookie's will be secretly exchanged for fakes). Then the offer of the bet will not change your probability for Heads, but the latter will still diverge from your betting rate. Your probability for Heads will still be ½, but you won't be willing to bet on Heads at these odds.

A less fanciful example that makes the same point is a case in which a bookie offers to make a bet with you on a given proposition, but with the proviso that you must be prepared to take each side of the bet. Only after you have accepted his offer, he will disclose whether you are to buy or to sell the bet. The bookie's offer in such a case doesn't change your probability for the proposition on which the bet is to be made, but it still makes you suspect that the bookie knows more than you do. Therefore, you wouldn't be willing to accept the offer at odds that correspond to your probability.

We think, however, that the problem for BI already arises in the original simple examples in which the bet offer does change your probability. In those cases, your *ex ante* credence for the proposition you are asked to bet on cannot be interpreted as your betting rate for that proposition. This is troublesome enough. BI is meant to be general and therefore it should also apply to a person's degrees of belief ex ante. In fact, to the extent that exposure to bet offers is a method to elicit the agent's probabilities, it is the ex ante probabilities that are meant to be the object of elicitation.

(ii) The second case is when information is revealed by the very fact that you accept the bet. Consider the following example: Your degree of belief that you are not a gambler is high. But then you are offered a bet on this proposition and you do accept the bet, which reveals something about yourself that is of relevance to the truth of the proposition you have betted on. So, in considering whether to accept the bet or not, it seems you should take into consideration that accepting it would be revealing in this way.

(iii) The third case is similar to the second: The fact that you make a bet offer can reveal something about yourself that is of relevance to the truth of the proposition betted on. For example, offering a bet on a proposition that you are a risk-lover (risk-hater) can increase (decrease) your probability for that proposition.

(iv) The fourth case, in which the information is provided by your bet offer being accepted, is similar to the first case: That your offer is accepted might be evidence that your opponent knows something you don't.

In these four kinds of cases, making a bet or being confronted with an opportunity to bet would have evidentiary (but not causal) bearings on the proposition betted on. This evidentiary effect explains why the rate at which you are willing to bet does not correspond to your degree of belief in the proposition under consideration.

Qualification: Causal versus evidential decision theory

At this point, however, we need to stop and re-consider. Cases of type (ii) and (iii) in the section about evidentiary relations are problematic. Our suggestion that they exemplify the Interference Problem for BI is incompatible with causal decision theory (CDT). If an action has purely evidentiary bearings on the relevant states of nature, but no causal bearings, CDT tells us that we should calculate the expected utility of the action in terms of the unconditional probabilities of states, rather than in terms of their conditional probabilities given the action under consideration, as evidentiary decision theory (EDT) would have it. In other words, on CDT, purely evidentiary bearings of actions should be disregarded in decision making. Now, suppose that the action I deliberate upon consists in placing a bet, and assume that this act has mere evidentiary bearings on the proposition on which the bet is to be made, as in the example in which that proposition states that I am not a gambler. Placing a single bet would not in itself cause me to become a gambler, but it would provide some evidence concerning my gambling dispositions. In such a case, CDT recommends that I should place the bet if my *un*conditional probability for the proposition under consideration is high. However, we have suggested above that, in my decision, I should take into consideration that placing the bet would provide evidence against the proposition on which the bet is to be made. This kind of counter-example against BI is incompatible with CDT. The same applies to type (iii) cases.

By the way, another seeming counterexample to BI in group (ii) is the well-known case of Sleeping Beauty. If a fair coin lands Heads, the Beauty will be awakened only once, while if it lands Tails, she will be awakened twice, with the memory of her first awakening erased. All causal links will thereby be removed between her actions on both occasions. Upon each awakening, she will be offered a bet on Heads. She knows all this. Now she wakes up and gets an offer of a bet on Heads. If she accepts the bet, this will increase her probability that she will accept another offer like this if she will be awakened again. These purely evidentiary bearings of her bet are bad: She will be awakened more than once only if the coin has actually landed Tails. If she for that reason abstains from betting, despite of the rate of the bet equaling her probability for Heads, her decision is in line with EDT, but not with CDT. For an analysis of Sleeping Beauty along these lines, see Arntzenius (2002). Bradley and Leitgeb's (2006) argue that betting odds and credences come apart in the case of Sleeping Beauty. Bovens & Rabinowicz (2010) suggest that this argument implicitly presupposes EDT.

If we stand by CDT, we have to reject such counterexamples. But other counterexamples we have presented still stand. Thus, CDT does imply that the agent's willingness to bet on a proposition A should depend on (a) the evidentiary bearings on A of the *opportunity to bet*, i.e. the evidentiary bearings on A of the agent's being in a position to place a bet on A if she so chooses, and (b) the expected causal effects that betting on A could have on that proposition.

Very roughly, then, a hypothetical bet b on A at a certain rate is acceptable to an agent iff this rate does not exceed the agent's current degree of belief that A *would* be the case if she were to make bet b, *conditioned* on the supposition that she will have an opportunity to make the bet in question. I.e., if we let P stand for the agent's current probability assignment and $\rightarrow$ for the subjunctive implication, b is acceptable to the agent if and only if b's price-stake rate does not exceed P(bet b on $A \rightarrow A$/opportunity to bet on A). Only if none of the interfering factors (a) and (b) mentioned above is present, this degree of belief is reducible to the agent's current unconditional degree of belief in A. In the absence of factor (b), P(bet on $A \rightarrow A$/opportunity to bet on A) = P(A/opportunity for bet on A), and in the absence of factor (a), P(A/opportunity to bet on A) = $P(A)$. Thus, it is only in the absence of both factors that BI does not encounter the interference difficulties we have raised.

Even this is an oversimplification. Even if an action of betting has purely evidentiary bearings on the proposition to be betted on and the opportunity to bet has no such bearings, the agent should still beware of fixing her betting rate in advance, in accordance with her current degree of belief in the proposition in question. Here's the reason why: When she starts to deliberate whether to accept the bet or not, she will often be able to gain introspective access to her psychological leanings – to her inclinations to decide one way rather than the other. Now, even on CDT, the information the agent will gain prior to her final decision is relevant to what she should decide. This means that the agent's ex ante probabilities aren't reliable betting guides in case like this, since they can be expected to change during deliberation leading to a decision on whether to bet or not. In what follows, we shall disregard this complication.

Other versions of the betting interpretation

There are other versions of BI apart from the simple operationalism we have been focusing on. Of special interest is the view that while degrees of belief can be *measured* by betting rates, they are distinct from the latter. Even de Finetti himself wavered between the operationalist account and the measurement view. One of the advocates of the measurement view was Richard Jeffrey. But Jeffrey certainly recognized that there were cases in which betting rates didn't seem to measure degrees of belief well (Jeffrey 1956, p. 238-9). In fact, all the standard problematic cases for the operationalist view are trouble-cases for the measurement view as well: For example, a person dislikes betting, her degrees of belief cannot be identified with her betting rates, since she doesn't have any, but neither can they be measured by these non-existent betting rates. However, the consequences are less drastic for the measurement view: Instead of constituting an outright refutation, the apparent counterexamples only restrict the range over which betting rates provide a reliable measure. It is standard that a particular measurement method only works under limited conditions. Thermometers only work within a certain range, but this doesn't compromise their reliability for the temperatures within that range. (cf. Jeffrey, ibid.)

It is arguable that Frank Ramsey also adhered to the measurement view. Cf. the following quotation from Ramsey (1931 [1926]), p. 172:

> The old-established way of measuring a person's belief is to propose a bet, and see what are the lowest odds which he will accept. This method I regard as fundamentally sound; but it suffers from being insufficiently general, and from being necessarily inexact. It is inexact partly because of the diminishing marginal utility of money, partly because the person may have a special eagerness or reluctance to bet, because he either enjoys or dislikes excitement or for any other reason, e.g. to make a book. The difficulty is like that of separating two different co-operating forces.

The quoted passage continues with Ramsey pointing out the interference problem that confronts this method of measurement.

The counter-examples considered in this paper can be treated as problems for measurement: Introducing bet opportunities sometimes creates an interference with what we want to measure. So does betting itself. One might conclude that this just shows yet another limitation of the betting approach viewed as a measurement procedure. However, this move considerably weakens the measurement view. The situation is not like the one in the thermometer analogy. There is no specific range of propositions for which the betting interpretation provides a reliable method. While some interference-based objections to this interpretation focus on particular classes of propositions (about future actions, about the agent's dispositions to bet, etc.), other such objections, like the one having to do with the evidentiary bearings of betting opportunities, apply – at least in principle – to *any* proposition.

A possible response to this problem might appeal to recent general discussions about dispositions. As has been pointed out by several authors, dispositions may well be subject to various interferences that falsify the simple stimulus-manifestation conditionals that they standardly are defined by. (For good up-to-date discussions, see Contessa 2012 and Fara & Choi 2012.) Thus, under some circumstances, the dispositions can be masked (as when a fragile object is carefully wrapped, so that it wouldn't break if it were dropped), or they might be overridden by antidotes (as when a poisonous food wouldn't kill if ingested provided we have taken an appropriate antidote), or they might be finkish, which means that the condition for an object losing (or acquiring) a disposition is the same as that disposition's stimulus condition. Here is an example of a fink, due to Martin (1994), who first introduced this idea: A wire is disposed to conduct electricity when touched by a conductor, but it is part of a circuit with a circuit-breaker that would open if a conductor were to touch the wire, which would hinder the wire from conducting electricity.

Now, an adherent of BI might conceivably argue that the dispositional interpretation of probabilities as betting rates an agent would be willing to accept if she got an opportunity is perfectly in order, but that the relevant dispositions are finkish: Their stimulus conditions, i.e. betting opportunities, would introduce new information and thereby cause the agent to lose these dispositions. In other words, betting opportunities have evidentiary bearings that interfere with the agent's dispositions to bet. A betting disposition exists as long as its stimulus condition is not actualized. That it does exist under those circumstances can be shown by a thought experiment: If the betting opportunity didn't have any evidentiary bearings, then the agent would accept the relevant bet if offered.

On this proposal, the evidentiary bearings of the betting opportunity in certain cases mask the disposition rather than make it disappear. This happens in the earlier mentioned example in which a bookie offers to make a bet with you on a given proposition, but only on condition that you would be willing to take each side of the bet. Only after you have accepted his offer, he will disclose which side you will have to take. The bookie's offer in such a case doesn't change your probability for the proposition in question, and thus, on BI, you still have the same disposition to bet. But the evidentiary bearings of the offer are such that you wouldn't be willing to accept it at the odds that correspond to your probability.

We are not sure what to say in response to this kind of proposal. It certainly is attractive, but there is something strange in dispositions that are supposed to be finkish or masked in such a pervasive and systematic way. It is one thing to accept finks and masks in rare cases, as exceptions, as in the example of a wire which is disposed to conduct electricity but would not do so because of the presence of a circuit-breaker. But if the interfering effect of the stimulus conditions on the purported dispositions is supposed to be a rule rather than an exception, then the very existence of such dispositions might be questioned. And even if we were to accept their existence, the whole point of identifying degrees of belief with behavioral dispositions – the underlying operationalist strategy – is undermined if these dispositions are so pervasively and systematically interfered with.

Some friends of BI respond to criticisms by taking a position that decision theory is not supposed to be about actual people anyway. Problems with betting rates as representations of degrees of belief that arise for actual people need not be equally serious when we consider agents viewed as *theoretical constructs*. In reply, the first thing to be noted is that this response leaves open the question why the degrees of belief of such theoretically constructed agents need to be defined, measured or otherwise elicited by betting behaviour. After all, the reason for introducing bets was to gain access to degrees of belief through observable phenomena. But theoretically constructed agents aren't observable anyway.

Talk of theoretical constructs, however, could also be used to justify various counterfactual or subjunctive moves as solutions to our problem. One such suggestion (due to Sebastian Enqvist) is that my degrees of belief at a given time *t* are given by the bets that my *hypothetical* alter ego would be willing to make at *t*, not about events in her hypothetical world, but about the *actual* world. This alter ego is exactly like me up to and including time *t*. The relevant bets, in other words, are those that we ourselves would be willing to accept about the actual world if we could take a time-out from reality. This allows us to avoid the problem that a bet on a proposition sometimes causally influences the proposition in question and that this influence needs to be taken into consideration by the bettor. Bets made in another world have no causal influence on what happens in the actual world.

While this counterfactual construction does take care of the interference phenomena related to the expected causal effects of bets, it still fails to attend to interferences generated by the evidential bearings of the opportunities to bet. But the most basic problem seems to be that the more idealized, hypothetical or counterfactual we go, the less it makes sense to look to bets for an interpretation of degrees of belief. What these cases at best show is that if we twist and turn enough, we might possibly construct hypothetical betting rates that would correspond to the agent's actual degrees of belief. *But it is our independent understanding of these degrees of belief that guides the hypothetical constructions.* Therefore, the former are not really illuminated by the latter.

What does this tell us?

It is not only the identification of the agent's degrees of belief proposition with her betting rates that is undermined by the cases we discuss. The very notion of a *betting rate* is put into question. As we remember, the betting rate for a proposition *A* is supposed to be the price-stake ratio common to all fair bets on *A*, where a bet is fair if and only if the agent is willing to take each of its sides. However, if a bookie's offer to sell a bet on *A* could decrease the agent's probability for *A*, while a bookie's offer to buy a bet on *A* could increase that probability, fair bets on *A* might not be available: There might be no ratio at which the agent would be willing both to buy and to sell the bet on *A*. On the other hand, if *A* is a proposition about some event that stands under agent's control (such as her current or future action), then there would be too many fair bets on *A*: As long as both the net gain $S - C$ and the price C are sufficiently large, the agent might be willing to both buy and sell bets with different C/S ratios. (If she buys this bet, it will be to her advantage to make *A* true if the net gain is sufficiently large, while if she sells it, she will profit by making *A* false if the price of the bet is sufficiently large.) Thus, no such ratio would be common to all fair bets on *A*. For both these reasons, the notion of a betting rate is problematic.

As we have seen, the interference problem creates difficulties not only for the view that identifies degrees of belief with dispositions to bet but also for other versions of BI: for the measurement view and for various counterfactual constructions. Can we do better?

An idea that seems right but would require much work to be properly developed is that beliefs should be given a *functional* interpretation: they are definable in terms of their double role of truth-trackers and guides of behaviour. On this view, beliefs are internal states of the agent that come in degrees and have propositional content. Their function is to adjust the agent's behaviour to the state of the world in such a way as to make the agent flourish, in some sense that would need to be spelled out. (This spelling out can be done in terms of desire satisfaction or in some more objective, Aristotelian way, by taking its point of departure from an appropriate theory of human nature. It could also be done on the evolutionary lines, in terms of the number of offspring.) This function accounts for their double role: On the input side, they track truth, which gives them a direction of fit different from desires, and their degrees are informed by the goal of truth approximation. (If we think of beliefs as coming in degrees, it is natural to expect that the truth-tracking aspect of beliefs should be accounted for in terms of some appropriate *accuracy* measure, such as the Brier score. See Joyce (1998), for a proposal along these lines.) On the output side, they guide the agent's behaviour and not least her betting behaviour: In one sense, all acting under uncertainty can be seen as betting. This means that there does exist, after all, a conceptual connection between degrees of belief and dispositions to bet. But the existence of such a conceptual connection does not imply any simple equation of the kind: the degree of belief in A = the betting rate for A. In the first place, the problem with this equation is that it wholly ignores the role of beliefs as truth-trackers. In the second place, as we have seen, the agent's betting dispositions with regard to A might well be determined by the degree to which the agent believes some other, more complex proposition – a proposition related to A but different from it (such as that A would be true if the agent made a bet on it). In addition, this complex belief, whose degree is relevant to the agent's disposition to bet, might well itself be conditioned on some suppositions – thus, in particular, it might be conditioned on the supposition that the agent will have an opportunity to bet on A.

Even this picture is extremely simplified. On the input side, the role of beliefs is not always to track truth and thereby enable one to act in an efficient way. At least sometimes, their presence is more directly tied to flourishing, as in the case of wishful thinking and ex post rationalization. And on the output side, the role of beliefs is not just to guide actions. They influence behaviour in other ways as well. Consider for example a person who is about to do a driving test: His confidence calms him down and makes him drive well, whereas a low confidence in one's driving skills would have the opposite effect. Furthermore, beliefs often affect other aspects of our functioning, apart from behaviour. Not least, they influence our emotions. Likewise, one's beliefs also guide various psychological processes, such as inferences, desire formation, etc.

All this shows that the functionalist approach of beliefs might turn out to be very complicated. But this is just to be expected. Correct theories about complicated creatures like us are not likely to be simple.

References

Arntzenius, Frank (2002), "Reflections on Sleeping Beauty," Analysis 62 (2): 53-62.

Bovens, Luc, and Wlodek Rabinowicz (2010), "The Puzzle of the Hats," *Synthese* 172: 57-78.

Bradley, Darren and Hannes Leitgeb (2006), "When Betting Odds and Credences Come Apart: More Worries for Dutch Book Arguments," *Analysis* 66 (2): 119-27.

Contessa, Gabriele (2012), "Dispositions and Interferences", *Philosophical Studies*, published online.

Fara, Michael, and Sungko Choi (2012), "Dispositions", Stanford Encyclopedia of Philosophy, http://plato.stanford.edu/entries/dispositions/

de Finetti, Bruno (1990), *Theory of Probability*, Vol. I (first edition 1974), New York: Wiley.

van Fraassen, Bas (1980), "Review of Brian Ellis, 'Rational Belief Systems'," *Canadian Journal of Philosophy* 10(3): 497-511.

Hájek, Alan (2009), "Interpretations of Probability", Stanford Encyclopedia of Philosophy, http://www.science.uva.nl/~seop/entries/probability-interpret/

Jeffrey, Richard (1956), "Valuation and Acceptance of Scientific Hypotheses," *Philosophy of Science* 23 (3): 237-246.

Joyce, James M. (1998), "A Nonpragmatic Vindication of Probabilism," *Philosophy of Science* 65: 575–603.

Levi, Isaac (1997), *The Covenant of Reason*, Cambridge: Cambridge University Press.

Martin, Charles B. (1994), "Dispositions and Conditionals", *The Philosophical Quarterly* 44: 1-8.

Rabinowicz, Wlodek (2002); "Does Practical Deliberation Crowd Out Self-Prediction?" *Erkenntnis* 57: 91-122.

Ramsey, Frank P. (1931 [1926]), "Truth and Probability," in Frank Ramsey, *Foundations of Mathematics and other Essays*, ed. By R. B. Braithwaite, London: Kegan, Paul, Trench, Trubner & Co., New York: Harcourt, Brace and Company: 156-98.

Spohn, Wolfgang (1977), "Where Luce and Krantz Do Really Generalize Savage's Decision Model," *Erkenntnis* 11: 113-34.

Wlodek Rabinowicz
Department of Philosophy
Lunds Universitet
222 22 Lund
Sweden
Wlodek.Rabinowicz@fil.lu.se

Lina Eriksson
Department of Philosophy
Flinders University
GPO Box 2100
Adelaide 5001, South Australia
Australia
lina.eriksson@flinders.edu.au

Ladislav Kvasz

Mathematics and Experience

Abstract

Mathematics was considered by the members of the Vienna Circle an *a priori* discipline consisting of analytic propositions. The aim of the present paper is to draw attention to the experiential dimension of mathematical knowledge. Following the interpretation of physical knowledge as knowledge constituted by the use of instruments, I am trying to interpret also mathematical knowledge as knowledge based on instrumental experience. This interpretation opens a new perspective on the place of the logicist program in philosophy of mathematics.

The aim of the present paper is to argue that mathematics is based on experience and that mathematical experience is instrumental. The difference between mathematics and physics is that in the case of mathematics the instruments are not measuring devices but *tools of symbolic representation*. In order to distinguish the mathematical experience from the physical one (i.e. from the experimental experience), I will call the instrumental experience in mathematics as *symbolic experience*. In analogy with the measuring devices of physics I will interpret the tools of symbolic representation as tools constituting symbolic experience.

1 The notion of symbolic experience

According to a widely held belief the instruments of symbolic representation can help us to find some mathematical propositions, but the justification of these propositions is purely logical, and therefore independent of any instruments. The decimal positional system *is not the reason* why it is that

$$135\,664 + 37\,863 = 173\,527,$$

to use the example from Frege's *Foundations of Arithmetic*. The role of the decimal system is only to compensate for our inability to perceive immediately this purely logical proposition. But here we have an analogy with the inclined plane that was used by Galileo to compensate for our inability to perceive the accelerated nature of free fall. Thus, already on the basis of the standard understanding of the role of instruments an analogy opens up between *the world of physics*, as the world which we access by means of measuring devices, and *the world of numbers*, as the world which we access by means of our tools of symbolic representation.

M.C. Galavotti et al. (eds.), *European Philosophy of Science – Philosophy of Science in Europe and the Viennese Heritage*, Vienna Circle Institute Yearbook 17,
DOI 10.1007/978-3-319-01899-7_9,

Nevertheless, this analogy does not reach far. Traditionally, the *physical world* is understood as independent of our representations and therefore its knowledge must be based on empirical experience. According to most philosophers of science the *world of physics* takes over that feature from the *physical world*, and thus the instrumental experience of physics, i.e. the *experimental experience*, is seen as a *kind of empirical experience*. On the other hand, many philosophers consider the propositions of mathematics as analytic and thus independent of empirical experience. So if we want to interpret mathematical knowledge as knowledge based on *symbolic experience*, this experience, it seems, has *non-empirical character*.

Thus between mathematics and physics, seen as two exact sciences, or between the experimental and the symbolic experience, seen as two kinds of instrumental experience, a basic asymmetry appears. Our goal is to *weaken* this asymmetry. To do so, we will radicalize the interpretation of the role of an instrument. Instead of interpreting an instrument as a "harmless" device, which only facilitates our access to some facts that exist independently of the instrument, we will stress the constitutive role of instruments. To achieve this we must first confront a widely held prejudice against mathematics.

a. *Chess as a false metaphor for mathematics*

Mathematics is often likened to the game of chess. This metaphor highlights a particular feature of mathematical symbols, namely that their meaning is fully determined by the rules of the particular calculus in which they function, just like the properties of a chess figure are determined by the rules of chess. A mathematician using a particular symbol resembles the chess player – the outcome of a calculation, just like the outcome of a chess game, is fully determined by the rules and does not depend on the state of the world. The following paragraphs are an attempt to show that this is a wrong metaphor for several reasons:

I. Infinitude. Chess is finite; it has a finite number of possible positions. Each representational tool, on the other hand, has a potentially infinite number of possible configurations. One could object that practically in the entire history of mankind only a finite number of such configurations (numbers, polynomials, or functions) have been used and this finite number is even much smaller than the number of possible positions in chess. Thus mathematics never used the infinitude of the possible configurations of its representational tools. This is true, but in order to create a potentially infinite number of configurations, the representational tools of mathematics must have an important feature, namely transparency.

*II. Transparency***.** The formulas of the various representational tools of mathematics are generated by a small number of simple rules that are constantly repeated. Therefore they are transparent and easy to comprehend. When one looks at a decimal number or an algebraic formula, it is easy to check whether it is properly constructed. In chess however, when we look at a complex situation, we cannot immediately say whether it is properly formed, thus whether it is an outcome of a possible game or not.

III. Standardness. For wide classes of tasks in mathematics there are standard methods to solve them. For the addition and multiplication of numbers of the decimal system we have standard algorithms, just like for solving linear or quadratic equations or for finding the integrals for wide classes of functions. In chess, on the other hand, we have to deal with each situation individually. There are no algorithms for solving classes of chess positions.

IV. *Approximativeness*. When we wish to calculate the integral of a function, we can expand it into an infinite series and integrate the series term by term. Taking the first few terms of the expansion we usually obtain a sufficiently good approximation of the result. Thus we approximate a difficult problem by an easy task, solve it and from the solution of the easy task we draw conclusions about the solution of the original problem. In chess there is no possibility of such approximation. There every situation is unique; it has its own peculiarities, which may be crucial for the outcome of the game. Every simplification is dangerous and approximation makes no sense.

V. Representativeness. The expressions of the representational tools (numbers, equations, integrals) can be put into a controllable relation to reality in such a way that from calculations with these expressions in the relevant calculus we can obtain information about some aspects of reality. When we count the number of coins in one pocket and in another one, we can find out how many coins there are by adding the two numbers. Of course, this is a banal example of the use of mathematics. Nevertheless, it is important to realize that by playing chess we achieve no such prediction. It is not possible to represent a slice of reality by a situation in a chess game and then, using only the rules of chess, to obtain some new knowledge about that slice of reality.

VI. Interpretativeness. Despite the freedom that forms the core of the chess metaphor, each representational tool of mathematics was created for a practical purpose, and thus its expressions naturally represent that aspect of reality, for the description of which the tool was originally created. Of course, the tools are not restricted to this particular aspect of reality, and can be used for purposes not foreseen by their creators. But still there is some aspect of reality, where the terms of the particular representational tool fit well. Chess does not allow any such thing. There is no segment of reality about which we could say that it allows a natural interpretation of chess.

VII. Mirroring. Another feature of representational tools is that they do not exist in isolation but are interrelated. There is a hierarchy of tools and when a tool encounters a situation which it cannot represent, sooner or later a new tool will be constructed that can do it. After it was discovered that the length of the diagonal of a unit square cannot be expressed as a ratio of integers, mathematicians created the real numbers for which this is no problem. The natural numbers are, of course, a part of the real numbers, so all calculations by natural numbers are mirrored in the system of the real numbers. Chess, on the contrary, exists in isolation. Nothing mirrors it and it does not mirror anything.

VIII. Syntheticity. Due to the fact that the calculi are interpreted, they can enrich our knowledge. When we show that solutions of a particular class of differential equations have some property (say stability), this can have deep consequences in physics, which is the area of natural interpretation of this particular tool. Chess is not capable of offering anything like this.

b. The horizon of representability as a permanent feature of instrumental experience

Now we can turn to the analysis of symbolic experience. First we must challenge the notion of reality (physical or mathematical) as something independent from instruments. Quantum mechanics has shown that measurement is not a passive registration of an independently existing phenomenon – the measurement enters the constitution of the phenomenon itself. Nevertheless, this does not violate the objectiveness of physical laws. The fact that the very notion of reality in quantum mechanics depends on the measurement process does not mean that we could measure what we wanted (in the sense that reason would prescribe laws to nature). The dependence of the phenomena on measurement (for instance in the double-slit experiment) does not open space to arbitrariness. Rather, Heisenberg's uncertainty principle determines the *boundaries of experimental experience* in an objective way.

When we interpret the world of mathematics as a world constituted by tools of symbolic representation, the example of quantum mechanics shows that the dependence of reality on the instruments does not mean the loss of objectivity. It does not open space for arbitrariness, but rather *puts limits to how far we can go.* So let us try to find such limits also in mathematics. All tools of symbolic representation are constructed so that

$$135\,664 + 37\,863 = 173\,527.$$

So the tools do not differ in content from the statements which can be derived by them. The point is rather that each tool of symbolic representation enables us to grasp only a fragment of the mathematical universe. If we use a non-positional system (e.g. the Roman numerals), it will be difficult to write numbers that in our positional system have more than 50 digits. To calculate, using the Roman numerals, the values of ln(0.8), ln(0.9), ln(1.1) and ln(1.2) *with an accuracy to 57 decimal places*, as Newton did in 1665, or to calculate π *with an accuracy to 60 decimal places*, as Euler did in 1736, is beyond human capabilities. It is likely that such values lie in the system of Roman numerals forever beyond the horizon. This does not mean that in this system these quantities would have other values. The choice of the tool of symbolic representation cannot affect the value of a particular quantity. It's just that if we use the Roman numerals, these values will remain beyond the horizon of possible experience.

The above arguments can be still challenged by saying that a mathematical truth (for instance the value of some quantity) is objectively given independently from any instrument. Thus we have not overthrown the position that mathematical truths are independent of our representational tools and that they are logically necessary. We have shown only that in some cases the full knowledge of mathematical truths is elusive. So far we have shown that the tools of symbolic representation determine the border of the accessible part of the realm of mathematical truths, but these truths may exist independently of any tools and may be necessary. In order to show that mathematical truths cannot be given independently of our representational tools and thus cannot be logically necessary, we have to turn to history.

c. Boundaries of language and the historical contingency of mathematical propositions

In the history of mathematics we can find a number of propositions that were proven by means of a particular tool of symbolic representation, but could not be proven by means of previous instruments. In (Kvasz 2008) I called this phenomenon the *logical boundaries of language*. The issue is not about difficulties of derivation, but about the objective non-existence of a derivation by means of a particular tool of symbolic representation. Similarly, a number of examples can be found when a mathematical object constructed by means of a particular tool could not be constructed by means of the previous ones. I have called this phenomenon the *expressive boundaries of language*. And again, the nonexistence of the object in the universe constituted by the particular tool of symbolic representation can be objectively proven. The issue is not that the object would lie far beyond the horizon. It simply does not exist in the universe constituted by the particular tool. What I want to point out is the objective existence of logical and expressive boundaries of every single tool of symbolic representation that mathematicians have invented so far.

The instrumental nature of experience imposes objective boundaries on knowledge. For numbers – elements of the universe of elementary arithmetic – all propositions, which can be proved by means of later representational tools (by means of algebra in algebraic number theory, by means of differential and integral calculus in analytic number theory), are valid. The problem is just that many of these propositions cannot be proven in elementary arithmetic. Thus the tension between the semantic notion of truth and the syntactic notion of provability has been present from the very beginnings of mathematics. Nevertheless, this tension is not easy to see by means of the framework itself. The unprovable propositions come usually to the fore only after a stronger instrument is invented.

I believe that the existence of the logical and the expressive boundaries is a *systematic feature of every tool of symbolic representation*. Every tool inevitably encounters such boundaries. I use history only for the reason that in the case of past representational tools we have at our disposal the newer tool, which allows us *to find explicitly the object or formula, which indicates the boundary of the former language*. This newer tool often allows us also *to prove* that the particular object is in the older language inexpressible, or that the formula is in the older language unprovable. But the very *fact of the existence of logical and expressive boundaries* of the particular language is independent of the existence of any later language. It is a systematic (and not only historical) feature of every mathematical language. The experiential aspect of mathematics consists in the fact that mathematics is able to break the boundaries of its present language, by creating a new representational tool, and so to open access to the study of new worlds that were inaccessible by means of the older tools. When mathematics thus *opens access* to a new universe of objects, and to new knowledge about these objects, it enriches our experience. *Breaking the logical and expressive boundaries of the language of mathematics can be interpreted as an opening of access to new kinds of experience*. In the present paper I will restrict myself to the discussion of the epistemological aspect of mathematics, leaving the ontological aspect for another occasion.

There is an analogy between the position described in this paper and the position of Saul Kripke concerning rigid designators. In both cases a kind of necessity is established by the introduction of a linguistic convention. This convention is arbitrary and thus, of course, a posteriori. In the linguistic community that accepts such a convention, there is a "causal link" connecting the actual users of the language with those, who introduced it. Nevertheless, there are also some differences between the two positions. Unlike Kripke, I am concerned not with names in ordinary language but rather with rules in artificial languages (the tools of symbolic representation). Further the theory I am trying to develop is concerned not so much with possible worlds semantics as with the history of mathematics. The necessities that I study are cultural and not metaphysical. Nevertheless, to spell out all these parallels and differences in detail would require much more space.

2 Logicism as the Thesis of Linguistic Pessimism

I see the experiential dimension of mathematical knowledge in the fact that the transition from the Roman to the Arabic numerals *opened access* to the value of the number π to 60 decimal places, or that the introduction of the algebraic symbolism *opened access* to the solution of cubic equations. Mathematics thus opened access to knowledge that was not possible to gain before. It got this access by breaking the boundaries of language, by creating a new tool of symbolic representation. This new tool allows us to create new representations, to experiment with these new representations and so to acquire new experience (for instance to accumulate sufficient experience with transformations of algebraic formulae, which is necessary for the discovery of the solution of cubic equations). The ability of mathematics to break the boundaries of every given tool of representation, which lies at the core of its ability to enrich our experience, challenges the traditional interpretations of mathematical knowledge.

Mathematical knowledge thus cannot be a priori. It depends on representational tools that are contingent cultural artifacts. Only when a certain tool is created, the corresponding area of symbolic experience becomes accessible. A characteristic feature of mathematical knowledge is that it deals with complex structures and subtle distinctions. Mathematics created its representational tools precisely in order to represent complex situations. The principles, on which mathematics is based, therefore concern not only the elements, but also the principles of construction of complex representations. And these principles, on which the tools of symbolic representation are based are conventions, and therefore cannot be a priori.

Nevertheless, *mathematical knowledge is not analytic either.* Knowledge gained by means of every particular representational tool can be made analytic (if we smuggle the conventions on which the particular representational tool is based into logic). *But to mathematics inseparably belongs also the activity of creating new representational tools.* Mathematical discoveries are often closely tied to linguistic innovations, which consist in creating a new representational tool. At first glance Hans Reichenbach and the logical positivists solved this problem by discriminating between the context of discovery and the context of justification. The construction of new representational tools falls within the context of discovery and thus it does not concern the program of logical positivism. But it is not so. Due to the fact that they defined the context of justification by means of formal logic, the logical positivists simply stripped one of the representational tools of mathematics of its historical context, and *out of a relative, contingent, historical mathematical tool* (of formal logic) *they made an absolute, universal, eternal framework of all knowledge*. However, during its long history, mathematics was never bound to any single framework and it managed to transcend each of them. The experiential aspect of mathematics is given precisely by its ability to create new representational frameworks that enable us to acquire new kinds of experience.

In hindsight, when all necessary linguistic innovations have already been completed, a particular piece of mathematical knowledge can be construed as analytic. However, this can be done only with knowledge for which we have a linguistic framework. In such a framework we can *discover* or *justify* that piece of knowledge. But mathematics can do much more than just discovering and justifying pieces of knowledge. It can *create new linguistic frameworks* within which the discoveries and the justifications of knowledge can take place.

The claim that mathematical knowledge is analytic can be interpreted as the thesis that *mathematics is forever tied to the representational tool of formal logic*, and contrary to all its historical experience, in future no new representational tool will be created. According to logicism mathematics has lost its creative power, by means of which it was able to create new representational tools and thus to overcome the logical and expressive boundaries of its past languages. Against the thesis of logicism we can argue that only (a part of) the *contemporary* mathematical knowledge is analytic, but not all *future* mathematical knowledge. It is likely that problems targeted by current mathematics will force it to create new representational tools, and some of the knowledge generated by means of these new tools will not be derivable by means of formal logic. The framework of formal logic has, just like all representational tools of the past, smaller expressive and logical power than later tools. A new tool will overcome the logical and expressive boundaries of its predecessors.

Therefore I will interpret the thesis that mathematical knowledge is analytic as the *thesis of linguistic pessimism*, that is, as the thesis that mathematics has already exhausted its creative force and that in the future there will be no radical breakthrough in mathematics, no overcoming of the logical and expressive boundaries of its language. All proofs and reasoning will forever be taking place in the linguistic framework created by Frege, Peano, Russell, Hilbert and Ackermann. Against this thesis, I would like to put the thesis of linguistic optimism, according to which the further development of mathematics will be in conformity with its history and mathematics will not remain trapped in its current framework.

The very fate of logicism indicates that Frege's pessimism was not justified. In the paper *Logicism reconsidered* (Rayo 2005) its author argues that logicism is refuted, and he uses as the main argument Gödel's incompleteness theorem. And Gödel's theorem is a result obtained using a new tool of symbolic representation. Thus despite Frege's pessimistic prognosis Gödel, Kleene, Church and Turing developed a new linguistic framework, by means of which it was possible to show that there are true sentences of arithmetic that cannot be proven. Thus the representational tool of Frege's logicism, just like the other tools of the past, has logical boundaries. Mathematics managed to create a more powerful tool of symbolic representation – recursive functions, lambda calculus, Turing machines – by means of which the boundaries of the previous framework became explicitly expressible.

3 Mathematics as an experiential science

Mathematics is not an a priori discipline, because it depends on the tools of symbolic representation, which are cultural artifacts, and therefore cannot be *a priori*. Mathematics, however, does not lose its character of objectivity and necessity, because the experience acquired by means of these tools is objective and necessary. But *mathematics is not analytic* in the sense of being reducible to logic or to any given linguistic framework. Thanks to its ability to create tools of symbolic representation, which open access to more and more complex patterns, *mathematics is an experiential science*.

Analytic geometry changed in a fundamental way our *experience of shape and of motion*. But it did so not in the sense that it would offer us some new facts (as experience is narrowly understood by the positivists) but that all the facts – those about the old curves, known since antiquity, and the new ones as well – were incorporated into a new linguistic framework. Against the background of this framework radically new and unexpected connections appeared and in the pile of data a pattern appeared. The fact that by means of a tool of symbolic representation mathematics opens access to phenomena and patterns that, without this tool would remain forever unknowable, characterizes the relation of mathematics to experience. Unless the particular representational tool was constructed, no one could perceive these patterns. These patterns could not be the subjects of our experience. Even if we would accidentally see them, we would not realize this. An essential feature of mathematics is its *ability to create* representational tools. These tools make possible the appearance of new kinds of experience, and this is *the experiential dimension of mathematics*. Analytic curves lay beyond the horizon of our ordinary experience. Not in the sense that the world around us would contain no objects having the shape of an analytic curve, but that without the appropriate representational tool *we cannot perceive them as such*. Mathematics by creating a representational tool opened our experience for these objects. Mathematics did not create any new fact (the facts have already lain before us), but it opened access to them.

Analytic geometry is not the only example when a representational tool fundamentally transformed our perception of reality. It seems that every representational tool brought a similar change. The discovery of the differential and integral calculus was constitutive for the emergence of physics in the same way analytic geometry was for the emergence of our perception of shape. That we understand some quantities as functions of other quantities is essential for all of modern physics, and at the same time it is utterly alien to the Ancient worldview. There is no doubt that physics is experiential science, but it became possible only after mathematics created the concept of a function and of a differential equation. Without these concepts we would perhaps still live in the universe of Aristotelian physics and would have very different experience with the world. Mathematics by creating tools, such as analytic geometry or differential calculus, which allow us to identify regularities in the pile of data and to make subtle distinctions (such as that between the first and the second derivative), enables us to *perceive meaningful information* where previous generations saw only meaningless chaos. The contribution of mathematics to experience is in creating tools for finding subtle patterns in what we perceive.

4 Logicism in the light of symbolic experience

In the 19th century a specific change occurred. One of the tools of symbolic representation, the predicate calculus, was identified with logic. This identification was a great mathematical achievement and led to the emergence of new disciplines such as proof theory or model theory. It was also a great incentive for philosophy, as it led to the emergence of analytic philosophy. In the philosophy of mathematics, however, the identification of the predicate calculus with logic concealed the fact that predicate calculus is one of the tools of symbolic representation which together constitute the world of mathematics. As a consequence of this identification predicate calculus started to be looked at as something more important than other mathematical theories; as something which can *serve as foundations* of mathematics. An attempt to reduce initially only arithmetic and then all of mathematics to logic was developed under the heading of logicism. Had it not been to formal logic but to "ordinary logic", logicism would be a remarkable program. But the proponents of logicism did not realize that as the result of the identification of formal logic with logic they did not reduce mathematics to logic, but to a formal calculus that is part of mathematics. If we put predicate calculus next to other tools of symbolic representation, we can see that its discovery fits into the overall scheme of development of mathematics. The predicate calculus is by no means exceptional and so *logicism is the reduction of mathematics to mathematics*. The logicist program of the reduction of arithmetic to logic becomes a mathematical problem of reducing one mathematical discipline to another mathematical discipline.

Logicism is parallel to a number of reductionist projects of the past. As the first we take "*arithmeticism*", as we may call the Pythagorean belief that all knowledge can be represented by (or reduced to) numbers and their ratios. According to Aristotle "The Pythagoreans believe that number is beginning of all being as its substance and also as its properties and states" (Metaphysics I, 5.985 b). Further we can take "*algebraicism*", as we may call Viète's belief, with which he concluded his *In Artem analytical isagoge* that by means of the analytical art (i.e. the formalism of algebra) he can solve all mathematical problems. He writes: "Finally, the analytical art that we put into threefold form of zetetic, poristic, and exegetic is entitled to appropriate the problem of all problems, which is: DO NOT LEAVE ANY PROBLEM UNSOLVED" (Viète 1591, p. 353). Viète's pronouncement expresses the belief that all of mathematics can be reduced to algebraic calculus – in full analogy with the way Frege wanted to reduce arithmetic to his (logical) calculus. A further parallel to logicism is "*analyticism*" as we can call the belief of Laplace from his *Philosophical Essay on Probability*:

> Given for one instant an intelligence which could comprehend all the forces by which nature is animated and the respective positions of the beings which compose it, if moreover this intelligence were vast enough to submit these data to analysis, it would embrace in the same formula both the movements of the largest bodies in the universe and those of the lightest atom; to it nothing would be uncertain, and the future as the past would be present to its eyes. (Laplace 1820, p. 4)

Laplace believed that calculus is able to solve all problems. The mathematical analysis, just like for Viète the analytical method, opens the gates for knowing everything. To this family belongs also "*logicism*", according to which by means of predicate calculus we can define all concepts of arithmetic and derive all of its basic propositions. The last member of this series is "*algoritmicism*", as we can call the belief that a Turing machine can simulate human thinking. In a textbook on recursion theory Odifreddi first formulates the AI thesis "Mental functions can be simulated by machines." and then comments that: "All work in artificial intelligence (pattern recognition, speech reproduction, problem solving, machine proving of theorems, playing games and learning with comprehension) constitute inductive evidence in favor of the AI thesis." (Odifreddi 1989, p. 117).

We see that whenever a new tool of symbolic representation was created, its creators succumbed to the fascination by the possibilities opened for mathematics by this new tool. And it must be admitted that it is understandable. Every new tool substantially improved the logical and expressive power of the language of mathematics, and thereby made it possible to solve many problems that were insoluble to previous generations. Therefore the hopes attached to the new tools are understandable. To their creators it appeared that they have found a tool that could solve absolutely everything. From this point of view we can understand logicism as a parallel to arithmeticism, algebraicism, analyticism and algoritmicism as an *interesting intra-mathematical project*: to reduce mathematics to a particular mathematical theory – mathematical logic. As in the case of other projects associated with the emergence of a new representational tool, also logicism substantially enriched mathematics with a number of new concepts, theorems and theories and thus fundamentally changed mathematics.

Similarly predictable, as the initial fascination with the possibilities of the new tool, was also the emergence of problems for the solution of which the new tool is not sufficient. Arithmetic was confronted with incommensurability, algebra with the insolubility of the equations of the fifth degree, analysis with the non-integrability of certain differential equations, and logic with the incompleteness of arithmetic. So the fate of logicism, from its initial plausibility till its final failure, is not different from the fates of the other programs. We can say that the project of logicism had the same fate as the similar projects of the past. Trying to save Frege's project by some form of neologicism, just like trying to save the other projects by means of some neopythagoreism, neoalgebraicism, or neoanalyticism is a mistake. Mathematics managed until now to surpass the boundaries of every tool of symbolic representation, and there is no reason why it should be otherwise in the case of logicism.

5 Conclusion

If we "trivialize" mathematics by closing it into the framework of logic (of intuition, or something else), we lose contact with its dynamic core, to which Cantor referred when he said that: "the essence of mathematics is its freedom". We feel this freedom when we read authors such as Descartes, Newton, or Riemann. *Descartes* liberated us from Euclid's narrow universe of shapes when he showed that besides the few curves that were known in Antiquity, there is a plethora of new ones. *Newton* freed us from the false Platonic identification of perfection with changelessness, which prevented the Ancients from understanding motion, when he showed that a dynamic system is a mathematical object and thus it is perfect, and yet it is subjected to change. *Riemann* freed us from the prison of the three-dimensional space by showing that we can create spaces of higher dimensions. We see that mathematics has an explosive power – it is able to break the limits into which human thought is imprisoned. It is only when we ignore this power that we can turn mathematics into a harmless discipline. This harmless discipline we can then put where we need according to our philosophical inclinations – into the realm of *a priori* intuition, into the realm of analytic propositions, or anywhere else.

Acknowledgements: The paper was written in the framework of the *Jan Evangelista Purkyně Fellowship* at the Institute of Philosophy of the Academy of Sciences of Czech Republic.

References

Ladislav Kvasz, *Patterns of Change, Linguistic Innovations in the Development of Classical Mathematics.* Basel: Birkhäuser 2008.

Pierre Simon Laplace, *A Philosophical Essay on Probabilities*. New York: John Wiley 1902.

Piergiorgio Odifreddi, *Classical Recursion Theory*. Amsterdam: Elsevier 1989.

Augustin Rayo, "Logicism Reconsidered", in: Stewart Shapiro (Ed.), *The Oxford Handbook of Philosophy of Mathematics and Logic*. Oxford: Oxford University Press 2005, pp. 203-235.

François Viète, "Introduction to the Analytical Art", in: Jacob Klein, *Greek Mathematical Thought and the Origin of Algebra*. Cambridge, Mass.: M.I.T. Press 1968, pp. 313-353.

Institute of Philosophy
Academy of Sciences
Jilska 1
110 00 Prague
Czech Republic
kvasz@fmph.uniba.sk

Eckehart Köhler

Gödel and Carnap[1] Platonism versus Conventionalism?

Introduction

Gödel emphatically opposed Carnap's "Syntax Program", based on a conviction that his Platonism was incompatible with Carnap's Conventionalism. These two doctrines were formulated much too sketchily by their authors to allow final conclusions. By paying closer attention to the real natures of these doctrines, i.e. by offering more exact explications, they can adequately be compared, with the surprising result that they are compatible, and that they differ only by a factor of "epistemological optimism".

What Are Foundations for Logic & Mathematics?

After publishing his *Aufbau* (1928), Carnap turned his attention to foundations of Logic and Mathematics. Carnap made these the most important problem for the Vienna Circle, using ideas of Hilbert (and of Wittgenstein) – and of Poincaré.

Especially in the 1920s, the foundations of Logic and Mathematics were poorly understood, particularly compared with those of the Natural Sciences. The main reason for this was that observation was shunted aside. In the Natural Sciences, measurement and other forms of **observation** had become rather sophisticated, and standards had become widely accepted, whereas no standards of observation for Logic and Mathematics had become generally accepted. It is crucial for a science to establish what the **content** of its knowledge is. Content is given by "data input". Without **data**, a science cannot pretend to constitute **knowledge**. And without observation, no data. Without data, we do not even have a science. This is the kernel of Gödel's objection to Carnap's "Syntax program". Unless Mathematics is subsumed under Natural Science – which Gödel strenuously denied – Mathematics has no subject domain and hence cannot exist.

What are the "observational data" for Logic and Mathematics? Hans Hahn (1933), for one, recognized the quandary into which the acceptance of Wittgen-

1 Presented on the occasion of the 20th anniversary of the Institute Vienna Circle, 6 Dec. 2011; earlier versions were presented at the meeting of the Gesellschaft für Analytische Philosophie, Berlin 2006, and at UC Irvine 2008. [In this text, important keywords are set **bold**; foreign and emphasized words are set *italic*; standard scientific subject areas and philosophical doctrines are capitalized, as in course catalogs.]

M.C. Galavotti et al. (eds.), *European Philosophy of Science – Philosophy of Science in Europe and the Viennese Heritage*, Vienna Circle Institute Yearbook 17,
DOI 10.1007/978-3-319-01899-7_10,

stein's position on the tautological character of Logic had put the Vienna Circle: Logic was meaningless (*sinnlos*), albeit not nonsense (*unsinnig*): Logic had no "Testability Criterion" of meaningfulness,[2] since no empirical observation could ever confirm Logic. All matters of (empirical) fact left it **invariant**. The *validity* of Logic, just like that of Ethics, seemed to be unobservable (Wittgenstein claimed it could only be "shown" and not written down in any observation statement – for one thing that would involve a metatheoretical claim, which Wittgenstein resisted). Since Logicists accepted the reduction of Mathematics to Logic, it followed in turn that all of Mathematics would be *sinnlos* as well. It followed that both Logic and Mathematics, like the allegedly non-cognitive Ethics, contained no knowledge. Then they were not sciences. Precisely this is what Hahn (1933) was so alarmed about.

This is absurd!! The battle against Metaphysics had gone too far! Many complaints were made. Perhaps the most sophisticated complaint was that of Beth (1963).[3]

The Conventionalist proposal offered by Carnap's *Logische Syntax* (1934) exacerbated the problem. Unfortunately, conventions were not analyzed in any detail.[4] But most particularly, Conventionalism tolerated so much freedom that *Syntax* was not unjustly accused of anarchism. At the time, Gödel was already a Platonist, hence disposed against Conventionalism; but he remained very discreet and lay low, discussing only his famous theorems. He only dared to write up his critique of Carnap's "Syntax Program" (so called by Gödel in his *1953) after being invited to contribute to Schilpp's (1963) volume on Carnap.

2 The Vienna Circle had originally been devoted to its beloved Verifiability Criterion of the meaningfulness of sentences. During the "Protokollsatzdebatte" at the latest, it was discovered that many legitimate sentences of science could not be verified by any finite number of observations. (The problem was that theoretical terms typically imply infinitely many sentences, only a finite subset of which are observable.) Carnap (1936/37) proposed his more practical Testability Criterion as the adequate resort. (It is clearly superior to Popper's Falsification Criterion, which arbitrarily ignores existential propositions.) See the section below on Empirical Tests for Platonism.

3 Beth (1963) argues in his §§V, VI that several passages of the *Logical Syntax* implicitly but clearly presuppose that *non-intuitive* (nonstandard) applications of Syntax-languages are *excluded*. This is partly hidden by Carnap's use of ordinary language as the meta-language for Syntax, but ordinary language automatically intends only *intuitive* (standard) interpretations. In a sense, Beth's argument is a variation of those of Quine (1936) and Gödel (*1953) dealt with below: all have in common that Syntax implicitly, but necessarily, presupposes content.

4 See the section "Decisions as Conventions" below for some reasons why the treatment was inadequate.

Is Carnap's Conventionalism Really Conventionalist?

In the 1980s, Hao Wang and I had many discussions about Gödel's thought and life. In discussing Gödel's critique of Carnap's Conventionalism,[5] I was stopped dead by a perceptive question. Wang pointed out the historical fact that Conventionalism is best known to philosophers of science as a version of neo-Kantianism concerning time and space, particularly in the form developed by Henri Poincaré (1902, 1905, 1908). Poincaré, as did many other Mathematicians since Gauss and Helmholtz, including Hilbert, agreed that Kant's epistemology based on intuition (Anschauung) would not work for Geometry. Poincaré proposed Conventionalism as a variety of neo-Kantianism.

Carnap (1934) abandoned neo-Kantianism: *Geometry* was all non-conventional, as it belonged to Physics since Einstein. Here Carnap made *all of Mathematics* and *all of Logic* be conventional instead. That doesn't look much like Poincaré. Wang asked: how could Carnap extend *Poincaré's* Conventionalism so far away from its originally relatively narrow area of application to part of Geometry? Logic had always been *univocal*, from Aristotle to Frege, to Russell, to (early) Poincaré. There had never been *alternative* theories to choose from; and without alternative *choices*, there can be no conventions.

Going beyond Poincaré's original position, I now treat the main stage of how Logic advanced from being regarded as clearly non-conventional, to become a candidate for conventionality (cf. Menger 1933), to Carnap's full-blown Conventionalism.

Carnap's Breakthrough

Transmogrification of Conventionalism

Especially after Brouwer introduced his Intuitionsistic Logic,[6] *Logic* (like Geometry) suddenly *also* allowed alternative choices, thus motivating Carnap to "extend" Conventionalism to Logic as well. Carnap was predisposed to do so, because he had always regarded Logic and Methodology from the standpoint of Instrumentalism, regarding them as *tools*. Leaving intuition now completely aside, Carnap (1934) made the newly available choices for Logic and Mathematics – **conventions**. Conventions are made through *our actions*; actors are now Nietzschean

5 Wang and I discussed various versions of Gödel's paper on Carnap, which we had obtained from his Nachlass at Princeton. Two versions ultimately appeared in Gödel (1995), edited by Solomon Feferman et al.

6 See van Stigt (1990, 1998) and Van Atten (2005).

Übermenschen, who feel no need to *justify* any actions by evanescent intuitions. (The entire Vienna Circle revered Nietzsche's decisionistic philosophy.)

Whereas Poincaré's Conventionalism said that empirically equivalent Mathematical structures of *Natural Science* could be chosen freely, the step now taken by Carnap encompassed all of *Logic and Mathematics*, which had previously been founded on *a priori* insight (Leibniz, Bolzano, Frege – all of them Logicists) and had been thought to be univocal. Carnap's breakthrough clearly goes way beyond what Poincaré's Conventionalism aimed at, so Wang's query to me was quite justified. Poincaré's conventionality of the Euclidean *metric* for space was held by the Vienna Circle to be *empirically* testable, like Newton's Mechanics. This was the position which Gauss, Helmholtz and Einstein had taken, *contra* Kant; and the only locus of conventionality remaining in Geometry was reduced to *coördinating definitions* (Reichenbach) for measuring distance, *i.e.* setting measurement standards.

In Logicism's new manifestation in Carnap (1934), Logic and Mathematics were now developed *together* in one system, as Hilbert had done it;[7] Mathematics now *appeared* to be as *empty* as Logic (Wittgenstein!?), no longer being based on intuition.

Decisions as Conventions

But what are conventions *really* ?? Neither Poincaré nor Carnap adequately explicated conventions, or discussed the politics of their negotiations.[8] E.g., no distinction was made between *individual* and *group* choices. Here are some points which the entire methodological literature has overlooked:

- According to (International) Law, conventions are *multilateral treaties*.
- If conventions had been acknowledged as simultaneous group choices, then methods of systematic consensus-formation might have been sought out to guide the search for usable principles and rules. But this required more sophistication in law and sociology than philosophical Methodologists or Mathematicians (usually) had before the 1950s.
- Decision and Game Theories had not yet matured enough to serve as a resource. They only became mature with von Neumann & Morgenstern (1944), especially (^{2}1947) with von Neumann's pioneering axiomatization of utility. Then in the 1950s, the great work of John Nash, Lloyd Shapley, Martin Shubick, John Harsanyi, and Jacob Marschak came along.

---

7 Carnap did not derive Number Theory from Type Theory, as orthodox Logicism required – thus sidestepping the controversies over the Axioms of Infinity and Reducibility. For a broader discussion of Logicism in the Vienna Circle, see Köhler (2006).

8 This had to wait for Lewis (1969), who uses Game Theory. This was a great initiative in what might be called "Behavioral Epistemics"; however, his work only treats the relatively trivial cases of self-enforcing rules (*e.g.* driving on the right); rather, for real-life conventions involving *both* conflict *and* coöperation, Bargaining Theory must be used. But Lewis doesn't mention "Nash-solution" anywhere.

What Carnap meant by Conventionalism arose mainly in opposition to dogmatism: if a decision can be made among several alternatives, Carnap insists on Tolerance. This attitude Carnap shared with the entire Vienna Circle: Hahn, Schlick, Menger, Neurath, all criticized Brouwer, preferring Hilbert's greater tolerance. And above all, Gödel wanted tolerance as well – though certainly not Carnap's Tolerance Principle.

A crucial feature of conventions left unmentioned by Poincaré and Carnap is their *inherent* "interpersonality". By their very meaning, they are never intended for individuals, but always for *all the people* in some group *simultaneously*! Their purpose is, *in principle*, to serve as universal standards for *everybody*. By nature, they are hence "intolerant" of idiosyncrasies impeding communication or mediation. Nothing much "subjective" about social institutions! Why has this not been taken more seriously? Understanding this "interpersonality" will play a key role below.

Vienna Circle Front against Dogmatism: Principle of Tolerance

In 1928, Brouwer (1929, 1930) delivered two lectures in Vienna, leaving an indelible impression on members of the Vienna Circle (especially on Gödel, but also on non-member Wittgenstein), who discussed them with great interest. But meanwhile, Gödel had been attending Carnap's new courses at the University of Vienna on foundations of Logic and Mathematics after his arrival there in 1928,[9] which quickly led to intensive discussions on foundations of Mathematics, on the Logicism of Frege (Carnap's teacher) and Russell, on Hilbert's Program, and Brouwer's Intuitionism. I reproduce here the most relevant of the discussions between Gödel and Carnap:

Discussion between Gödel and Carnap 1931 concerning Brouwer's restrictions on Classical Analysis[10]
Finitism. It is arbitrary where one wants to draw the line:
1) only concrete sentences
2) number variables; complete induction
3) Hilbert (in metamathematics)
4) Brouwer
5) classical mathematics
[…]
Gödel: It is entirely a matter of free decision (as Menger also thinks) which formulas and rules one first admits (in mathematics) as "meaningful" on the basis of associating some ideas with them. Hence one might just as well accept classical mathematics right away.

9 For a description of Gödel's relations with Carnap and the Vienna Circle, see Köhler (1991, 2002).

10 My translation from Carnap's discussion notes of 10 June 1931 (now in the University of Pittsburgh Library). All of Carnap's discussion notes with Gödel were first published in Wang (1987, p. 51), Köhler (1991), and also in Köhler et al. (2002).

There is no plausible distinction; even though one can draw a well-defined boundary line at various places.

Thus Carnap and Gödel in 1931 agreed to oppose dogmatism. What was omitted from this discussion, however, was an understanding of what *objects* the theory is really talking about: The problem was whether Tolerance implies denying that theory has any content (the Instrumentalism of Duhem, perhaps Hilbert, and Carnap), or whether Tolerance just gives the researcher freedom to work on his own theory, "agreeing to disagree" about principles, but agreeing that content exists. Clearly Gödel's using the term "meaningful" in his discussion made little impression on Carnap's work in pure Syntax, which aimed at eliminating all talk of meanings.

Carnap's new position culminated in the notorious **Principle of Tolerance** (1934, §17), the core of the "Syntax Program" (following the translation of 1937):

In the foregoing we have discussed several examples of negative requirements (especially those of Brouwer, Kaufmann, and Wittgenstein) by which certain common forms of language – methods of expression and of inference – would be excluded. Our attitude to requirements of this kind is given a general formulation in the *Principle of Tolerance: It is not our business to set up prohibitions, but to arrive at conventions [Festsetzungen].*

...

In Logic, there are no morals. Everyone is at liberty to build up his own logic, i.e. his own form of language, as he wishes. All that is required of him is that, if he wishes to discuss it, he must state his methods clearly, and give syntactical rules instead of philosophical arguments.

Karl Menger, in an exchange of letters, claimed priority for the Principle of Tolerance, since Menger (1928, 1930, 1933) had written on Intuitionism in a similar vein; so Carnap (1934, §17) gave *him* credit for Tolerance.[11] Not a whisper about Gödel's discussion. To be sure, Gödel would want no credit for Carnap's Principle anyway: he could not possibly subscribe to its rather anarchic formulation.

Carnap (1934) drove his Syntax Program to an extreme: he emphatically urged all right-minded Methodologists to, please, drop all contentual language (*inhaltliche Sprache*, called "the material mode" in 1937) and to replace it everywhere by the "formal mode", in order to avoid confusion – and above all, to avoid Metaphysics.[12] This Gödel could not abide. But he waited several decades to

11 Menger's (1928) "dictionary" of Intuitionistic terms was a precursor to Gödel's (1933, 1933a) interpretation of Intuitionistic Arithmetic and Logic, two of the most important contributions to Intuitionism, and influential in Kripke's later work on Modal Logic. Unfortunately, the otherwise very helpful introductory notes in Gödel (1995) by A.S. Troelstra missed Menger's precedence. Menger had been a docent with Brouwer in Amsterdam (1925–1927) and had arranged his invitation to Vienna in 1928. *Cf.* Menger (1979).

12 Carnap (1934, §81) *did* permit use of the "material mode" – so long as it was always clear that any apparent metaphysical references could be eliminated by a simple shift to formal mode. Carnap's later transition to Semantics relaxed the restriction much

counter Carnap. Gödel waited until Paul Arthur Schilpp began preparing a volume in honor of Carnap in the early 1950s. The Carnap volume appeared as Schilpp (1963); but, alas, without Gödel.

Gödel's Refutation of Carnap's "Syntax Program"

Unfortunately, readers had to wait until after Gödel's death to find out his views. In his article, Gödel proved that Syntax can never eliminate content (Inhalt). Here are the steps of Gödel's argument:

In every system of conventions, their mutual consistency must be assumed.

- Without *intuitions* about correctness, or even plausibility, we are clueless about correctness, and so we are forced to *prove* consistency.
- But the (meta-)Mathematics used for such proofs has to be "contentual Mathematics" in Hilbert's sense, not merely "formal".[13]
- Hence the very attempt to found Mathematics by avoiding reference to its contents forces one right back to the contents themselves. QED.

It was the essence of Hilbert's Program that consistency proofs for mathematics must be undertaken to "secure" mathematics, which Hilbert held to be an adequate foundation. But Carnap instead apparently assumed that advances in Logic succeeded in overcoming inconsistencies, so consistency required no proof.

Gödel's argument is unnecessarily prolix. Quine (1936) long before pointed out that a lot of Set Theory ("contentual Mathematics" *par excellence*) must be used in order to *merely display* conventions in the first place – even *before* consistency proofs are started. E.g. the famous definition of truth by Tarski (1936) uses all the types of Type Theory. And Carnap (1934) himself used all of Type Theory

more, but as Gödel complained, Semantics was just as conventional and non-intuitive as Syntax, so nothing had changed. Cf. the section "Carnap's Semantics" below.

13 The proofs must persuade us, so we must "see" their point. Content can only be given by intuition, which observes normative validity *immediately*, *i.e.* without inference from other evidence: what Russell (1911) called "acquaintance". (Of course, content can also be extended to include all the theory-mediated inferences from directly intuited data as well.) In the "Introductory Note" to Gödel (*1953), Warren Goldfarb (1995) overlooks that Evert Beth preempted Gödel's refutation of *Logical Syntax*'s rejection of intuition. Indeed, the high point of Beth (1963, §VI) was that, in Theorem 36.6 of the *Logical Syntax*, the sentence W_{II} was called analytic (*and* true), where W_{II} says precisely that Language II is *consistent*, using Gödel-numbering. Beth notes that this theorem is convincing *only if* Language II is interpreted in the standard (intuitive) way, which of course was just Gödel's point that consistency proofs presuppose intuition. Carnap's reply to Beth in Schilpp (1963, *pp.* 927*ff.*) concedes the whole game: "And perhaps I would also further agree with his further statement: 'I think that it is even possible to show that, in the absense of such non-formal, intutitive interpretation, the whole edifice of Logical Syntax would miss its purpose'." More Gödel could not wish for. But Gödel took no note of Beth, nor of Carnap's reply, in his discussions with Wang in the 1970s; *cf.* Wang (1996).

contentually, particularly in Carnap (1935), where his criterion of validity for sentences of Mathematics resorts to *all higher types* of Russell's Type Theory (Gödel helped Carnap with this).

Explication of Intuition

My explication of intuition is simple:[14] **intuitions are nothing other than value judgments**. They tell us what propositions and what rules are correct / good / valid. Now there are two ways to represent value judgments:

- as **preference rankings** (or as **utility functions** derived from them; *cf.* von Neumann 1947), and
- as **heuristics**, which prioritize principles or rules in hierarchical rankings (Herbert Simon).

Either of these make intuition *easily and immediately distinguishable* from sensory observation. *This*, I claim, is *just what* distinguishes Mathematics from Natural Science. The **values** of Mathematics are "orthogonal" to the **facts** of Natural Science. Hume already saw the analogous point for Ethics in his famous discovery that "Oughts" are never derivable from "Is's".[15]

Putnam (2002) claims to have proved the opposite in his *Collapse of the Fact / Value Dichotomy*, following the orthodoxy meanwhile established by Quine. But he proved no such thing. Putnam "demonstrated" that many colloquial value declarations are *ambiguous* ("thick", "entangled"). They are disambiguatable! Putnam reneges on the supreme obligation on Analytic Philosophers to strive for explications in order to solve philosophical problems. He forgot Carnap's great chapter on the topic of Explication.[16]

All the Explication we need can be gotten from Decision Theory, the most prominent model of Rationality ever developed, which absolutely requires that

---

14 This explication just reformulates the explication in Carnap (1968), quoted below in the section "Conventions Connected with Intuitions": "the ability to discriminate between valid and invalid reasoning". This means we make a *value judgment* about *correctness* of sentences or rules. Long before I (re)read Carnap (1968), I had realized that my explication was the only feasible one, even though no one, except perhaps White (1956), (1981), (1986) had proposed it before. (White had proposed that logic & math are branches of *ethics*; however, he did not cleanly separate empirical observation from ethical intuition, which many naturalistic pragmatists such as Dewey, Lewis and Quine are loath to do.)

15 In Köhler (2002b), I emphasize this feature of Logic and Mathematics unnoticed by Gödel (*1953), and I quoted there the famous passage in Hume's *Treatise* arguing for the dichotomy. The two approaches to declare value judgments are partly interchangeable; *cf. e.g.* Cornides (1974).

16 Carnap (1962, Ch. I "On Explication"). Putnam (1951) wrote his dissertation, "The Meaning of the Concept of Probability in Application to Finite Sequences", under Hans Reichenbach at UCLA; but he didn't list Carnap (1945), a famous paper explicating two different concepts of probability, which led in consequence to a deeper treatment of Explication in (1950) at the beginning of that treatise.

facts and values be declared independently. Otherwise *neither* descriptive *nor* prescriptive applications of Decision Theory can work. This is *my* Transcendental Deduction!

- **Empirical applications** of Decision Theory require empirical psychologists to distinguish between *sensory stimulation* ("cues") and *enforcement*, the latter addressing *motives* of behavior (values behind action), in order the explain or predict *actual* behavior at all.
- **Normative applications** of Decision Theory require utility functions (value distributions) to be independent of an actor's probability functions (representing the facts an actor estimates) – otherwise an optimal action (which maximizes expected utility) cannot be computed.

Putnam knows Decision Theory rather well, and has worked in it. Why did he not use its conceptual framework?

Carnap's Semantics

Soon after the publication of the Logische Syntax (1934), Carnap was influenced by Tarski (1932, 1936) to extend Syntax to Semantics. Prima facie, it would seem that Semantics explicitly acknowledges a contentual Mathematics, so that Carnap (1936, 1942, 1943, 1947) may seem to be abandoning his "Syntax Program". Shouldn't the turn to Semantics require Gödel to rethink his critique of Carnap's Syntax Program?

Gödel thought nothing of the sort.[17] Carnap himself said Semantic rules are also conventions! True, Carnap now used the ontological *terminology* of reference, and explicitly uses object terms – but all of that can be interpreted purely Instrumentalistically! Intuition is *still* not accepted, so we still have no real content for Logic or Mathematics, because Carnap still had not acknowledged that thinkers need to **perceive** the objects of reference with their minds. To show how right Gödel was, we look to Carnap's *Introduction to Semantics* (1942) and find the following: **Conventionalism *remained* the basis for Semantics:**

> The principle of tolerance (perhaps better called "principle of conventionality"), as explained in [1934/37, §17], is still maintained. On the other hand, the construction of a system of logic … within a given semantical system is not a matter of mere convention; here the choice is essentially limited if the concepts are to be adequate … (Carnap (1942, p. 247)

Such "limits" were not enough to assuage Gödel. Despite superficial appearances of Realism (which had infuriated Neurath into thinking Carnap had become a Metaphysician), a form of Instrumentalism seems to have been retained:

- Insofar as conventions are used to *circumvent* (or eliminate) intuition, they do not represent any *knowledge*, irrespective of the fine talk of meanings.

---

17 Gödel (*1953, §45) analyzes Carnap (1950), where "internal questions" have no "cognitive content".

- To be sure, Conventionalism is at least "relativized" to certain well-known nomenclature.
- "Content" is now allowed to be explicitly *mentioned*, but it has no *backing* – human beings have no access to it. The words are just a *façon de parler*.

Of course, it is obviously absurd to *mention* objects in a large theoretical treatise without admitting any epistemic, intuitive *access* to them. Few people noticed.

Gödel's Definition of Platonism

What was Gödel's position really? At the end of his so-called Gibbs Lecture (*1951, p. 38), Gödel offered his one and only precise, and I think also *adequate*, definition of Platonism, a passage which I quote in its entirety because of its importance:[18]

> Assuming that mathematical knowledge exists, … the Platonistic view is the only one tenable. Thereby I mean the view that mathematics describes a non-sensual reality, which exists independently both of the acts and [of] the dispositions of the human mind and is only perceived, and probably perceived very incompletely, by the human mind. This view is rather unpopular among mathematicians; there exist, however, some great mathematicians who have adhered to it. For example, Hermite once wrote the following sentence [Darboux: *Éloges académique et discours*, 1912, p. 142]:
> [Gödel quotes from the French:] 'There exists, unless I am mistaken, an entire world consisting of the totality of mathematical truths, which is accessible to us only through our intelligence, just as there exists the world of physical realities; the one, just as the other, is independent of us, both of them divinely created, …'

Analysis of Gödel's Platonism

According to the analysis I gave in my "Gödels Platonismus" (2002b), there are *three aspects* to Platonism implied by Gödel's definition (ultimately reducible to *one*):

i. [**Intuition**] Humans have a faculty (an "access" / "our intelligence" / Mathematical intuition) enabling them to view the "*other world*".
ii. [**Dualism**] The "other world" *differs* from "Physical reality".
iii. [**Objectivity**] Our Intuition *reveals* the "other world" – albeit only "partially".

Notice that sensory observation employs the *same* criterion of objectivity as Gödel requires for intuition: if we know whether we are hallucinating or suffering from illusions, we can work to *correct* them; and gradually we achieve objectivity.

18 The AMS talk is informally called the "Gibbs Lecture" by Gödel scholars. The translation from Darboux (1912) is mine (E.K.). In Gödel's later discussions with Wang, Gödel claims there is a "weak Platonism" which "everyone can accept"; I think Hermite's definition is so general that it is probably what Gödel had in mind here.

Why Intuition Fell into Disrepute

By the 1920s already, intuition had seriously fallen into disrepute – especially within the Vienna Circle. Intuition seemed unsalvageable in the wake of the "crises" of reason or intuition: Kant thought intuition to be apodictic, whereas the intuition actually used in Mathematics seemed highly unreliable, as Bolzano (1810) emphasized long ago. Hahn (1933) expresses the exasperation of an experienced practitioner of higher Analysis, a field which in the earlier 19th century had become notorious for its many antinomies and contradictions. Neurath and Neider agreed (1929).

Making one drastic break with older Platonism and with the traditional Rationalism which had evolved from it, Gödel found intuition to be incontrovertibly *fallible*! Fallibility was "completely obvious" to Gödel, due to the *antinomies of set theory*: the paradoxes of Russell, Burali-Forti, Richard *et al.* And Gödel was insightful enough to put the Fallibility of intuition to work, in two remarkable ways (Köhler 2002b):

- Gödel noticed that, due to its very fallibility, intuition could be seen to be objective (point iii. above). Objectivity follows, because, when we perceive an error, we conclude that we cannot see all things the way we *want* to see them: our *will* alone cannot guarantee freedom from error. The existence errors proves that something must exist outside our minds, and this forces us to correct our intuition.
- Gödel allowed intuitions to function as data for an *experimental*, *inductive* approach to Logic and Mathematics – *fully analogous to sensory observation* in the Natural Sciences! (Here Gödel's analogy between intuition and sensory observation culminates.)

But Gödel was shy and didn't press the issue in the Vienna Circle, and his vision lay fallow. Carnap (1968) finally acknowledged intuition, but Gödel took no note of that. Even after 1968, no one picked up the ball because the prejudices against intuition had become so entrenched, particularly among "Naturalists" such as Quine. However, ignoring philosophical dogmas, experienced senior Mathematicians were in effect resurrecting intuition, specifically the methodology of Heuristics of Pólya (1954). Pólya was highly regarded by Carnap, who had visited him in Zürich in the early '30s and later in Stanford, where he had become University Professor.[19] Just such heuristics correspond exactly to my general explication of intuition.

19 It was not Pólya, however, who influenced Carnap's acknowledgment of intuition, but rather work Carnap was doing on probability with Abner Shimony and Richard Jeffrey, especially on de Finetti. I report on this and many other discoveries in a book, *Intuition Regained* (forthcoming). Pólya ultimately influenced Artificial Intelligence and Cognitive Science, especially through Herbert Simon, Carnap's student at Chicago, who based his whole approach of ("boundedly rational") Heuristics on Pólya's methodology for Mathematical discovery and inference. Also influenced by Pólya, Douglas Lenat (1977), a doctoral student at Stanford who had previously studied with Simon at

Conventions Connected with Intuitions

Does Intuition exist as a distinct faculty of observation? (Explication of Intuition continued):

None other than Carnap himself provides incontrovertible support for my explication of intuition. In 1965, at a large conference organized by Imre Lakatos at Bedford College in London, Carnap "came out of the closet" and publicly accepted intuition as the correct epistemic foundation for not only *inductive* Logic, but also *deductive Logic* as well! (For Carnap as Logicist, Logic included Mathematics as well.) Among the most engaged discussants the next day was Karl Popper, who (not realizing this would destroy Mathematical knowledge) rejected intuition *tout court*.[20] In replying to Popper (1968), the endlessly patient Carnap (1968, *p.* 310) assumed that, as so often in philosophy, terminological problems are causing the trouble, and tried finessing "intuition" by proposing to relabel it:

> Since Popper seems allergic to the terms 'inductive' and 'intuition', let us for the moment use neutral terms. Instead of 'inductive' we shall say 'probabilistic' (understood either in the logical or personalistic sense of 'probability', not the frequency sense); and instead of 'intuition': 'the ability to **discriminate** (at least in simple cases) **between valid and invalid** reasoning', or briefly 'discriminatory ability'. I can hardly see reasons for serious doubt when I make the following assertion, which indeed seems to me rather trivial: we cannot hope to teach someone to use a language and, furthermore, to speak and think in a logically correct way unless he has from the beginning certain basic discriminative abilities; and this holds for the whole field of reasoning, including deductive and probabilistic reasoning. But what I would like to emphasize is the following point, which is not trivial. I believe that deductive discriminatory ability, even of a very high degree, is not sufficient as a basis for the whole field of reasoning, including probabilistic reasoning. … [my emphasis, E.K.]

Here Carnap reverts to the earlier terminology of his definition (1935) of a general criterion of *validity* for classical Mathematics, whose sentences he, in this case, did *not* call "L-true", but rather "valid" (gültig"). The term "valid" is very interesting, because it is obviously chosen to avoid attributing (factual) truth to Mathematics. The term makes it explicit that we have here to do with a *value judgment* rather than a factual judgment; it emphasizes the modal distinction between two types of judgment. To find a Logical principle "valid" is a *value judgment, i.e.* judging a norm to be correct. And, as I said before, the way we declare this is by establishing a preference ranking or by prioritizing norms. (See also Köhler 1985,

---

Carnegie-Melon in Pittsburgh, developed a program called AM, which featured hundreds of heuristics. (AM conjectures and derives theorems of *e.g.* prime number theory starting from simple axioms of set theory and a large suite of heuristics covering set theoretical concept formation, and finding proofs.)

20 In consequence, Popper not only deprived himself of knowing truth in Natural Science (since he rejected all acceptance methods of statistics), he deprived himself of knowing any normative principles, too.

2002b.) On the other hand, the term "discrimination" is a standard notion from the Psychology of Perception, *i.e.* of observation – but clearly *not* of sensory observation; rather of registering whether something is felt to be rational or not, which is based on … feelings.[21]

We are now prepared to further explicate Gödel's Platonism towards my goal of showing it to be compatible with Carnap's Conventionalism. First I repair a misconception about Plato's Heaven; then I look at conventions.

Correction of Old Errors about Plato's Heaven

We need an *improved explication* of Plato's "second world" (Hermite's "other world" in Gödel's citation). Hermite, Gödel and other Platonists fall into the same trap as Plato himself if they assume that the domain of Ideas comprise a *second domain of objects* (Aristotelian "substances"), existing factually somehow together with empirical objects. (I refer to the traditional picture of Socrates's real bed and ideal bed somehow posted side-by-side in the same bedroom.) Plato himself knew perfectly well the absurdities that follow from his evident failure to explicate his Ideal world adequately: I refer to the contortions he puts Socrates through in the second part of the *Parmenides*.

The correct explication should make the ideality of Plato's Heaven not that of ideal substances or objects, but rather that of *ideal propositions*. I claim that Plato[22] and his student Aristotle were using the *wrong logical categories*. (They did not yet have a Propositional Logic, which the Stoics were shortly to introduce, nor a Deontic Logic.) The distinction which Pythagoras and Plato want is the *deontic* one between real situations and optimal situations.[23] NB: the objects (Aristotle's substances) are *the same* for both. Leibniz knew the correct categories, because Leibniz had Deontic modalities. Aristotle was better known for distinguishing the related *epistemic modalities* of contingency and necessity, but that's another story. We are all acquainted with Wittgenstein's definition of the (real) world as "*alles, was der Fall ist*", *i.e.* the set of all facts. We now place another world, the ideal

21 Better: rational inclinations, or sentiments, the term used by Adam Smith (1759).

22 Plato got his Heaven from Pythagoras, and hence from Persian Theology. In the Persian, and especially the Zoroastrian tradition, Heaven was "outside" our world, not of it; and the gods Ahura Mazda and Ariman, etc., were in *this* Heaven, and not material beings. Pythagoras's numbers shared this feature of abstractness, and Plato's ideas were just generalizations of Pythagoras's numbers.

23 Most influential on Deontic Logic has been von Wright (1963). Rose Rand (1939), a member of the Vienna Circle, was one of the first to publish on it, with Hans Kelsen's help. Very interesting is Cornides (1974), who connects Deontic Logic with Preference Logic. Protagoras, the law teacher of Plato's dialogue, presented an early deontic paradox: *cf.* Åqvist (1984) for a most authoritative overview on Deontic Logic. Carnap (1963) wrote on "optatives", which he got from Menger (1939); but Carnap did not connect these with the epistemology of Logic or intuition. The best general introduction to values is Rescher (1969), and a classic reference on the good is von Wright (1963a).

world, next to it for comparison – something which every businessman and householder can do, *e.g.* when comparing actual *vs.* budgeted costs. (In German: *Ist kontra Soll.*) We must now learn to compare not bedrooms, but possible worlds!

The Real and the Ideal Worlds:

- The **real world** is the set of all facts, *i.e.* of all true propositions (Wittgenstein): Real World ≡ $\bigcup_i$ {facts$_i$}
- The **ideal world** is the set of all valid/correct/good propositions (Brentano): Ideal World ≡ $\bigcup_i$ {max$_j$ preferred$_{ij}$}; or alternatively:
  Ideal World ≡ $\bigcup_j$ {valid propositions$_j$} [all *optimally preferred* propositions$_i$.]

Advantages: This explication avoids the notorious "*Verdoppelung*", because we have no separate domains of *objects* (*e.g.* between concrete and abstract ones, involving issues of ontology quite distinct from modes of propositions). The above explication also has the great advantage of ensuring a *causal connection* between human perception and its objects (the "causal theory of perception"), for there is no causal gap hindering our recognizing a proposition as valid. Calling a situation optimal is easy if we observe it and verify its properties. We thus are able to avoid the two main objections long lodged against Platonic Idealism, the seriousness of which were *recognized by Plato himself*, who had no answer to them.)

Normative ideality is not an empirical property like 'red' or 'square'. It is an *operator* applied to propositions (states of affairs), not directly to objects. When objects are called 'ideal', they are so only with respect to specified standards, and this involves normative propositions. Decision Theory expresses such things by assigning *utilities* to propositions (at the same time it *independently* assigns *probabilities*).[24]

It is also useless to try defining Mathematics as the science of "abstract structures", *etc.*, which even Gödel was wont to do, as did many other Mathematicians.[25] Instead, Mathematics is the science of valid calculations and associated valid propositions, things that real computers and people can do or be. (Calculations can get infinite, but that doesn't make them any less physical; *cf.* Köhler 2000, where I investigate von Neumann's idea of "physicalizing" all of Mathematics and Logic.) Gödel opposed Aristotle's doctrine of "matter and form", thinking it made Dualism impossible, assuming Aristotle was trying to explicate ideality. But abstractness avails nothing here.

---

24 Richard Jeffrey does this in his well-known *Logic of Decision* (1965). Brentano had the corresponding ontology for propositions, where he allowed separate and simultaneous value assignments for factuality and goodness (true/false, and good/bad). A classic source for Decision Theory is Luce & Raiffa (1957).

25 Problem: many perfectly good *physical entities* are also abstract, *e.g.* electromagnetic fields, which are sets of *n*-tuples of points with their field-strength vectors.

Conventions Reveal Intuitions

Now I must destroy Gödel's misconception that Conventionalism always excludes intuition, a mistake which Poincaré, Hilbert and Carnap all fomented. None of them realized that conventions are connected *nolens volens* with intuitions – because (as Gödel noticed without noticing) they're connected to will, therefore to interest and therefore to value. But no one had an adequate explication for intuition, so the connection could not be made.

But now we *have* an adequate explication. Carnap had it already, but he did not live long enough to work through all the consequences, especially with respect to his old Syntax Program. To do so, we need mainly to apply Decision Theory to connect actions with value judgments. Conventions are **group decisions** or (multilateral simultaneous) **choices** of norms: **social institutions**. Now there are standard ways to connect values with choices, since our common sense tells us that our choices must follow our values (assuming rationality, of course!).

Very simply: if we choose something, we can infer that we value it. So our choices *reveal* our values. Since intuitions are evaluations, as I say, our choices reveal our intuitions! Just this relation was "discovered" by economists back in the 1930s and made to be the foundation of Consumer Choice Theory: every purchase reveals the preference of a consumer for one product over another. Paul Samuelson (1938, 1947) used this "discovery" in his campaign against the previously dominating introspectionist psychology of the Marginal Utility school (Carl Menger, Jevons, Walras), which had previously dominated the literature on the economic value concept. Of course Samuelson would not want to associate his "discovery" of "revealed preference" with intuition, which is always *introspective*; and Samuelson was not thinking about Epistemology of Mathematics. Nowadays, Economists have disengaged from Samuelson's campaign, and are willing to allow *both* introspective *and* choice-based data for valuations. We can now easily judge the connection between choices and opinion surveys concerning values.

But are **values** not quite subjective, and therefore worthless for epistemology? Nonsense, subjective judgments are perfectly valid, when done right. Rescher (1969, Ch. X) surveys evaluation of values; Jeffrey (1974) discusses "Preferences among Preferences". Brouwer's Intuitionism assumes that subjective calculations of mathematicians' minds are objectively perfectly valid. All judgments begin subjective, but "become objective" under suitable conditions. Here is how.

Making Conventions Objective

As Carnap (1968) indicated, training in Statistics, Logic and Mathematics all require subjective "abilities of discrimination". Assuming professional levels of competence, the performance of well-trained experts will rise to quite good standards of objectivity. Assuming also that the value judgments of experts are declared using preference rankings (or their derived utility functions) we may formulate this statement as a theorem holding for all norms that the experts arrive at:

> **Representation Theorem for Objective Intuitions** (sketch, E.K.):
> If the choices of norms made by an individual (or by groups of individuals) are **sufficiently reliable and stable**, then we may infer *preference rankings* and *utility functions* from them, assignable to the individuals (or to the groups). Since the utility functions represent value assignments, they constitute all of the intuitions of the individuals (or groups). Since they are reliable and stable, they are at least **coherent** (de Finetti 1937). Hence they will be useable in statistical reasoning, hence objectively rational by that criterion.

This is a standard type of theorem about measurement scales, showing that a measurement does in fact observe what is intended. It is also called a general proof of **validity** in Measurement Theory. It is proved using well-known Decision-Theoretic methods of measuring utilities, such as those of von Neumann (1947), or those of Davidson & Suppes (1957).[26] We may therefore infer that (introspections about) intuitions are objective when they are *sufficiently reliable and stable*. Rawls (1971, *p*. 20) explicitly made equivalent claims for the objectivity of moral intuitions: assuming these achieve "**Reflective Equilibrium**" (a standard of reliability and stability), then they are regarded as objective.[27] Intuition works the same way with all norms.

The Role of Pragmatic Criteria Backing Conventions

In his *Logische Syntax* (1934), Carnap explicitly used "pragmatic criteria" to guide him in choosing conventions (but without obligating himself *epistemically*: i.e. without taking such criteria to be evidence justifying belief). Even in Carnap's *Aufbau* (1928), we find pragmatic criteria. Already Poincaré pointed out that, however "free" conventions may be, they are often not "arbitrary" – because we always tend to prefer particular ones, depending on our own training, circumstances and predilections, or even just our tastes. Later on, in Carnap's famous "Empiricism, Semantics, and Ontology" (1950), these "pragmatic criteria" play an especially *prominent role* in connection with settling "external ontological" questions.

Quine refused to assign pragmatic criteria any role distinct from ordinary empirical evidence. (Quine is a "Holist" here: all evidence is "one-sorted".) But I now ask: To what extent may such pragmatic criteria nevertheless be regarded as *epistemic justifications* for belief, and hence as foundations for knowledge of

---

26 Such representation theorems specify the exact conditions of rationality (orderliness) which preference rankings must have in order to be measurable by utility functions. Such conditions typically include transitivity, completeness and so-called independence. On the relations between preference and utility, see the classic survey by Chipman (1960), the classic handbook article by Luce & Suppes (1965), and the more recent standard reference article by Mehta (1998).

27 Rawls (1971, *pp*. 20*f*.) in fact adapted Reflective Equilibrium from the approach that Goodman (1955, *pp*. 65–68) used to approximate, not moral objectivity, but reasonable standards for inductive inferences. Goodman was apparently thinking along the same lines as Carnap (1968) on inductive intuition.

norms? Bryan Norton (1977) claims that they *are* distinguishable from empirical evidence, and that precisely *this* fact is what distinguishes Carnap's "external ontological" questions about language systems or conceptual frameworks from "internal" ones. For pragmatic criteria are *value judgments* – that's just what distinguishes them from *factual judgments*. Though they are as subjective as any other judgment, they can be "made objective" when correctly applied – just as empirical observation can be objective!

Since pragmatic criteria have now been unclothed as *value judgments*, they're no different from intuitions! If this is so, then we may say that, in a certain sense, Carnap had (implicitly) acknowledged the faculty of intuition all along, ever since his (1928) – only he didn't find the right word for it. Why? Because accepted Mathematical terminology, such as Gödel consistently used, always had intuition giving us "Mathematical truth" or "facts", whereas there are no "truths" in value judgments.

Thus we see what chaos terminological confusion brings.

"Independence of the Human Mind"

Gödel's requirement that Plato's Heaven be "independent of acts and dispositions of the human mind" seems to put Heaven way beyond Conventionalism, which seems to be bound to (finite sequences of) human behavior. But conventions can be made arbitrarily strong! Conventions, being multilateral agreements valid for all minds, are already "independent of [all individual] human minds" in any case. And Carnap (1934) had already promoted conventions about transfinite procedures! Carnap was the first to make use of Hilbert's ω-rule, in order to define Logical-Mathematical validity [Gültigkeit] for his "indefinite" (Fraenkel) Language II. Clearly Carnap wasn't thinking the ω-rule-convention was for human operations, but rather for ideal, infinitely capable thinkers.

But a "Weak Platonist" working with only within finite bounds doesn't need Unbounded Rationality to indulge in Dualism, as does the "Strong Platonist" who accepts Cantor's Transfinite Hierarchy of numbers. He just needs to be perfect.

Conventionalism Is Compatible with Platonism

Since conventions *nolens volens* "reveal" implicit intuitions, we must see to what extent they confirm the aspects of Gödel's Platonism *à la* Hermite. The answer is yes, if such conventions are "well behaved". If they behave right, then Conventionalism "leads to" Platonism – just as Gödel said that "subjective Mathematics" somehow "leads to objective Mathematics" (Wang 1996, 8.3.2).

If so, then Conventionalism and Platonism are *compatible*. They do *not* really contradict each other at all, as Gödel (and presumably also Carnap) thought they did. This is not to say that the two "isms" are *the same*; but they do *not* struggle in mortal combat with each other. And since they *are* compatible, Platonism should hold for all conventions which "behave nicely".

The only real difference between Conventionalism and (Gödel's) Platonism is that the latter affirms decisively that there *will be* a Reflective Equilibrium for *every* open question concerning norms; whereas the Conventionalism (officially) doesn't care and remains agnostic. Gödel is convinced that human Reason is so powerful, it will find a way to decide them all. (This is why Hilbert's famous denial of du Bois-Reymond's "Ignorabimus" at his festival lecture at the meeting of the Gesellschaft für Naturforscher und Ärzte in Königsberg in 1930 makes him a Platonist just as Gödel would want.)

To salve the consciences of the rock-ribbed Logical Empiricists, we ask: do conventions satisfy the legendary Testability Criterion? Of course! Carnap (1936/37) first proposed a Testability Criterion for Empirical meaningfulness, and we need only extend it to allow the additional "sixth sense" of intuition to cover all judgments of Rationality. Since even opponents of intuition admit that it exists empirically, they cannot stop anyone from using its data to test the objectivity of "Norm-Workers".

Empirical Tests for Platonism

Recall that Gödel's Platonism, as defined in his "Gibbs Lecture", is actually an *empirical hypothesis*: Gödel (Hermite) says that *we humans* really & actually, in *this* world, *have* a "sixth sense". To empirically confirm this hypothesis, three hurdles must be cleared: we must derive the three elements of Gödel's (Hermite's) definition: i. Intuition, then ii. Dualism, and then iii. Objectivity.

- First, the very existence of intuitions is shown by pointing out that conventions, *i.e.* research choices of working Mathematicians, "reveal" the intuitions they're using. Samuelson's terminology was methodologically misleading at the time, but the term is very nice here.
- Dualism is (surprise!) *quite easy* to confirm: it simply follows from "Hume's Law". (How ironic: the arch-Empiricist among philosophers providing the critical argument in favor of Platonism!)
- The *most difficult* hurdle is Objectivity. (Other authors might use "Realism" here, but I dislike this term almost as much as Otto Neurath did.) Reflective Equilibrium is achieved if we have intersubjectivity and robustness.

Once again **Gödel's Analogy Argument**: Mathematical intuition is objective *in good analogy* with the objectivity of *sensory observation*. Empirical Science has done a good job convincing us of *its* objectivity. If Reflective Equilibrium presumably is an adequate test for the objectivity of inductive inference (Goodman) and that of moral intuition (Rawls), it should suffice for all other norms as well. Charles Parsons (2008, *p*. 324) endorses "wide reflective equilibrium" as a test for reasonableness.

Prognosis for Platonism

We seem to have approached a Reflective Equilibrium for *many topic areas* in Normative Sciences, especially in Logic and Mathematics; but also in Statistics, Decision Theory, Accounting, Contract Law, Rules of Order, etc. Hence Platonism seems to be reasonably confirmed for *some* normative areas. (Dualism is *trivially* verified, Objectivity with a *little bit* of work.) But in other areas (fashion, literature, punk rock) it may very well be that continual random fluctuations could be expected … forever! Fashion seems to wander *essentially randomly*, like the hemline levels of dresses over time. In contrast, science is allegedly cumulative and "progressive". Students of the Vienna Circle, however, are familiar with the warnings of Neurath and Frank against naïveté here. Whether Platonism is confirmable for all normative questions about the Pythagorean categories of Truth, Beauty, Goodness, is a vast expectation.

I once conjectured to Hao Wang that the Japanese and Chinese "discovery" of European classical music seems to confirm Platonism in art, and he liked that idea. It is an *empirical matter*, but we can conjecture some answers even now. Such conjectures may be influenced by our wishes for its success, however. Hence, there are also *normative aspects* to consider. Gödel both expected *and* wished very strongly that human reason and many sorts of rationality theories would grow spectacularly in the future.

Clearly Gödel's wish motivated his passionate belief in Platonism.

Carnap's Qualified Optimism

The general prospects for achieving equilibrium were explicitly discussed by Carnap in his letter joyfully acknowledging receipt of Thomas Kuhn's manuscript of "The Structure of Scientific Revolutions" (1962) for the *Encyclopedia of Unified Science*. Because of its importance and direct relevance, I quote the letter at length:[28]

Simultaneously I am returning your manuscript "The Structure of Scientific Revolution" [*sic*]. I am happy that it is now in final form and that the U. of Chicago Press has found a

28 Two letters from Carnap to Kuhn were excerpted by Reisch (1991, *pp.* 266*f.*); this citation is from the second of them. Carnap was predisposed in favor of Kuhn's approach from long acquaintance with Otto Neurath and Philip Frank, both of whom emphasized sociological-historical motives in scientific research. Frank founded the Institute for Unified Science in Cambridge MA, whose meetings were attended by Kuhn as a student. The Institute's archives relocated to Chicago after Frank's death. Reisch's discussion about whether "Kuhn Kill[ed] Logical Empiricism" takes no notice of Carnap's relations with Neurath and Frank; neither does it refer to work on scientific discovery methodologically much more important than that of Kuhn, Toulmin or Hanson, namely that of Carnap's student, Herbert Simon (1973), (1977); see especially the *magnum opus* by Langley, Simon *et al.* (1987).

way of publishing it in its full length. I am especially gratified by the fact that we can incorporate this work into the Encyclopedia.

I am convinced that your ideas will be very stimulating for all those who are interested in the nature of scientific theories and especially the causes and forms of their changes. I found very illuminating the parallel you draw with Darwinian evolution: just as Darwin gave up the earlier idea that the evolution was directed toward a predetermined goal, men as the perfect organism, and saw it as a process of improvement by natural selection, you emphasize that the development of theories is not directed toward the perfect true theory, but is a process of improvement of an instrument. In my own work on inductive logic in recent years I have come to a similar idea: that my work and that of a few friends in the step for step solution of problems should not be regarded as leading to "the ideal system", but rather as a step for step improvement of an instrument. Before I read your manuscript I would not have put it in just those words. But your formulations and clarifications by examples and also your analogy with Darwin's theory helped me to see clearer what I had in mind.

If the "instrument", of course, continually improves, it could be approaching equilibrium, hence an optimum; Carnap's point here, in reference to Darwin, is obviously that *local* improvements don't necessarily achieve a *global* optimum ("the ideal system"). Carnap had originally strongly believed in the possibility of finding a universally valid (globally "ideal") Inductive Logic, *i.e.*, that a single system of rules existed to compute the best theories for every set of data, constituting a global "logic of discovery".

Most working Statisticians, in particular Bruno de Finetti and Jimmy Savage, and even Dick Jeffrey (Carnap's own student and friend), held Carnap's original goal to be hopelessly over-optimistic; hence Carnap's letter seems to indicate a loss of optimism. But late in life, Carnap asserted he revered Leibniz more than all other philosophers. That means Carnap preferred optimism despite everything! Again in reference to Darwin: Cultural institutions are likely not chained in their evolution the way living beings are to their chemistry and to the chances of helpful mutations; Gödel's belief was that man's theoretical imagination can manage more radical mutations than DNA-based genes can. In any case, memes seem to vary much more quickly than genes.

Gödel loved Leibniz even more than Carnap did. Gödel was the greatest optimist of all, and Gödel would want his optimum to be *global.*

References

Lennart Åqvist (1984): "Deontic Logic", in Dov Gabay & Franz Guenther (eds.): *Handbook of Philosophical Logic* II, Extensions of Classical Logic, Reidel Publishing Co., Dordrecht.

Magdalena Aebi (1947): *Kants Begründung der "Deutschen Philosophie"*, Verlag für Recht und Gesellschaft, Basel.

Salvador Barberà, Peter J. Hammond & Christian Seidl (eds.) (1998): *Handbook of Utility Theory* 1, *Principles*, Kluwer Academic Publishers, Dordrecht.

Paul Benacerraf & Hilary Putnam (1964): *Philosophy of Mathematics: Selected Readings*, Prentice-Hall, New York; 2nd edition (with the section on Wittgenstein entirely replaced by a section on set theory), Cambridge Univ. Press, Cambridge 1983.

Bernard Bolzano (1810): "Über die Kantische Lehre von der Konstruktion der Begriffe durch Anschauungen", appendix to his *Beiträge zu einer begründeteren Darstellung der Mathematik*, Prague; reprinted by the Wissenschaftliche Buchgesellschaft, Darmstadt 1971.

Luitzen Egbertus Jan Brouwer (1929): "Mathematik, Wissenschaft und Sprache", *Monatshefte für Mathematik und Physik* 36, 153-164; translated as "Mathematics, Science, and Language", in Mancosu (1998).

—— (1930): *Die Struktur des Kontinuums*, brochure, Vienna; translated as "The Structure of the Continuum", in Mancosu (1998).

Evert W. Beth (1963): "Carnap's Views on the Advantages of Constructed Systems over Natural Languages in the Philoosphy of Science", in Schilpp (1963).

Bernd Buldt, Eckehart Köhler, Peter Weibel, Michael Stöltzner, Carsten Klein & Werner DePauli-Schimanovich-Göttig (2002): *Kurt Gödel, Wahrheit & Beweisbarkeit* II, *Kompendium zum Werk*, öbv&hpt, Vienna.

Rudolf Carnap (1922): *Der Raum. Ein Beitrag zur Wissenschaftslehre*, supplementary number 56 of *Kant-Studien*, Reuther & Reichard, Berlin.

—— (1923): "Über die Aufgabe der Physik und die Anwendung des Grundsatzes der Einfachstheit", *Kant-Studien* 28, 90-107.

—— (1928): *Der logische Aufbau der Welt*, Weltkreis-Verlag, Berlin; 2nd edition Felix Meiner, Hamburg 1961; translation by Rolf George as *The Logical Structure of the World*, Univ. of California Press, Berkeley 1967.

—— (1934): *Logische Syntax der Sprache*, Julius Springer, Vienna; English translation, including Carnap (1935), as *The Logical Syntax of Language*, by Amethe Smeaton, Countess von Zeppelin, Kegan Paul, London 1937.

—— (1935): "Ein Gültigkeitskriterium für Sätze der klassischen Mathematik", *Monatshefte für Mathematik und Physik* 42, 163-190.

—— (1936): "Wahrheit und Bewährung", *Actes du Congrès international de philosophie scientifique, Sorbonne, Paris 1935,* 4. *Induction et probabilité*, Actualités scientifique et industrielles 391, Hermann & Cie, Paris; translated and adapted as "Truth and Confirmation", in Feigl & Sellars (1949).

—— (1936/37): "Testability and Meaning", *Philosophy of Science* III (1936), 419-471; IV (1937), 1-40; reprinted in entirety in Robert Ammerman: *Classics of Analytic Philosophy*, McGraw-Hill, New York 1965.

—— (1942): *Introduction to Semantics*, Harvard Univ. Press, Cambridge MA.

—— (1945): "The Two Concepts of Probability", *Philosophy and Phenomenological Research*, 5, 513-532; reprinted in Feigl & Sellars (1949), and in Feigl & Brodbeck (1953).

—— (1947): *Meaning and Necessity: A Study in Semantics and Modal Logic*, Univ. of Chicago Press, Chicago; 2nd edition with supplements, Chicago 1956.

—— (1950): "Empiricism, Semantics, and Ontology", *Revue internationale de philosophie* 4, 20-40; reprinted in 2nd edition of Carnap (1947), and reprinted in Benacerraf & Putnam (1964), among many other places.

—— (1950a): *Logical Foundations of Probability*, Univ. of Chicago Press, Chicago; 2nd revised edition, Chicago 1962.

—— (1955): "On Some Concepts of Pragmatics", *Philosophical Studies* 6, 89-91; reprinted in the 2nd edition of Carnap (1947).

—— (1963): "Reply to Abraham Kaplan", in Schilpp (1963).

—— (1968): "Inductive Logic and Inductive Intuition", in Lakatos (1968).

John S. Chipman (1960): "The Foundations of Utility", *Econometrica* 28, 193-224; reprinted in R. Duncan Luce, Robert Bush & Eugene Galanter (eds.): *Readings in Mathematical Psychology* II, John Wiley, New York 1965.

Thomas Cornides (1974): *Ordinale Deontik: Zusammenhänge zwischen Präferenztheorie, Normlogik und Rechtstheorie*, Springer-Verlag, Vienna.

Louis Couturat (1905): *Les Principes de Mathématiques, avec un appendice sur la philosophie des mathématiques de Kant*, Alcan, Paris; translated as *Die philosophischen Prinzipien der Mathematik* by Carl Siegel, Verlag von Dr. Werner Klinkhart, Leipzig 1908.

—— (1906): "Pour la logistique (Réponse à M. Poincaré)", *Revue de Métaphysique et de Morale*, 208–250.

Donald Davidson & Patrick Suppes (1957): *Decision Making: An Experimental Approach*, in collaboration with Sidney Siegel, Stanford Univ. Press, Stanford.

Herbert Feigl & Wilfrid Sellars (eds.) (1949): *Readings in Philosophical Analysis*, Appleton-Century-Crofts, New York.

Herbert Feigl & May Brodbeck (eds.) (1953): *Readings in the Philosophy of Science*, Appleton-Century-Crofts, New York.

Bruno de Finetti (1937): "La prévision: Ses lois logique, ses sources subjectives", *Annales de l'Institut Henri Poincaré* 7, 1-68; translated as "Foresight: Its Logical Laws, Its Subjective Sources", in Henry Kyburg, Jr. & Howard Smokler (eds.): *Studies in Subjective Probability*, John Wiley & Sons, New York 1964.

Warren Goldfarb (1995): "Introductory Note to *1953/9" [= Gödel's Carnap-paper], in Gödel (1995).

Kurt Gödel (1933): "Zur intuitionistischen Arithmetik und Zahlenlehre", *Ergebnisse eines mathematischen Kolloquiums*, 4, 34-38; translated as "On intuitionistic arithmetic and number theory" in Gödel (1986).

—— (1933a): "Eine Interpretation des intuitionistischen Aussagenkalküls", ibid., 4, 39-40; translated as "An interpretation of the intuititionistic propositional calculus" in Gödel (1986).

—— (*1951): "Some Basic Theorems on the Foundations of Mathematics and Their Implications", in Gödel (1995); a lecture presented at a meeting of the American Mathematical Society specifically designated to honor Willard Gibbs [* indicates: previously unpublished].

—— (*1953): "Is Mathematics Syntax of Language?", in Gödel (1995).

—— (*1970): "Ontological Proof", in Gödel (1995).

—— (1986): *Collected Works* I, *Publications 1929–1936*, edited by Solomon Feferman, John Dawson, Jr., Stephen C. Kleene, Gregory H. Moore, Robert Solovay & Jean van Heijenoort, Oxford University Press, Oxford.

—— (1995): *Collected Works* III, *Unpublished Essays and Lectures*, edited by Solomon Feferman, John Dawson, Jr., Warren Goldfarb, Charles Parsons & Robert Solovay, Oxford University Press, Oxford.

Nelson Goodman (1955): *Fact, Fiction, & Forecast*, Harvard Univ. Press, Cambridge MA.

Hans Hahn (1933): "Die Krise der Anschauung", in *Krise und Neuaufbau in den exakten Wissenschaften. Fünf Wiener Vorträge*, [edited by Karl Menger, who did not want his name on the publication,] Franz Deuticke, Leipzig and Vienna; translated as "The Crisis in Intuition" in Hans Hahn: *Empiricism, Logic and Mathematics: Philosophical Papers*, Reidel Publishing Co., Dordrecht 1980.

Peter J. Hammond (1998): "Objective Expected Utility", in Barberà *et al.* (1998).

—— (1998a): "Subjective Expected Utility", in Barberà *et al.* (1998).

Daniel Hausman (1992): *The Inexact and Separate Science of Economics*, Cambridge Univ. Press, Cambridge.

David Hilbert (1899): *Grundlagen der Geometrie*, Teubner, Leipzig; translation of the 10th German edition as *Foundations of Geometry*, Open Court, LaSalle IL 1990.

—— (1905): "Über die Grundlagen der Logik und der Arithmetik", *Verhandlungen des Dritten Internationalen Mathematiker-Kongresses, Heidelberg, August 1904*, Teubner, Leipzig; translated as "On the Foundations of Logic and Arithmetic" by Jean van Heijenoort (ed.): *From Frege to Gödel: A Source Book in Mathematical Logic, 1879–1931*, Harvard University Press, Cambridge MA 1967.

Richard Jeffrey (1965): *The Logic of Decision*, McGraw-Hill, New York.

—— (1974): "Preference among Preferences", *Journal of Philosophy* LXXI, 377-379; reprinted in R. Jeffrey: *Probability and the Art of Judgment*, Cambridge University Press, Cambridge 1992.

Eckehart Köhler (1985): "On the Unity of All Normative Sciences as a Hierarchy of Rationality Principles", in *Proceedings of the 9th International Wittgenstein Symposium, Kirchberg am Wechsel, Aug. 1984*, Hölder-Pichler-Tempsky, *Vienna.*

—— (1991): "Gödel und der Wiener Kreis", in Paul Kruntorad (ed.): *Jour fixe der Vernunft: Der Wiener Kreis und die Folgen*, Verlag Hölder-Pichler-Tempsky, Vienna; revised version issued as Köhler (2002).

—— (2000): "Why von Neumann Rejected Carnap's Dualism of Information Concepts", in Miklós Rédei & Michael Stöltzner (ed.): *John von Neumann and the Foundations of Quantum Physics*, Vienna Circle Institute Yearbook 8, Springer, Dordrecht.

—— (2002): "Gödel und der Wiener Kreis", in Köhler et al. (2002).

—— (2002a): "Gödels Jahre in Princeton", in Köhler et al. (2002).

—— (2002b): "Gödels Platonismus", in Buldt, Köhler, et al. (2002).

—— (2006): "Ramsey and the Vienna Circle on Logicism", in Maria Carla Galavotti (ed.): *Cambridge and Vienna: Frank P. Ramsey and the Vienna Circle*, Vienna Circle Institute Yearbook **12**, Springer, Dordrecht.

—— (to appear): "Poincaré, Carnap and Gödel on Conventionalism and Platonism".

——, Peter Weibel, Michael Stöltzner, Bernd Buldt, Carsten Klein & Werner DePauli-Schimanovich-Göttig (2002): *Kurt Gödel, Wahrheit & Beweisbarkeit* I, *Dokumente und historische Analysen*, öbv&hpt, Vienna.

Georg Kreisel (1958): "Hilbert's Programme", *Dialectica* 12, 346-372; reprinted in Benacerraf & Putnam (1964).

Thomas Kuhn (1962): *The Structure of Scientific Revolutions*, brochure in the *International Encyclopedia of Unified Science* II, 2, edited by Otto Neurath,

Rudolf Carnap & Charles Morris, University of Chicago Press, Chicago; issued in various formats, finally enlarged with a postscript and included in the bound collection *Foundations of the Unity of Science* II, same publisher, 1970.

Imre Lakatos (ed.) (1968): *The Problem of Inductive Logic, Proceedings of the International Colloquium in the Philosophy of Science, London, 1965*, Vol. 2, North-Holland Publishing Co., Amsterdam.

Pat Langley, Herbert A. Simon, Gary L. Bradshaw & Jan M. Zytkow (1987): *Scientific Discovery: Computational Exploration of the Creative Process*, MIT Press, Cambridge MA.

Douglas Lenat (1982): *AM: Discovery in Mathematics as Heuristic Search*, Part 1 of Randall Davis & Douglas Lenat: *Knowledge-Based Systems in Artificial Intelligence*, McGraw-Hill, New York.

Sophus Lie (1871, 1959): "On a Class of Geometric Transformations", in D. E. Smith (ed.): *A Sourcebook in Mathematics*, Dover Publishing Co., New York.

R. Duncan Luce & Howard Raiffa (1957): *Games and Decisions: Introduction and Critical Survey*, John Wiley, New York; reprinted by Dover Publishing, New York.

R. Duncan Luce & Patrick Suppes (1965): "Preference, Utility, and Subjective Probability", in R. Duncan Luce, Robert Bush & Eugene Galanter (eds.): *Handbook of Mathematical Psychology* III, John Wiley, New York 1965.

Paolo Mancosu (1998): *From Brouwer to Hilbert: The Debate on the Foundations of Mathematics in the 1920s*, Oxford University Press, Oxford 1998.

—— (1998a): "Hilbert and Bernays on Metamathematics", in Mancosu (1998).

Ghanshyam B. Mehta (1998): "Preference and Utility", in Barberà *et al.* (1998).

Karl Menger (1928): "Bemerkungen zu Grundlagenfragen I: Über Verzweigungsmengen", *Jahrbuch der Deutschen Mathematikervereinigung* 37, 213-226; translated as "An Intuitionistic–Formalistic Dictionary of Set Theory", in Menger (1979).

—— (1930): "Der Intuitionismus", *Blätter für Deutsche Philosophie* 4, 311-325; reprinted in Menger (1979).

—— (1933): "Die neue Logik", in *Krise und Neuaufbau in den Exakten Wissenschaften. Fünf Wiener Vorträge* [organized by Karl Menger, who did not want his name on the publication]; translated as "The New Logic", *Philosophy of Science* IV, 299-336; reprinted in Menger (1979).

—— (1939): "A Logic of the Doubtful. On Optative and Imperative Logic", *Reports of a Mathematical Colloquium* 1, 53-64; reprinted in Menger (1979).

—— (1979): *Selected Papers in Logic and Foundations, Didactics, Economics*, Reidel Publishing Co., Dordrecht.

—— (1979a): "Logical Tolerance in the Vienna Circle", in Menger (1979).

John von Neumann (1947): "The Axiomatic Treatment of Utility", in John von Neumann and Oskar Morgenstern: *Theory of Games and Economic Behavior*, Princeton Univ. Press, 2nd edition (the 1st edition of 1944 had been *without* this famous appendix).

Bryan Norton (1977): *Linguistic Frameworks and Ontology: A Re-examination of Carnap's Metaphilosophy*, Janua Linguarum, Series Minor 145, Mouton Publishers, The Hague.

Charles Parsons (2008): *Mathematical Thought and Its Objects*, Cambridge Univ. Press.

Henri Poincaré (1902): *La Science et l'Hypothèse*, E. Flammarion, Paris; translated by F. Lindemann as *Wissenschaft und Hypothese*, Teubner, Leipzig 1904; translated as *Science and Hypothesis*, Dover Publications, New York 1952.

—— (1905): *La Valeur de la Science*, E. Flammarion, Paris; translated by E. Weber as *Der Wert der Wissenschaft*, Teubner, Leipzig 1906; translated as *The Value of Science*, Dover Publications, New York 1958.

—— (1908): *Science et Méthode*, E. Flammarion, Paris; translated by F. Lindemann as *Wissenschaft und Methode*, Teubner, Leipzig 1914; translated as *Science and Method*, Dover Publications, New York 1956.

—— (1908): *Dernier Pensées*, E. Flammarion, Paris 1913; translated by Karl Lichtenecker as *Letzte Gedanken*, Akademische Verlagsgesellschaft, Leipzig 1913; translated as *The Last Essays*, Dover Publications, New York 1963.

George Polya (1954): *How to Solve It*, Princeton University Press; several further editions, one a Doubleday Anchor Book, Garden City, New York 1957.

—— (1954a): *Induction and Analogy in Mathematics: Mathematics and Plausible Reasoning* I, Princeton University Press.

—— (1954b): *Patterns of Plausible Inference: Mathematics and Plausible Reasoning* II, Princeton University Press.

Karl Popper (1968): "Theories, Experience, and Probabilistic Intuitions", in Lakatos (1968).

Hilary Putnam (2002): The Collapse of the Fact/Value Dichotomy,

Willard Van Orman Quine (1936): "Truth by Convention", in O. H. Lee (ed.): *Philosophical Essays for A.N. Whitehead*, Longmans, New York; reprinted in W. V. O. Quine: *The Ways of Paradox and Other Essays*, Random House, New York 1966.

—— (1951): "Two Dogmas of Empiricism", *Philosophical Review* 60, 20-43; reprinted many times, first in W. V. O. Quine: *From a Logical Point of View*, Harvard University Press, Cambridge MA.

—— & Joseph Ullian (1951): *The Web of Belief*, Random House, New York.

John Rawls (1971): *A Theory of Justice*, Harvard Univ. Press, Cambridge MA.

Rose Rand (1939): "Logik der Forderungssätze", *Revue internationale de la Théorie du Droit*, Nouvelle Serie I, 308-322.

George A. Reisch (1991): "Did Kuhn Kill Logical Empiricism", *Philosophy of Science*, 58, 264-277.

Bertrand Russell (1900): *The Philosophy of Leibniz*, Allen & Unwin, London.

—— (1903): *The Principles of Mathematics*, Allen & Unwin, London.

—— (1911): "Knowledge by Acquaintance and Knowledge by Description", *Proceedings of the Aristotelian Society*, New Series XI, 108-128.

Paul Samuelson (1938): "A Note on the Pure Theory of Consumer's Behaviour", *Economica* 5, 61-71.

—— (1947): *Foundations of Economic Analysis*, Harvard Univ. Pr., Cambridge MA.

Paul Arthur Schilpp (ed.) (1963): *The Philosophy of Rudolf Carnap*, Open Court, LaSalle IL.

Adam Smith (1759): *The Theory of Moral Sentiments*, A. Miller, London; Glasgow Edition edited by D. D. Raphael & A. L. MacFie, Oxford University Press, Oxford 1976.

Herbert Simon (1972): "Theories of Bounded Rationality", *Decision and Organization*, ed. by C.B. Radner & R. Radner, North-Holland Publishing Co., Amsterdam; reprinted in H.A. Simon (1982): *Models of Bounded Rationality* II, *Behavioral Economics and Business Organization*, MIT Press, Cambridge MA.

—— (1973): "Does Scientific Discovery Have a Logic?", Philosophy of Science, 40, 471-480; reprinted in Simon (1977).

—— (1977): *Models of Discovery, and Other Topics in the Methodology of Science*, Reidel Publisching Co., Dordrecht.

Adam Smith (1759): *The Theory of Moral Sentiments*, London.

Alfred Tarski (1932): "Der Wahrheitsbegriff in den Sprachen der deduktiven Disziplinen", *Akademie der Wissenschaften in Wien, Mathematisch-Naturwissenschaftliche Klasse, Akademischer Anzeiger* LXIX, 23-25.

—— (1936): "Der Wahrheitsbegriff in den formalisierten Sprachen", *Studia Philosophica* I, 261-405; originally in Polish, 1933; translated by Joseph Woodger as "The Concept of Truth in Formalized Languages", in Alfred Tarski: *Logic, Semantics, Metamathematics: Papers from 1923 to 1938*, Oxford University Press, Oxford 1956.

Louis L. Thurstone (1954): "The Measurement of Values", *Psychological Revue* LXI, 47-58; reprinted in the posthumous collection L. L. Thurstone: *The Measurement of Values*, University of Chicago Press, Chicago 1959.

Mark Van Atten (2004): *On Brouwer*, Thomson/Wadsworth, London.

Dirk van Dalen (1999): *Mystic, Geometer, and Intuitionist: The Life of L. E. J. Brouwer*, 1: *The Dawning Revolution*, Oxford University Press, Oxford.

—— (2005): *Mystic, Geometer, and Intuitionist: The Life of L. E. J. Brouwer 1881–1966*, 2: *Hope and Disillusion*, Oxford University Press, Oxford.

Walter P. van Stigt (1990): *Brouwer's Intuitionism*, North-Holland Publishing Co., Amsterdam.

—— (1998): "Brouwer's Intuitionist Programme", in Mancosu (1998).

Morton White (1956): *Toward Reunion in Philosophy*, Harvard University Press, Cambridge, MA.

—— (1981): *What Is and What Ought to Be Done: An Essay on Ethics and Epistemology*, Oxford University Press, New York.

—— (1986): "Normative Ethics, Normative Epistemology, and Quine's Holism", in Lewis Edwin Hahn & Paul Arthur Schilpp (eds.) (1986): *The Philosophy of W. V. Quine*, The Library of Living Philosophers XXVIII, Open Court Publishing Co., La Salle IL.

Hao Wang (1987): *Reflections on Kurt Gödel*, MIT Prss, Cambridge MA.

—— (1996): *A Logical Journey: From Gödel to Philosophy*, MIT Press, Cambridge MA.

Stanley Wong (1978): *The Foundations of Paul Samuelson's Revealed Preference Theory: A Study by the Method of Rational Reconstruction*, Routledge & Kegan Paul, London.

Georg Henrik von Wright (1963): *Norm and Action*, Routledge & Kegan Paul, London.

—— (1963a): *Varieties of Goodness*, Routledge & Kegan Paul, London.

University of Vienna and Lauder Business School, Vienna
Lichtenauergasse 9/6
A-1020 Wien
Austria
eckehart.koehler@univie.ac.at

Pablo Lorenzano

What is the Status of the Hardy-Weinberg Law within Population Genetics?

1. Introduction

Some philosophers of biology argue that laws or principles in biology are non-empirical, but a priori or analytical, mentioning the Hardy-Weinberg law as an example.[1] Sandra Mitchell summarizes this view, though without accepting it: "That the mathematical laws used in biology (like Hardy-Weinberg equilibrium) are logical truths is neither controversial nor particularly pertinent".[2]

On the other hand, for biologists who analyze and present population genetics by means of models,[3] as well as for philosophers of biology, who work within the semantic conception of theories and therefore analyze biological theories (population genetics in particular) in terms of models,[4] the Hardy-Weinberg law

1 See Robert N. Brandon, "Adaptation and Evolutionary Theory", *Studies in History and Philosophy of Science* 9, 1978, pp. 181-206; Robert N. Brandon, "A Structural Description of Evolutionary Biology", in: Peter D. Asquith/Thomas Nickles (Eds.), *PSA 1980*. vol. 2, East Lansing, Michigan: Philosophy of Science Association 1982, pp. 427-439; Robert N. Brandon, "Does Biology Have Laws? The Experimental Evidence", *Philosophy of Science*, 64 (Proceedings), 1997, S444-S457; Elliott Sober, *The Nature of Selection: Evolutionary Theory in Philosophical Focus*. Cambridge, Massachusetts: MIT Press 1984; Elliott Sober, *Philosophy of Biology*. Boulder: Westview Press 1993; Elliott Sober, "Two Outbreaks of Lawlessness in Recent Philosophy of Biology", *Philosophy of Science*, 64 (Proceedings), 1997, S458-S467; and Mehmet Elgin, "Biology and A Priori Laws", *Philosophy of Science*, 70, 2003, pp. 1380-1389.

2 Sandra D. Mitchell, "Pragmatic Laws", *Philosophy of Science*, 64 (Proceedings), 1997, S471.

3 See Daniel L. Hartl, *A Primer of Population Genetics*. 3rd ed., Sunderland, MA: Sinauer Associates 2000; Daniel L. Hartl/Andrew G. Clark, *Principles of Population Genetics*. 3rd ed., Sunderland, MA: Sinauer Associates 1997; Albert Jacquard, *Structures Génétiques des Populations*. Paris: Masson & Cie, Editeurs, 1970; Richard C. Lewontin, *The Genetic Basis of Evolutionary Change*. New York: Columbia University Press 1974; Bruce Wallace, *Topics in Population Genetics*. New York: W. W. Norton and Co. 1968.

4 See John Beatty, "What's Wrong with the Received View of Evolutionary Theory?", in: Peter D. Asquith/Thomas Nickles (Eds.), *PSA 1980*. vol. 2, East Lansing, Michigan: Philosophy of Science Association 1982, pp. 397-426; John Beatty, "On Behalf of the Semantic View", *Biology and Philosophy*, 2, 1987, pp. 17-23; Elisabeth Lloyd, "A Semantic Approach to the Structure of Population Genetics", *Philosophy of Science*, 51, 1984, pp. 242-264; Elisabeth Lloyd, *The Structure and Confirmation of Evolutionary Theory*. New York: Greenwood Press 1988; Paul Thompson, "The Structure of

M.C. Galavotti et al. (eds.), *European Philosophy of Science – Philosophy of Science in Europe and the Viennese Heritage*, Vienna Circle Institute Yearbook 17,
DOI 10.1007/978-3-319-01899-7_11,

is a central law, but not the fundamental law, of (classical) population genetics: They consider it to be a simplified "(mathematical) model" (the most simple case in classical population genetics), since in (classical) population genetics there is an heterogeneous multiplicity of models that are obtained by violating and/or by eliminating the conditions assumed by the Hardy-Weinberg law (and by introducing new parameters).

A summary of such a position can be found in Bas van Fraassen:

> The scientific literature on a theory makes it relatively easy to identify and isolate classes of structures to be included in the class of theoretical models. It is on the contrary usually quite hard to find laws which could be used as axioms for the theory as a whole. Apparent laws which frequently appear are often partial descriptions of special subclasses of models, their generalization being left vague and often shading off into logical vacuity. Let me give two examples. The first is from quantum mechanics: *Schrödinger's equation.* [...] The second is the Hardy-Weinberg law in population genetics. [...] it appears in any foundational discussion of the subject. But it could hardly be an axiom of the theory, since it holds only under certain special conditions. If we look into the general case, we find a logical fact: that certain assumptions imply that it describes an equilibrium which can be reached in a single generation, and maintained. The assumptions are very special, and more complex variants of the law can be deduced for more realistic assumptions – in an open and indefinite sequence of sophistications.[5]

The connection between the two different positions of philosophers of biology on the Hardy-Weinberg law is evident; since this law is "just a bit of fairly simple mathematics", a "(mathematical) model", it is considered to be *a priori* or analytical.

The aim of this paper is to further develop van Fraassen's diagnosis, expanding a previous analysis of the fundamental law of classical genetics and the status of the so-called 'Mendel's laws'.[6] According to this diagnosis the Hardy-Weinberg law: 1) cannot be considered as axiom (or fundamental law) for classical population genetics, since it is a law that describes an equilibrium that 2) holds only

---

Evolutionary Theory: A Semantic Approach", *Studies in History and Philosophy of Science* 14, 1983, pp. 215-229; Paul Thompson, *The Structure of Biological Theories*. Albany, N.Y.: State University of New York Press 1989; Paul Thompson, "Formalisations of Evolutionary Biology", in: Mohan Matthen/Cristopher Stephens (Eds.), *Handbook of the Philosophy of Science: Philosophy of Biology*. Amsterdam: Elsevier, 2007, pp. 485-523.

5 Bas van Fraassen, "The Semantic Approach to Scientific Theories", in: Nancy Nersessian (Ed.), *The Process of Science*. Dordrecht: Nijhoff, 1987, p. 110.

6 Pablo Lorenzano, "Fundamental Laws and Laws of Biology", in: Gerhard Ernst/ Karl-Georg Niebergall (Eds.), *Philosophie der Wissenschaft – Wissenschaft der Philosophie. Festschrift für C.Ulises Moulines zum 60. Geburstag*. Paderborn: Mentis-Verlag 2006, pp. 129-155; Pablo Lorenzano, "The Influence of Genetics on Philosophy of Science: Classical Genetics and the Structuralist View of Theories", in: Anne Fagot-Largeault/Juan Manuel Torres/Shahid Rahman (Eds.), *The Influence of Genetics on Contemporary Thinking*. Dordrecht: Springer 2007, pp. 99-115.

under certain special conditions, and 3) only determines a subclass of models, 4) whose generalized form (and fundamental law) being shading off into logical vacuity, and 5) more complex variants of the fundamental law (and of the Hardy-Weinberg law) can be "deduced" for more realistic assumptions.

In order to achieve this, I will use notions of the structuralist view of theories,[7] a version that is related to but different from that of van Fraassen's. These are the notions of fundamental law, specialization, and special law.

Having as a background a structuralist reconstruction of classical population genetics, view of theories, I will show why the Hardy-Weinberg law should not be in fact considered the fundamental law of such a theory, but a special law (and not even a "terminal" specialization, i.e. a "non-terminal" specialization).[8]

2. The Hardy-Weinberg Law (or Principle or Equation)

This law was known in the English-speaking world as 'Hardy's Law' until 1943, only rarely called the 'Hardy-Weinberg-Castle Law', and even more rarely 'Castle-Pearson-Hardy-Weinberg Law' (even though we cannot find it literally and in a strict sense neither in Hardy's nor Weinberg's (or Castle's or Pearson's) work).[9] It states that, if only one gene with just two alleles, A and a, is considered it can be mathematically proved that given that certain conditions are fulfilled, the frequen-

7 For a comprehensive presentation of this metatheory, see Wolfgang Balzer/C. Ulises Moulines/Joseph D. Sneed, *An Architectonic for Science. The Structuralist Program*. Dordrecht: Reidel 1987.

8 Due to limitations of space we have to left for other works the presentation of a complete structuralist reconstruction of (classical) population genetics as well as its comparison with other attempts made within some informal variant of the received-view of scientific theories (such as Michael Ruse, *The Philosophy of Biology*. London: Hutchinson 1973) and within some other version of the semantic view of theories (such as those mentioned in footnote 4).

9 For instance, Hardy states the law in terms of characteristics and not in terms of alleles, and Weinberg states a more general version of it in terms of '*Anlagen*'; see Godfrey Harold Hardy, "Mendelian proportions in a mixed population", *Science*, 28, 1908, pp. 49-50; Wilhelm Weinberg, "Über den Nachweis der Vererbung beim Menschen", *Jahreshefte des Vereins für vaterländische Naturkunde in Württemberg*, 64, 1908, pp. 368-382. See also George Udny Yule, "Mendel's laws and their probable relation to intra-racial heredity", *New Phytologist*, 1, 1902, pp. 193-207, 222-238; William E. Castle, "The laws of Galton and Mendel and some laws governing race improvement by selection", *Proceedings of the American Academy of Arts and Sciences*, 35, 1903, pp. 233-242; Karl Pearson, "Mathematical contributions to the theory of evolution. XI. On the influence of natural selection on the variability and correlation of organs", *Philosophical Transactions of the Royal Society of London, Ser. A*, 200, 1904, pp. 1-66; and Curt Stern, "The Hardy-Weinberg law", *Science*, 97, 1943, pp. 137-138; James F. Crow, "Hardy, Weinberg and language impediments", *Genetics*, 152, 1999, pp. 821-825.

cies or relative proportions of the alleles *A* and *a* in the population will not change from one generation to the next, after the second generation. Thus, equilibrium will be reached in one generation and will remain unchanged after the second generation, meaning the same that the gene pool will be in a stationary state – equilibrium – with respect to those alleles. This equilibrium is expressed by the following equation: $p^2 + 2pq + q^2 = 1$, where p denotes the frequency of one allele, q denotes the frequency of the other allele, p^2 denotes the frequency of homozygous individuals for an allele, q^2 is the frequency of homozygous individuals for the other allele, $2pq$ is the frequency of heterozygous, and the sum p and q should always equal l, since $p + q$ represents the 100% of the alleles of that particular gene in the gene pool, from which it could be obtained $q = 1 - p$.

The conditions that should be met in order to hold the equilibrium stated by the law are the following: 1) organisms within population are diploid; 2) reproduction is sexual; 3) generations are non-overlapping, i.e. individuals from different generations do not breed together; 4) mating is random; 5) population size is very large (large enough for probability laws to be applied, i.e. it can be treated as infinite, or as sufficiently large for effects of genetic drift to be minimized); 6) migration is negligible; 7) mutation can be ignored; and 8) natural selection does not affect the alleles under consideration.

3. The notions of fundamental law or guiding-principle, of specialization and of special law

It can be claimed that, despite sustained efforts, a complete satisfactory concept of scientific law does not yet exist, since we still lack an adequate set of necessary and sufficient conditions serving as criteria for a statement to be considered a "(scientific) law". Moreover, "[i]t is likely that no such set of conditions can ever be found that would appear satisfactory to everybody since the notion of a law is a strongly historical, discipline-dependent kind of notion".[10]

Within the structuralist tradition, discussion of the notion of a law have usually focused on what Stegmüller called the 'fundamental law of a theory'[11] (in opposition to Fodor's sense,[12] where the expression 'fundamental laws' refers to laws of what he thinks are basic or fundamental sciences). When the criteria for a statement to be a fundamental law or guiding-principle of a theory are discussed, there is a tendency to speak about "necessary conditions", "*weak* necessary conditions"

10 Wolfgang Balzer / C. Ulises Moulines / Joseph D. Sneed, *op. cit.*, p. 15.

11 Wolfgang Stegmüller, *Theorienstrukturen und Theoriendynamik.* Heidelberg: Springer 1973.

12 Jerry Fodor, "Special Sciences (or: the Disunity of Science as a Working Hypothesis)", Synthese, 28, 1974, 97-116; Jerry Fodor, "Hedged Laws and Psychological Explanations", Mind, 100, 1991, 19-33.

or about "'symptoms', some of which can be even formalizable".[13] These are: 1) having a cluster or synoptic character (to include "all the fundamental or basic concepts that characterizes the theory", "several of the magnitudes", "diverse functions", "possibly many theoretical and non-theoretical concepts", "almost all" or "at least two"); 2) possessing "validity in *every* intended application" of the theory, but not an unrestricted scope nor applying every time and everywhere; 3) having a quasi-vacuous character (they are highly abstract, schematic, and contain essential occurrences of *T*-theoretical terms, which in structuralist sense are terms whose extension can only be determined through the application of a theory's fundamental(s) law(s) so that they can resist possible refutations, but which nevertheless acquire specific empirical content through a non-deductive process known as 'specialization'; this process, which provides more specific laws (the so-called 'special laws'), consists in the introduction of further restrictions, or specifications of (some of the components of) the fundamental law(s), in such a way that they become progressively concrete in diverse directions until we finally obtain the so-called 'terminal specializations' in which all components are specified; 4) playing a systematizing role, which allows to include diverse applications within the same theory since it provides a guide to and a conceptual frame for the formulation of other laws (the so-called 'special laws'), which are introduced to impose restrictions on the fundamental laws and thus apply to particular empirical systems.[14]

Most scientific theories have laws of very different degrees of generality within the same conceptual setting. We may say that all axioms of a theory are axiomatic but some are more axiomatic than others. A theory is not a 'democratic' sort of entity. Rather, it is a strongly hierarchical system. Usually there is a single fundamental law – which connects every fundamental term of the theory in only one 'big' formula, which is accepted as valid in all of the theory's applications by the scientific community and whose primary role is to provide a frame for the

---

13 C. Ulises Moulines, *Pluralidad y recursión*. Madrid: Alianza 1991, p. 233.

14 For a more extended treatment of these conditions, and the manner in which a better way of conceiving of fundamental laws is not as being mere definitional or analytical statements (or a priori as opposed to empirical) but as being 'quasi-analytical' or – by using a label with a more traditional flavour – 'synthetic a priori', and the relationships of such a proposal with the notion of a relativized a priori as first introduced by Reichenbach (Hans Reichenbach, Relativitätstheorie und Erkenntnis A Priori. Berlin: Springer 1920) and further developed by Friedman (in several works, among others in: Michael Friedman, Dynamics of Reason. Stanford: CSLI Publications 1999), see Pablo Lorenzano, op. cit. and "Lo a priori constitutivo y las leyes (y teorías) científicas", Revista de Filosofía, 33, 2008, 21-48. On the other hand, we think that this characterization of laws is compatible with recent proposals about laws in biology, such as the "paradigmatic" (Martin Carrier, "Evolutionary Change and Lawlikeness: Beatty on Biological generalizations", in: Gereon Wolters and James Lennox (Eds.), *Concepts, Theories and Rationality in the Biological Sciences*. Konstanz: Konstanz University Press and Pittsburgh University Press 1995, pp. 82-97) and "pragmatic" (Sandra Mitchell, *op. cit.*) ones (which we take as complementary).

formulation of other laws – 'on the top' of the hierarchy and a vast array of more special laws – which apply to more restricted domains.

The picture that emerges here is close to Kuhn's conception of disciplinary matrix, nomic generalizations, and exemplars/applications. According to Kuhn, in highly unified and developed theories, like Classical Particle Mechanics, there are some generalizations that are not "specific" but "schemes" that take specific forms for specific problems/applications:

> […] generalizations [like $f = ma$…] are not so much generalizations as generalization-sketches, schematic forms whose detailed symbolic expression varies from one application to the next. For the problem of free fall, $f = ma$ becomes $mg = md^2s/dt^2$. For the simple pendulum, it becomes $mg \,\mathrm{Sin}\, \theta = - md^2s/dt^2$. For coupled harmonic oscillators it becomes two equations, the first of which may be written $m_1 d^2s_1/dt^2 + k_1 s_1 = k_2(d + s_2 - s_1)$. More interesting mechanical problems, for example the motion of a gyroscope, would display still greater disparity $f = ma$ and the actual symbolic generalization to which logic and mathematics are applied.[15]

Note that the top-bottom relation is *not* one of *implication*, but – in structuralist terms – of *specialization*, i.e. bottom laws are specific versions of top ones.

4. The Structure of Classical Population Genetics and its Fundamental Law

Population genetics is usually characterized as the study of origins and dynamics of genetic variation in populations or, equivalently, as the study of changes in the genetic composition (or structure) of populations from generation to generation. The genetic composition of a population could be affected by mating its individuals and of the subsequent offspring in the next generation.

Several factors could affect the transmission and the change of the genetic structure of a population. Such changes constitute the so-called "evolution" of a population. In the literature four evolutionary factors or forces (the "causes" of changes) are often mentioned: migration, mutation, selection and genetic drift. On the other hand, individuals of a population have certain traits, characters or characteristics, by which they can be identified and that, by mating of the individuals in the population, are distributed in the next generation in a certain way – given in the form of ratios or relative frequencies. These ratios or relative frequencies in the distribution of the traits, characters or characteristics in the successive generations constitute what allows us to test (classical) population genetics (that is, its "(empirical) basis of testing"), allowing at the same time to express those empirical phenomena that the theory intends to account for. Finally, a certain kind of

15 Thomas S. Kuhn, "Second Thoughts on Paradigms", in: Frederick Suppe (Ed.), *The Structure of Scientific Theories*. Urbana, Ill.: University of Illinois Press 1974, p. 465.

relationship is assumed – that we call 'determining' – between the genetic composition of individuals of populations and their traits, characters or characteristics. We refer to this relationship as 'determining', and could depend not just on the genetic composition, but also on the environment in which the individuals of the populations develop (to include this in our account requires just a couple of easy modifications).

Now, we have all we need in order to characterize classical population genetics in a more precise way. Its basic concepts are: (1) A set of *populations* (*POP*) defined in terms of their *individuals* (*i* is an individual and *I* is the set of individuals): **pop** $\in$ *POP* and **pop** $\subseteq$ *Pot*(*I*). (2) A set of *types of characteristics* ($(C_i)_{i\leq k}$) that constitute the different *phenotypes* ($\pi_1, \ldots, \pi_k$); each phenotype $\pi \in P$ has the form $\langle c_1, \ldots, c_k \rangle$, where $c_1 \in C_1, \ldots, c_k \in C_k$. (3) A set of *types of (allelic) factors* ($(F_i)_{i\leq s}$) that constitute the different *genotypes* ($\gamma_1, \ldots, \gamma_s$); for every $i \leq s$ and any two factors f_1 and f_2, that f_1 and f_2 belong to the same set F_i means f_1 and f_2 are alleles; each genotype $\gamma \in G$ has the form of a finite list of *pairs of allelic factors* $\langle\langle f_{i1}, f_{i2}\rangle, \ldots, \langle f_{s1}, f_{s2}\rangle\rangle$, where $i \leq s$ and f_{i1} and f_{i2} are members of F_i. (4) A set of parameters, the *evolutionary factors, forces* or "causes" of changes in the distributions of gene frequencies (genotypes) in the offspring (*FEV*): migration, mutation, selection and genetic drift (symbolized by *m*, *n*, *s*, *F*, respectively). (5) The *time,* represented by a linear order $\langle T, < \rangle$, where it is assumed that *T* is a finite, non-empty, set, of indexes for generations (of populations), which also can be considered as representing points or periods of time *t*, and where $< \subseteq T \times T$ is transitive, anti-reflexive and connected; $\langle T, < \rangle$ is isomorphic with the pair $\langle \mathbb{N}, < \rangle$, consisting in the set of natural numbers $\mathbb{N}$ and in the relation-less-than on the natural numbers <. (6) The *appearance* of the populations – being parental or progeny –, given by distributions of phenotypes in relative frequencies (*APP*); we symbolize parental appearance by $APP(\mathbf{pop}_1) = \langle \pi_{11}, \ldots, \pi_{1k} \rangle$, and $APP(\mathbf{pop}_2) = \langle \pi_{21}, \ldots, \pi_{2k} \rangle$, and progeny appearance by $APP(\mathbf{pop}_n) = \langle \pi_{11}, \ldots, \pi_{ij}, \ldots, \pi_{nk} \rangle$ (where $i \leq n, j \leq k$). (7) The *mating* of the populations (*MAT*), that is, a partial function which represents the transition from the parents to their progeny, symbolized by equations of the form $MAT(\mathbf{pop}_1, \mathbf{pop}_2) = \langle \mathbf{pop}_1, \ldots, \mathbf{pop}_n \rangle$, where the number *n* varies according to the parental populations. (8) The *transition from a distribution of phenotypes* of one generation *to another distribution of phenotypes in the next generation* (*DIST*) (given by relative frequencies);[16] formally, a distribution is a function that assigns to each component of a given set a real number which indicates that element's "weight" or "probability of occurrence"; the set of all *genetic* distributions or Γ-distributions over the set *P* are described by means of *D*(*P*); formally, members **p** of *D*(*P*) are *stochastic genetic processes* over *P* with respect to *T*; this means that $\mathbf{p}: T \times P \rightarrow \mathbb{R}$; it is a function that assigns to points of time

16 For a phenotype $\pi \in P$ and a set of populations $X \subseteq Pot(I)$, the *relative frequency of* π *in X*, $FR(\pi/X)$ is defined in the following manner:
$FR(\pi/X)$ = (the number of elements in the sets $i \in X$, for which $APP(i) = \pi$) over
(the number of elements of elements of *X*).

and phenotypes non-negative real numbers, such that, for a fixed index of time t, $\mathbf{p}_t$ is a genetic distribution, where $\mathbf{p}_t(\pi) \geq 0$ and, for every $\pi \in P$, $\sum_{\pi \in P} \mathbf{p}_t(\pi) = 1$; *DIST* takes a set of such functions $\mathbf{p}_t(\pi) = \mathbf{p}(t, \pi)$ as arguments and assumes sets of functions in the next generation as values; such distributions can be written in the more convenient way $\sum r_i \pi_i$, where π_i varies in the set P and where r_i is the value of the corresponding function of the distribution, i.e. $r_i = \mathbf{p}_t(\pi_i)$; actually, *DIST* is not a real primitive concept, then it can be defined through *MAT* and *APP*; beginning with two parental populations $\mathbf{pop}_1$, $\mathbf{pop}_2$, we see the value of $MAT(\mathbf{pop}_1, \mathbf{pop}_2)$, i.e., the set of progeny $\{\mathbf{pop}_1, \ldots, \mathbf{pop}_n\}$; we determine the value of $APP(\mathbf{pop}_n)$ for $i \leq n$, i.e, phenotypes that occur in the progeny; we count the total number n of progeny as well as the number m_i of progeny that exhibits a given phenotype and calculate the relative frequency $r_i = m_i/n$ of that phenotype; the list of all relative frequencies obtained for the different progenies is thus the desired distribution of phenotypes in the progeny of $\mathbf{pop}_1$ and $\mathbf{pop}_2$, that is, the value of $DIST(\mathbf{pop}_1, \mathbf{pop}_2)$. (9) The *determination* of phenotypes by genotypes (and eventually also the *environment*) (*DET*) represented as a function which assigns phenotypes to genotypes; the genotypes *determine* phenotypes, but the inverse does not occur; a given phenotype may happen to be determined by several different genotypes; that is why we have chosen different numbers: k for the number of phenotypes and s for the number of genotypes, being k normally equals or less than s; for the case of individuals, we have equations of the form $DET(\gamma_i) = \pi_j$, and, for the case of populations, equations of the form $DET(\langle \gamma_1, \ldots, \gamma_s \rangle) = \langle \pi_1, \ldots, \pi_k \rangle$ (where $i \leq s$, $j \leq k$).[17] (10) The *transition from parental population genotypes to genotypes in the progeny population* (*COMB*); this is a function that represents the transition from parental genotypes to genotypes in the progeny, assigning to any distributions of particular genotypes a combination or a mixture (distributions) of genotypes of progeny. A distribution of genotypes could be considered as a genetic distribution, i.e. as a function that assigns numbers ("weights") to genotypes; this transition of genotypes is a kind of theoretical analogue of the transition of phenotypes (the difference between them is, in certain respects, one between experimental probabilities and expected or theoretical probabilities); the set of all *genetic* distributions or Γ-distributions of genotypes are described by means of $D(G)$ ("theoretical probabilities"); formally, members $\mathbf{p}^*$ of $D(G)$ are *stochastic genetic processes* over G with respect to T and of the set of parameters *FEV*; this means that $\mathbf{p}^*$: $T \times G \times FEV \rightarrow \mathbb{R}$; it is a function that assigns to points of time, genotypes and evolutionary factors non-negative real numbers, such that, for a fixed index of time t, and fixed evolutionary factors *fev* (which could be one or some of the above

17 If it is considered convenient to take account of the environment, assuming that the determination of the phenotypes by the genotypes depends on it as well, this could be carried out through the following modifications: 1) introducing the non-empty finite set *ENV* of "environments" (variable ε: $\{\varepsilon_1, \ldots, \varepsilon_r\}$), and 2) redefining the function *DET* in the following way: DET: $G \times ENV \rightarrow P$ is surjective (with $DET(\gamma, \varepsilon) = \pi$, for individuals, or $DET(\langle \gamma_1, \ldots, \gamma_s \rangle, \varepsilon) = \langle \pi_1, \ldots, \pi_k \rangle$, for populations).

mentioned ones: migration m, mutation n, selection s or genetic drift F), $\mathbf{p}^*_{t,fev}$ is a genetic distribution, where $\mathbf{p}^*_{t,fev}(\gamma) \geq 0$ and, for every $\gamma \in G$, $\sum_{\gamma \in G} \mathbf{p}^*_{t,fev}(\gamma) = 1$; $\mathbf{p}^*_{t,fev}$ is formally defined by establishing $\mathbf{p}^*_{t,fev}(\gamma) = \mathbf{p}^*(t, \gamma, fev)$, such that $\mathbf{p}^*_{t,fev}$: $G \rightarrow \mathbb{R}$ has the right kind for being a Γ-distribution; *COMB* takes a set of such functions $\mathbf{p}^*_{t,fev}(\gamma) = \mathbf{p}^*(t, \gamma, fev)$ as arguments and assumes sets of functions in the next generation as values; we have equations of the form $COMB(t, \langle \alpha_1\gamma_{11}, \ldots, \alpha_s\gamma_{1s} \rangle, \langle \alpha_1\gamma_{21}, \ldots, \alpha_s\gamma_{2s} \rangle, m, n, s, F) = \langle \alpha_1\gamma_{11}, \ldots, \alpha_j\gamma_{ij}, \ldots, \alpha_s\gamma_{ns} \rangle$ (where all α_i are positive real numbers, such that $\sum_{1 \leq i \leq s} \alpha_i = 1$).

We can we put together all basic concepts in one structure x, which thus contains the "conceptual framework" of the theory: $x = \langle I, (C_i)_{i \leq k}, (F_i)_{i \leq s}, FEV, T, <, APP, MAT, DIST, DET, COMB \rangle$. Its different objects and functions can be graphically represented in the following way (see Fig. 1), where the objects are represented by rectangles and the functions by arrows.

Fig. 1

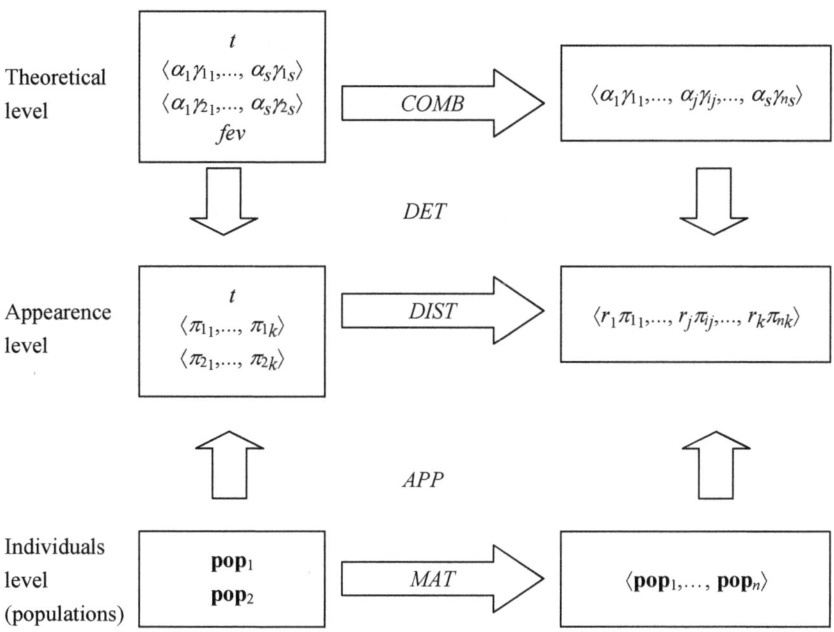

The set of all structures of type x, interpreted in the indicated way, constitute the set of *potential models* of (classical) population genetics. However, this theory does not simply propose concepts, but rather it claims uses them to make claims with regard to the "world", that is, it imposes certain restrictions of a nomic kind over these structures. For these structures it is reasonable to ask whether or not they are actual *models* of the theory. In the case the nomic restrictions are satisfied, especially the so-called fundamental law(s), these structures would be considered models. Informally speaking, what we have identified as the fundamental law of (classical) population genetics, which we can call "law of population matching", states that

- in given populations,
- that mate and produce progeny,
- a certain *fit* or *match* takes place between the distributions of the characteristics (relative frequencies) and the distributions of genes theoretically postulated (expected or theoretical probabilities), given certain relationships between genes (genotypes) and characteristics (phenotypes).

Formally, if $x = \langle I, (C_i)_{i \leq k}, (F_i)_{i \leq s}, FEV, T, <, APP, MAT, DIST, DET, COMB \rangle$ is a potential model, then x is a *model of classical populations genetics* iff

(1) for every **pop**, **pop'** $\subseteq Pot(I)$ such that MAT is defined for $\langle$**pop**, **pop'**$\rangle$, for every $t \in T$, for every $\langle \gamma_1, \ldots, \gamma_s \rangle, \langle \gamma_1', \ldots, \gamma_s' \rangle \in G$, for every $fev \in FEV$, such that $DET(\langle \gamma_1, \ldots, \gamma_s \rangle) = APP(\mathbf{pop})$ y $DET(\langle \gamma_1', \ldots, \gamma_s' \rangle) = APP(\mathbf{pop'})$:

$$COMB(t, \langle \alpha_1 \gamma_{11}, \ldots, \alpha_s \gamma_{1s} \rangle, \langle \alpha_1 \gamma_{21}, \ldots, \alpha_s \gamma_{2s} \rangle, fev) = DIST(DET(\langle \alpha_1, \ldots, \gamma_s \rangle), DET(\gamma_1', \ldots, \gamma_s' \rangle))$$ [18]

In order to achieve a thorough understanding of this law, consider two parental populations **pop**, **pop'**, with phenotypes $\langle \pi_1, \ldots, \pi_n \rangle, \langle \pi_1', \ldots, \pi_n' \rangle$, genotypes $\langle \gamma_1, \ldots, \gamma_s \rangle, \langle \gamma_1', \ldots, \gamma_s' \rangle$, evolutionary factors *fev* and corresponding genetic distributions over phenotypes and genotypes in their progeny: $D(P) = \langle r_1 \pi_{11}, \ldots, r_j \pi_{ij}, \ldots, r_k \pi_{nk} \rangle$, $D(G) = \langle \alpha_1 \gamma_{11}, \ldots, \alpha_j \gamma_{ij}, \ldots, \alpha_s \gamma_{ns} \rangle$. Consider first the simple case in which DET is one-one. In this case each phenotype π_j comes from exactly one of the genotypes $\gamma_1, \ldots, \gamma_s$. So $k = s$ and we may assume that each π_j is produced by γ_j. The natural notion of fit between the two distributions $\langle r_1 \pi_{11}, \ldots, r_j \pi_{ij}, \ldots, r_k \pi_{nk} \rangle, \langle \alpha_1 \gamma_{11}, \ldots, \alpha_j \gamma_{ij}, \ldots, \alpha_s \gamma_{ns} \rangle$ in this case is following. We say that $D(P)$ and $D(G)$ *ideally fit* iff for all $j \leq s$: $r_j = \alpha_j$.

In general, the situation is more complex, since a phenotype may be produced by a combination of several different genotypes, i.e. $k < s$. In such cases we have to compare the probabilities of all these genotypes with the relative frequency of the phenotype they produce. Formally, let us introduce, for given parental genotypes $\langle \gamma_1, \ldots, \gamma_s \rangle, \langle \gamma_1', \ldots, \gamma_s' \rangle$, and given index $j \leq k$, the set $C(\langle \gamma_1, \ldots, \gamma_s \rangle, \langle \gamma_1', \ldots, \gamma_s' \rangle, j)$ of all probabilities a_i occurring in $D(G)$ such that the corresponding genotype γ_i produces phenotype π_j (see Fig. 2). Moreover, let us write $c_j = \sum \alpha_i$, $i \in C(\langle \gamma_1, \ldots, \gamma_s \rangle, \langle \gamma_1', \ldots, \gamma_s' \rangle, j)$, as the sum of all those probabilities α_i whose corresponding genotypes γ_i give rise to the same phenotype π_j with relative frequency r_j.

In order to fit $D(P)$ and $D(G)$, we have to compare then each relative frequency π_j with the sum c_j. In the general case, $D(P)$ and $D(G)$ *ideally fit* iff for all $j \leq k$: $r_j = c_j$. Two genetic distributions of genotypes $\langle \alpha_1 \gamma_{11}, \ldots, \alpha_j \gamma_{ij}, \ldots, \alpha_s \gamma_{ns} \rangle$ and of phenotypes $\langle r_1 \pi_{11}, \ldots, r_j \pi_{ij}, \ldots, r_k \pi_{nk} \rangle$ *ideally fit* to each other iff: i) $k \leq s$; ii) each phenotype π_j comes, through DET, from a genotype γ_i or from more different

18 Or, alternatively, recalling that *DIST* can be defined through *MAT* and *APP*: $COMB(t, \langle \alpha_1 \gamma_{11}, \ldots, \alpha_s \gamma_{1s} \rangle, \langle \alpha_1 \gamma_{21}, \ldots, \alpha_s \gamma_{2s} \rangle, fev) = APP(MAT(\langle \mathbf{pop}, \mathbf{pop'} \rangle))$.

genotypes γ_i; and iii) the probability coefficients of the items related in ii) fit each other.

Fig. 2

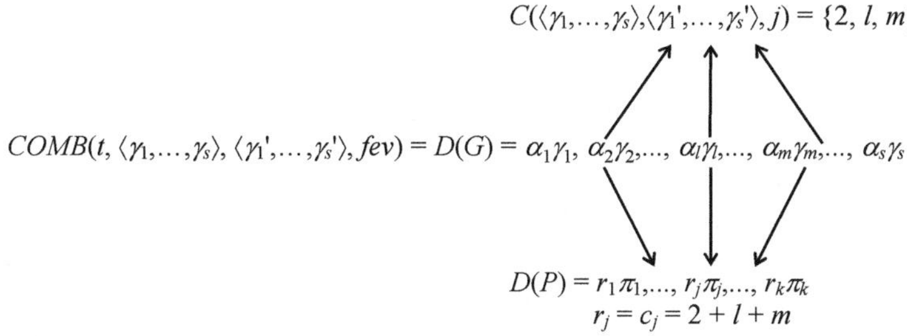

In the law of population matching we can easily identify the characteristics or "symptoms" of fundamental laws indicated in the previous section. First, the law of matching can be seen as a *synoptic* law because it establishes a substantial connection between the most important terms of classical population genetics in a "big" formula. It contains all the important terms that occur in population genetics. Second, the law of matching is highly schematic and general and it possesses so little empirical content that it is irrefutable on its own and with no consideration of further restrictions (i.e. it has a "quasi-vacuous" character). In order to examine the empirically determined relative frequency of characteristics and the theoretically postulated distribution of genes and set out to test what the law claims – namely, that the coefficients in the distribution of characteristics and of genes in the progeny are (approximately) equal –, without introducing any kind of further restrictions, amounts to a "pencil and paper" exercise that does not involve any empirical work. Nevertheless, as we would expect in the case of any fundamental law, despite being irrefutable in the explicated sense it provides a conceptual frame in which all special laws can be formulated; that is, special laws with an increasing degree of specificity and with an even more limited domain of application until we reach "terminal" specializations, whose associated empirical claims can be seen as particular, testable and, eventually, refutable hypotheses.

In addition, it is important to observe, and that is a central point, that this law has *implicitly* been *accepted as valid in every application of the theory* because the scientific community has been using it as a general background assumption, that provides a starting point for the analysis of different distributions of characteristics and serves as *a guide for dealing* with the plethora of empirical situations that population geneticists face (systematizing character). Consequently, the primary role of the law of matching is to guide the process of specialization, determining the ways in which it must be specified to obtain special laws.

4.1. The specializations of classical population genetics

In order to account for specific situations, the following parameters must be specified in the fundamental law: a) the number of pairs of alleles involved, b) the way in which the factors or genes (genotypes) are related to the characteristics (phenotypes) (either complete or incomplete dominance, codominance or epistasis), c) the number, and d) the nature of evolutionary factors (either migration, mutation, selection and/or genetic drift) to be considered, and e) how the parental genes are distributed in the progeny (the way in which the genetic constitution of the parental genotypes changes). When these five types of specifications are made, terminal special laws can be obtained, and the empirical claims associated with these laws are what are capable of direct empirical test. In case these laws "survive" a test, which means that the introduced specifications turn out to be the appropriate ones, one can say that the intended applications have become "successful" and that the empirical systems become "models" of the theory.

5. The Hardy-Weinberg Law as a Non-Terminal Specialization

In particular, the Hardy-Weinberg Law, inasmuch as it imposes additional restrictions on the law of matching, thereby adding information that is not already contained in its highly schematic formulation and restricting its area of application, can be obtained from the fundamental law through specialization and hence must be considered a "special law" of classical population genetics even though not a "terminal specialization".

This is because the Hardy-Weinberg law is a kind of "(pure) theoretical law" (a specific version of the theoretical analogue of the transition of phenotypes), which only establishes what happens at the level of the alleles (or (allelic) factors) or genotypes, but doesn't say anything about how to connect such a level with that of "the experience", "the empirical", that is, with the characteristics or phenotypes.

If we carry out suitable specifications of indicated type a), c), d) and e), we obtain the Hardy-Weinberg law:

So, if 1) the number of pairs of factors (or genes) is limited to one (with just two alleles *A* and *a* – f_{i1}, f_{i2} or a_1 and a_2) (specification of type a), 2) there are no evolutionary factors acting ($FEV = \varnothing$) (specification of type c and d), 3) the distributions of parental population genotypes in the offspring are equally probable, i.e. $COMBINATOR(\langle \mathbf{p}^*_t a_1 a_2\rangle, \langle \mathbf{p}^*_t a_2 a_1\rangle, \varnothing\rangle) = (\langle \mathbf{p}^*_{t+1} a_1 a_1 + \mathbf{p}^*_{t+1} a_1 a_2 + \mathbf{p}^*_{t+1} a_2 a_1 + \mathbf{p}^*_{t+1} a_2 a_2\rangle)$ (specification of type e), then (conclusion) the reached equilibrium will remain unchanged in the next generations, i.e. for every $a_1 a_2$ and for every $t + n$: $\mathbf{p}^*_{t+1} = \mathbf{p}^*_{t+n}$.[19]

19 Where $a_1 a_2$ represent alleles *A* and *a*, $\mathbf{p}^*_{t+1}$ represents the frequencies, or relative proportions, of alleles reached in the first generation next to the moment in which mating takes place, and $\mathbf{p}^*_{t+n}$ represents the frequencies, or relative proportions, of alleles in

Its "deduction" takes place only if a series of additional assumptions (premises) are added. In such "deduction" not all the specifications are carried out, in particular the way in which the factors (genotypes) are related to the characteristics (phenotypes) (specification of type b) is not specified. This is the reason why, even when it is a specialization, the Hardy-Weinberg law cannot be considered a "terminal specialization". Besides, because we do not find any of the necessary conditions or "symptoms" pointed out above (1. the synoptic character, 2. validity in *every* intended application, 3. the quasi-vacuous character, and 4. the systematizing role), it is clear that the Hardy-Weinberg law could not be considered as the fundamental law of classical population genetics.[20]

6. Conclusions

Having as a background a structuralist reconstruction of classical population genetics, we have identified the fundamental law of such a theory, the "law of population matching", and characterized the Hardy-Weinberg law as a non-terminal specialization. The analysis carried out throughout this paper allows to capture the elements stressed by Bas van Fraassen in his diagnosis, understanding better in which sense such a law: 1) cannot be considered as axiom (or fundamental law) for classical population genetics, since, 2) it is a law that describes an equilibrium that holds only under certain special conditions, 3) that only determines a subclass of models, 4) whose generalized form (and fundamental law) being shading off into logical vacuity, and 5) more complex variants of the fundamental law (and of the Hardy-Weinberg law) can be "deduced" for more realistic assumptions. However, our analysis also allows 6) a better understanding of the status of the Hardy-Weinberg law in classical population genetics, and 7) to have a unified picture of the different, really heterogeneous, models of classical population genetics as well, and provides conceptual tools to advance in the analysis of the relationships

---

next generations.

20 Although there are general similarities between the approaches in terms of 'fundamental law / specialization relation' and in terms of 'theoretical idealizational/concretization relation' (see Martti Kuokkanen (Ed.), *Idealization VII: Structuralism, Idealization and Approximation.* Amsterdam: Rodopi 1994), our analysis of the Hardy-Weinberg law using the former differs from the analyses carried out within the latter (see Krzyszlof Łastowski, "The Method of Idealization in Population Genetics", *Poznan Studies*, 3, 1-4, 1977, pp. 199-212; Krzyszlof Łastowski, "The Idealizational Status of Theoretical Biology", in: Władysław Krajewski (Ed.), *Polish essays in the philosophy of the natural sciences.* Dordrecht, Reidel 1982, pp. 353-364; Matti Sintonen/Mikka Kiikeri, "Idealization in Evolutionary Biology", in: Martti Kuokkanen (Ed.), *op. cit.*, pp. 201-216) basically in considering that the idealizational character in general of that law doesn't explain the peculiar status of it within classical population genetics (being neither the fundamental law nor a terminal specialization, or a full concretized version, of the law of population matching).

between classical population genetics and evolutionary theory, and thus in the analysis of the last one. This is a future direction that exceeds, however, the limits of our present article.

Acknowledgments: I would like to thank Theo Kuipers, Gereon Wolters, Matti Sintonen, Kenneth Waters, John Beatty and an anonymous referee for their helpful comments on an earlier draft of this paper. This reasearch was supported by the research project PICT2007 N° 1558 (ANPCyT, Argentina).

National University of Quilmes
National Scientific and Technical Research Council
Roque Sáenz Peña 352 (B1876BXD) Bernal
Prov. Buenos Aires
Argentina
pablol@unq.edu.ar

Jan Woleński

Kazimierz Twardowski and the Development of Philosophy of Science in Poland

In 1930 Otto Neurath asked Alfred Tarski for information about Polish philosophers of science and their work. Tarski answered on 24 April, 1930, saying in one of the first sentences (see Tarski 1992, p. 31) that:

> Almost all researchers, who pursue the philosophy of exact sciences in Poland, are indirectly or directly the disciples of Twardowski, although his own works could hardly be counted within this domain.

Then, Tarski mentions several names of mathematical logicians and philosophers of science who worked on general methodology (I use the terms "general methodology" and "general philosophy of science" as equivalents) of natural and/or social sciences. Tarski's quoted statement shows the enormous role of Twardowski in the development of philosophy in Poland in the first half of the 20th century. Twardowski became a professor of philosophy in Lvov in 1895. He was a very successful teacher and very soon trained a group of outstanding philosophers, among others (I give names in alphabetical order and restrict to philosophers of exact sciences) Kazimierz Ajdukiewicz, Tadeusz Czeżowski, Tadeusz Kotarbiński, Stanisław Leśniewski, Jan Łukasiewicz and Zygmunt Zawirski. They became professors of Polish universities in the interwar period and educated the next generations of students. Teaching activities of Twardowski and his students constituted the Lvov-Warsaw School, especially famous for its logical part formed by Leśniewski and Łukasiewicz (the story of this movement is described in Woleński 1989). This brief information indicates what Tarski intended by words "Almost all researchers [...] are indirectly or directly the disciples of Twardowski". Tarski himself, a student of Leśniewski and Łukasiewicz, belonged to indirect disciples. Let me add that the word "almost" cannot be erased or replaced by "all", as other circles in the philosophy of science arose in Poland as well, although they were much less important than the Lvov-Warsaw School.

Yet Tarski's statement should be modified. It is true that Twardowski's "own work could hardly be counted within this domain" if the domain were identified with mathematical logic and the foundations of mathematics. Twardowski did not work in these fields and was even skeptical about the relevance of purely logical methods for philosophical analysis. Although Twardowski considered general logic as an indispensable and valuable tool of philosophical analysis, he warned strongly against so-called symbolomania and pragmatophobia as dangers of an ex-

M.C. Galavotti et al. (eds.), *European Philosophy of Science – Philosophy of Science in Europe and the Viennese Heritage*, Vienna Circle Institute Yearbook 17,
DOI 10.1007/978-3-319-01899-7_12,

cessive use of formal methods (see Twardowski 1912; dates of papers always refer to first original editions; page reference to English originals or translations if available). He probably directed these reservations to Leśniewski and Łukasiewicz, although their names were not mentioned in this context. On the other hand, Twardowski did some work which can be „counted within this domain", if the domain includes general problems of the philosophy of science, that is, the philosophy of exact sciences. This paper discusses some of Twardowski's ideas related to general methodology, their origin and their significance for the work of his students.

Twardowski studied in Vienna in 1885–1889 and belonged to the last generation of Brentano's students. He is, however, usually qualified as an old Brentanist as opposed to so-called young Brentanists, for instance, Oskar Kraus or Alfred Kastil. Twardowski essentially contributed to Brentanism by his famous work on the concept and object of presentation (see Twardowski 1894). This book is of significant importance to descriptive psychology as it tries to clarify the essential distinction between content and object of mental acts, particularly presentations. Twardowski shared Brentano's methaphilosophical postulates, especially the one saying that that philosophy should be clear, precise, anti-speculative and that it should consist of justified statements. He also shared Brentano's negative attitude towards German idealistic metaphysics. The famous fourth Brentano's *Habilitation* thesis stating that the true (correct) method of philosophy is the same as of natural sciences, was fully accepted by Twardowski, who propagated the ideas of his master among Polish students of philosophy. Twardowski's main educational task consisted in creating scientific philosophy in Poland and building firm foundations for its subsequent development. His success in this respect was enormous. He trained many philosophers and representatives of other fields (psychologists, philologists or historians of art), including over 30 full university professors.

Twardowski's interest in general methodology was motivated to a great extent by his very serious attitude towards the mentioned Brentano's *Habilitation* thesis on the identity of philosophical and (natural) scientific method and his entire project regarding scientific philosophy as based on descriptive psychology. Recall that this kind of psychological research, although sharply contrasted with so-called genetic psychology, was considered as empirical by the Brentanists, even more than by Brentano himself. Hence, application of empirical methods to philosophy was evident in Brentano's circle. Twardowski, guided by this view, tried to elaborate various methodological problems, related to the concept of science, its structure and methods. Yet, he did not produce any major systematic work about the philosophy of science. Twardowski expressed his methodological views in several works of a different character, in elementary textbooks (see Twardowski 1901), short pieces devoted to teaching (for example, Twardowski 1920) and fairly sophisticated papers (see for instance, Twardowski 1923). His methodological views, opinions, claims and remarks (there are many occasional notes on the philosophy of science in Twardowski's writings on psychology, ethics or history of philosophy) were motivated by practical needs of education at all levels of teaching as well as deep theoretical problems that needed to be solved or at least elucidated.

Before I pass to Twardowski's methodological views, I would like to say something more about his metaphilosophical and general philosophical views as well as about reactions of his students to his views. Moreover, I will give an example of Twardowski's analytic technique, which is also very instructive for understanding philosophical analysis in the entire Lvov-Warsaw School. This section is intended as a brief report on a broader context of Twardowski's more specific methodological views and their reception in the Lvov-Warsaw School. Twardowski employed two kinds of understanding philosophy. Firstly, philosophy investigates objects, which are present in internal and/or external experience. Secondly, philosophy is a collection of disciplines including psychology, logic, metaphysics, epistemology, ethics and aesthetics. The former understanding closely followed Brentano and immediately resulted in psychologism, both ontological and methodological. Twardowski favored this account of philosophy at the beginning of his academic career, but he was later more inclined towards the collective meaning of philosophy. Although he still considered psychology as the most fundamental philosophical field, being under the influence of Husserl, he rejected ontological psychologism, accepting however, its methodological version. According to the collective approach to philosophy, it should not be defined via a uniform property, for instance that philosophy is the most general or the most fundamental science. Every part of philosophy provides a set of problems, which can be scientifically investigated and eventually solved. No traditional philosophical problem should be a priori or definitely excluded from scientific research On the other hand, because it seems unrealistic to effectively build universal philosophical systems, scientific philosophy must restrict itself to concrete and well-stated questions. For Twardowski, traditional metaphysics did not consider scientific problems, but issues related to world-views, for instance, religious beliefs. The requirement that philosophical problems have to be well-stated is of the utmost importance, as vagueness and the lack of precision make it impossible to deliver scientific solutions of related problems. One could say that there is no solvability without conceptual and linguistic clarity.

Twardowski's students, notably logicians, followed him in adapting the collective understanding of philosophy. They rejected psychologism and, according to the academic classification of sciences, excluded psychology from the domain of philosophy. Logic *sensu largo*, covering semantics, formal logic (logic *sensu stricto*) and the methodology of science, became the basic philosophical field and the ultimate source of methodological principles necessary for performing scientific philosophy. In other words, analysis in terms of descriptive psychology was replaced by logical analysis focused on properties of language. This tendency was also present in other philosophical centers, particularly in Vienna, however in Poland it had some important consequences. Roughly speaking, intentionality as the central property of mental acts, was interpreted as a semantic attribute of nominal words (see Woleński 2006, Woleński 2009 for a fuller account of the role of Brentanist ideas in the rise of semantics within Polish philosophy). Moreover, the irreducibility of intentionality to non-intentional phenomena, was interpreted in Poland as the priority of meaning over syntactic properties of language. One aspect of this issue is very delicate. Polish logicians considered intensionality of linguistic contexts as derivable from the phenomenon of mental intentionality. It is remarkable that in some Polish works intensionality was was labeled as intentionality of language. On the other hand, from the logical point of view intensionality was counted as a defective property. Hence, according to the standards of Polish logicians, all correct logical systems must be extensional. The priority of meaning was a by-product of this ideology.

Although Polish analytic philosophers active after Twardowski became much more skeptical about metaphysics than their master, they did not reject metaphysical problems as meaningless *a limine*. For example, they considered such issues as the causal structure of the world, determinism, indeterminism or the question of the infinity of space and time as genuine and not only verbal matters, solvable by adopting some conventions. In consequence, the Lvov-Warsaw School never accepted the principle of empirical verifiability or, speaking more generally, empirical testability as the exact and simple criterion for distinguishing between meaningful science and meaningless metaphysics. For Polish philosophers, testability was a criterion for the scientific character of problems and statements, but not a measure of meaning. Independently of these details, the philosophy of science, being a part of logic *sensu stricto*, functioned in the Lvov-Warsaw School as one of the most important philosophical fields. Although Twardowski's students radicalized his views in many respects, their work remained within the general frameworks created by their teacher. In particular, they believed that scientific philosophy should be done step by step, that it must be performed clearly and precisely and that the research conducted within scientific philosophy should deal with problems rather than systems in the traditional sense. German idealistic metaphysics was for Polish analytic philosophers equally unacceptable as it was for Brentano.

Twardowski strongly defended absolutism in the theory of truth and epistemology (see Twardowski 1900, Twardowski 1973; the second paper is based on Twardowski lectures delivered in 1924/1925). Twardowski also defended absolutism in ethics, however it is less relevant in the context of this paper, I will focus only on epistemological matters. According to Twardowski, arguments of epistemological relativists are mistaken and stem from overlooking differences between incomplete sentences and complete propositions. Sentences defined as uttered linguistic items are often incomplete for they frequently contain indexical words. Since they can always be eliminated, sentences are easily converted into propositions. Other relativist argument comes from a correct observation that scientific hypotheses are only partially confirmed. Twardowski argued that this argument confuses truth and its criteria. Twardowski's views on truth very strongly influenced Polish philosophy. He initiated the tradition in Poland, which culminated in Tarski's semantic theory of truth frequently used in discussing various methodological questions. It is worth noting that this theory was usually interpreted in Poland as classical and absolutist. In general, Polish philosophers shared realistic views regarding scientific theories although some representatives of the Lvov-Warsaw School expressed other views in this respect (see Giedymin 1984 for a extensive treatment of anti-realism in Polish philosophy of science). For instance, Łukasiewicz (see Łukasiewicz 1912) considered scientific theories as relatively free constructions of human mind intended not to give an accurate description of empirical phenomena, but to grasp them via conceptual resources.

Perhaps one example, related to Twardowski's distinction between truth and truth criteria would be interesting in the present context. Polish philosophers were interested in French conventionalism. In particular, Ajdukiewicz proposed radical conventionalism (see Ajdukiewicz 1934, Ajdukiewicz 1934a) as an improvement of philosophical ideas of Henri Poincaré and Pierre Duhem. Roughly speaking, according to radical conventionalism not only theoretical schemes but even empirical protocols, can be accepted or rejected depending on what conceptual apparatus we choose. In other words, we can keep empirical data and change observation sentences by replacing one conceptual apparatus with a different one. The earlier conventionalism restricted changeability to theoretical systems and postulated the stability of empirical protocols. The immediate problem with Ajdukiewicz's account consisted with the question whether radical conventionalism was coherent with the classical theory of truth. Ajdukiewicz himself pointed out that the problem in question disappears because he discussed acceptance and rejection, not truth. Izydora Dąmbska (see Dąmbska 1938; she was Twardowski's student who belonged to the second generation of the Lvov-Warsaw School) argued that conventionalism, moderate or radical, does not imply relativism, because it concerns solvability of problems and not the truth. or falsity of statements. It is easily to see the close link between Twardowski's argument that we should distinguish truth and its criteria and the standpoints of Ajdukiewicz and Dąmbska.

Twardowski discussed the problem of objectless presentations (see Twardowski 1894, p. 18-26, especially p. 20). He argues that there are no such presentations. One of his arguments runs as follows. Defenders of the opposite view, saying that objectless presentations exist point out that we use sentences of the form 'Nothing is *X*', for example (*) 'Nothing is eternal'. Thus, it seems that the subject term 'Nothing' does not refer to something. Twardowski says that 'Nothing' functions as a syncategorematic part of a sentence. Consequently, one cannot consider 'Nothing' as a subject-term in (*), because it would be at odds with its syncategorematicity (any subject term must be categorematic). The problem is solved by translating (*) into (**) 'There is not something which is eternal'. Since Twardowski correlated presentations with names, his reasoning has a clear semantic import. Although Twardowski's argument is quite elementary and uses only traditional resources, we can easily transform it into the language of modern predicate logic. In fact, (**) is translatable into (***) 'for no *x*, *x* is eternal' (or 'there is no object which is eternal'). So 'nothing' is a combination of quantification and predication. Twardowski's approach to objectless presentations consists in translating a philosophical problem into the language of logic under the assumption that names and presentations coincide. This kind of analysis was widely used in Poland as well as in other centers of analytic philosophy. For example, Carnap's famous criticism of Heidegger's dictum *Das Nichts nichtet* used the same strategy. Tarski's semantic theory of truth, Leśniewski's definition of being an object (*a* is an object if and only if *a* is something) or Łukasiewicz's analysis of determinism and indeterminism via three-valued logic are further examples of the same analytic tactic and show the continuity of Polish analytic philosophy from Twardowski to mature mathematical logicians.

As I have earlier noted Twardowski's research in the philosophy of science was not systematic and exposed in a relatively small number of works. He pondered such issues as: the concept of science, classification of reasoning and the distinction between deductive sciences vs. inductive sciences. Twardowski's account of science is very simple and elementary, namely he defined science as an ordered body of propositions. Twardowski's more important idea concerning science comes from his distinction between actions and products (see Twardowski 1912). The need to draw such distinction was recognized within the Brentano circle before Twardowski, for instance, by Carl Stumpf. However, Twardowski was the first who investigated this distinction in detail. Twardowski observed that we frequently use pairs of expressions in which the first element expresses an action and the second a product of the action in question. As examples we can take the following pairs (a) <to jump, jump>, (b) <to think, thought> or (c) <to write, writing>. Actions are expressed by verbs, but their products by related verbal nouns. The given examples illustrate various possible situations. In (a) the action and its products have physical nature, in (b) – both elements are psychical, but (c) represents the situation in which both elements are psycho-physical. Moreover, products can be enduring (existing after the end of a related action) or non-enduring (they disappear when the associated action comes to an end). An important application of the discussed distinction is that judgments (propositions) are conceived in the psychological sense as actions and in the logical sense as products. Thus, the pair <judgment in the psychological sense, judgment in the logical sense> is equivalent to the pair <judging, judgment>. Although Twardowski himself did not directly apply the distinction between actions and products to an analysis of science, his students made this step. It was mainly Ajdukiewicz who used Twardowski's distinction in the philosophy of science (see Ajdukiewicz 1960). Ajdukiewicz correlated science as an action and science as a product with two kinds of approach to scientific theories. Science as an action is subjected to so-called pragmatic methodology which takes into account various psychological and sociological aspects of scientific work. Since scientific activity involves beliefs, expectations, hopes, etc, methodology of science understood as action cannot ignore such factors. On the other hand, science understood as a product is subjected to apragmatic methodology (metascience) similar to metamathematics. Metascience consists in employing formal logical tools. According to Ajdukiewicz, both kinds of methodology are indispensable in order to form an adequate theory of science.

Twardowski was firmly convinced that justification of a proposition is a logically regular process and consists in explicitly analyzable reasoning. Hence, he wanted to apply general logical categories to analysis of arguments. Twardowski defined reasoning as a relation between a logical antecedent and a logical consequent (see Twardowski 1901, p. 29). We can distinguish two kinds of reasoning, namely deduction and induction. Whereas the former searches for logical consequences of given assumptions, the latter consists in looking for logical antecedents of the given assertions. Twardowski seems to be the first, at least in elementary textbooks, who characterized reasoning (inference) not via psychological factors, but appealing to logical connections. Since most Polish philosophers shared logicism in methodological analysis of science, the logical characterization of reasoning became very popular in Poland (see Woleński 1991 for an extensive survey). Let me mention a few facts. Łukasiewicz (see Łukasiewicz 1912) distinguished between reasoning from premises to a conclusion and reasoning from a logical antecedent to a logical consequent. He observed that the actual direction of a reasoning (an inference) as going from chosen premises to a conclusion does not necessarily coincide with its logical direction from the logical antecedent to the logical consequent. It allowed for a better discrimination of logical and psychological elements in reasoning. Moreover, Łukasiewicz was able to define more specific kinds of reasoning, namely proof, testing, inference and explanation. Although not all Polish methodologists were inductionists (for example, Łukasiewicz argued that induction does not play any significant role in the mature science), the conceptual framework introduced by Twardowski immediately led to the problem of inductive logic. Finally, Tarski's famous definition of logical consequence belongs to the line of thought about inference which originated with Twardowski, although Tarski himself did not refer to this tradition.

Discussing the distinction between deductive (a priori, rational) sciences and inductive (a posteriori, empirical) sciences, Twardowski rejected the view that the former are not based on experience, whereas the latter are constructed on empirical data, because, as he put it, such explanation is not satisfactory. In particular, we cannot say that deductive arguments produce theorems exclusively by means of reasoning, and that the empirical procedure appeals only to experience. Twardowski points out that the term 'experience' is ambiguous. If one says that propositions refer to experience, he or she means concrete singular protocols. On the other hand, when Mill considered axioms of mathematics as empirical, he understood them as experiential generalizations, because axioms are general by definition. Now, inductive disciplines also refer to generalization from experience, because they deduce their laws from other laws. Thus, deduction is present in a posteriori sciences. On the other hand, induction can also be present in a priori sciences as a heuristic method. Twardowski agreed that deduction is the correct method of justification in sciences called deductive, and that induction performs the justification game in inductive sciences. On the other hand, rejected two following identifications: (i) deductive sciences = a priori (independent of experience) sciences; (ii) inductive sciences = a posteriori (independent of reasoning) sciences. Moreover, Twardowski sharply distinguished between the context of discovery and the context of justification. He pointed out that logical regularity can only be attributed to the latter. This view was accepted by the entire Lvov-Warsaw School and essentially contributed to its logical methodology.

Twardowski's own philosophical work as well as his teaching activities are entirely immersed in the philosophical heritage created by Brentano in Vienna. It does not mean that Twardowski's students could be included in the Brentanist philosophical movement. In fact, although they inherited many ideas of Brentano via Twardowski (for example, the thesis that mental phenomena are intentional) in many respects they became much closer to the Vienna Circle than to traditional or even modified Brentanism. Schematically speaking, the development of the Lvov-Warsaw School can be considered as a path from Brentanism to the style of philosophy recommended by logical empiricism. This is, however, a separate issue which cannot be discussed in this paper. Although some tenets represented in the Lvov-Warsaw School, similar to tendencies of Moritz Schlick and his followers (Kotarbiński's idea of the reistic language as suitable for science including the humanities anticipated physicalism), appeared in Poland before inception of the Vienna Circle, there were developed and radicalized under the direct influence of logical empiricism, especially by philosophers who belonged to the second generation of the Lvov-Warsaw School, such as Maria Kokoszyńska and Henryk Mehlberg. The role of the Viennese heritage for the development of Polish analytic philosophy until 1939 was crucial. Thus, this paper fits well with the main topic of conference "Philosophy of Science in Europe – European Philosophy of Science and the Viennese Heritage".

References

Ajdukiewicz, K. 1934, "Sprache und Sinn", *Erkenntnis* IV, pp. 100-138. Translated as "Language and Meaning". In Ajdukiewicz 1978, pp. 35-69.

Ajdukiewicz, K. 1934a, "Das Weltbild Und Begriffsapparatur", *Erkenntnis* VI, pp. 259-287. Translated as "The World Picture and the Conceptual Apparatus". In Ajdukiewicz 1978, pp. 67-89.

Ajdukiewicz, K. 1960, "Axiomatic Systems from the Methodological Point of View". *Studia Logica* 9, pp. 205-218. Reprinted in Ajdukiewicz 1978, pp. 282-294.

Ajdukiewicz, K. 1978, *The Scientific World–Perspective and Other Essays, 1931–1963*. Dordrecht: Reidel.

Dąmbska, I. 1938, "Konwencjonalizm a relatywizm". *Kwartalnik Filozoficzny* 15, pp. 328-337. Translated as "Conventionalism and Relativism". *Organon* 41, 2009, pp. 9-15.

Giedymin, J. 1986, "Polish Philosophy of the Inter-War Period and Ludwik's Fleck's Theory of Thought-Styles and Thought-Collectives". In *Cognition and Fact. Materials on Ludwik Fleck*. Ed. by R. S. Cohen and Th. Schnelle. Dordrecht: Reidel, pp. 179–216.

Łukasiewicz, J. 1912, "O twórczości w nauce". In *Księga Pamiątkowa ku uczczeniu 250-tej Rocznicy założenia Uniwersytetu Lwowskiego przez króla Jana Kazimierza*, Lwów: Uniwersytet Lwowski, t. 1, pp. 1–15. Translated as "On Creative Element in Science". In J. Łukasiewicz, *Selected Works*, Amsterdam: North-Holland 1970, pp. 1–15.

Tarski, A. 1992, "Drei Briefe an Otto Neurath". *Grazer Philosophische Studien* 43, pp. 1–32.

Twardowski, K. 1894, *Zur Lehre vom Inhalt und Gegenstand der Vorstellungen*. Wien: Hölder. Translated as *On the Concept and Object of Presentation*. The Hague: Nijhoff 1977.

Twardowski, K. 1900, "O tak zwanych prawdach względnych". In *Księga Pamiątkowa Uniwersytetu Lwowskiego ku uczczeniu pięćsetnej rocznicy Fundacji Jagiellońskiej Uni-wsersytetu Krakowskiego*, Lwów: Senat Akademicki Uniwersytetu Lwowskiego, pp. 1-25. Translated as "On So–Called Relative Truths". In Twardowski 1999, pp. 147–169.

Twardowski, K. 1901, *Zasadnicze pojęcia dydaktyki i logiki* (The Fundamental Concepts of Didactics and Logic). Lwów: Towarzystwo Pedagogiczne.

Twardowski, K. 1912, "O czynnościach i wytworach. Kilka uwag z pogranicza psychologii, gramatyki i logiki. In *Księga Pamiątkowa ku uczczeniu 250-tej rocznicy założenia Uniwersytetu Lwowskiego przez króla Jana Kazimierza*, Lwów: Uniwersytet Lwowski, t. 2, pp. 1-33. Translated as "Actions and Products. Some Remarks from the Borderline of Psychology, Grammar and Logic". In Twardowski 1999, pp. 103-132.

Twardowski, K. 1920, "O wykształcenie logiczne" (In Favour of Logical Teaching). *Ruch Filozoficzny* V, pp. 65-71.

Twardowski, K. 1921, "Symbolomania i pragmatofobia", *Ruch Filozoficzny* VI, pp. 1–10. Translated as "Symbolomania and Pragmatophobia". In Twardowski 1999, pp. 261-270.

Twardowski, K. 1923, "O naukach apriorycznych, czyli racjonalnych (dedukcyjnych) i naukach aposteriorycznych, czyli empirycznych (indukcyjnych). In *Główne kierunki filozofii w wypisach z dzieł ich klasycznych przedstawicieli*

(Teoria poznania – logika –metafizyka). Lwów: Jakubowski, pp. 180–190. Translated as „A Priori, or Rational (Deductive) Sciences and A Posteriori, or Empirical (Inductive) Sciences". In Twardowski 1999, pp. 170-179.

Twardowski, K. 1975, "Wykłady z teorii poznania w r. a. 1924–25". *Archiwum Historii Filozofii i Myśli Społecznej*, pp. 241–299. Translated as „Theory of Knowledge. A Lecture Course (1924/25". In Twardowski 1999, pp. 181-239.

Twardowski, K. 1999, *On Actions, Products and Other Topics in Philosophy*, Amsterdam: Rodopi.

Woleński, J. 1989, *Logic and Philosophy in the Lvov-Warsaw School*. Dodrrecht: Kluwer.

Woleński, J. 1991, "Theories of Reasonings in the Lvov-Warsaw School". In *Topics in Philosophy and Artificial Intelligence*. Ed. by L. Albertazzi and R. Poli, Bozen: Kulturinstitute der Zentral Europe, Bozen, pp. 101-112. Reprinted in J. Woleński, *Essays in the History of Logic and Logical Philosophy.* Jagiellonian University Press, Kraków 1999, pp. 52-58.

Woleński, J. 2006, "Brentanism and the Rise of Formal Semantics". In *Actions, Products and Things. Brentano and Polish Philosophy*. Ed. by A. Chrudzimski and D. Łukasiewicz, Ontos Verlag, Heustenstamm: Ontos, pp. 217-232.

Woleński, J. 2009, "The Rise and Development of Logical Semantics in Poland". In *The Golden Age of Polish Philosophy. Kazimierz Twardowski's Philosophica l Legacy*, Berlin: Springer. Ed. by S. Lapointe, J. Woleński, M. Marion and W. Miśkiewicz, pp. 43-59.

Department of Epistemology
Jagiellonian University Kraków
ul. Grodzka 52
31-044 Kraków
Poland
wolenski@if.uj.edu.pl

TOMASZ PLACEK

VIENNA CIRCLE ON DETERMINISM

ABSTRACT

Members of Vienna Circle explicated determinism in terms of predictability in principle, or calculability. This paper attempts to uncover the rationale for this explication. It argues that the explication was an attempt to escape trivialization arguments; another important factor was the Circle's views on meaning as testability.

1. INTRODUCTION

My aim in this paper is to investigate what concept of determinism and indeterminism was held by the Vienna Circle by examining what were the views of the prominent members of the Circle on this subject: Moritz Schlick, Rudolf Carnap, Herbert Feigl, and Philipp Frank.[1] A handbook's answer to this question is that the group understood (in)determinism along Laplacean lines and further identified it with predictability in principle, or calculability in principle. The present paper confirms this diagnosis, adding perhaps a few nuances. My concern, however, is not so much to uncover what concept of determinism was advocated by the philosophers mentioned, but rather, what line of thoughts drew them to accept a rather implausible view of determinism as predictability in principle. After all, the world might be deterministic in the sense that each initial segment of it has a single possible continuation, but not calculable (even in principle), since the mathematical functions representing forces are not computable. To say that the evolution of this world is calculable (or predictable) in principle one needs to stretch the meaning of the modifier "in principle" beyond any reasonable limit.

To put the Circle's view on indeterminism in a historic perspective, the beginning of the twentieth century saw a revival of interest in future contingents and attempts to provide a rigorous framework for the discussion of (in)determinism and alethic modalities.[2] One development of this kind, Łukasiewicz's multi-valued logic,[3] is of particular note since it was developed in the Lvov-Warsaw school

1 I am grateful to an anonymous referee for useful comments on this paper. The support of the research grant 668/N-RNP-ESF/2010/0 of the (Polish) Ministry of Science and Higher Education is kindly acknowledged.

2 For an overview of history of modal logic, cf. Goldblatt (2006).

3 Łukasiewicz's first publication on three-valued logic was in his (1920).

M.C. Galavotti et al. (eds.), *European Philosophy of Science – Philosophy of Science in Europe and the Viennese Heritage*, Vienna Circle Institute Yearbook 17,
DOI 10.1007/978-3-319-01899-7_13,

which the Viennese were in contact with, and Łukasiewicz's logic was known to members of the Vienna Circle. Łukasiewicz's aim was to provide a response to Aristotle's sea battle problem, a response that would allow for the freedom of actions of human agents. Underlying his project is a modal reading of (in)determinism, clearly present in Aristotle's "coat story" or his sea battle (*De Interpretatione* 19a and 9, respectively). On this reading, a basic thought behind indeterminism is that the course of events until some time has more than a single *possible* continuation. In Aristotle's example, in one continuation of today's events there is a sea battle tomorrow, and in another continuation of these events, there is no sea battle. The Aristotelian modal approach and Laplacean epistemic approach express two distinct intuitions concerning (in)determinism. Learning the Viennese's reaction to Łukasiewicz's project, apart from being interested in itself, may cast some light on the Circle's choice of the Laplacean concept.

Another observation relevant to the Vienna Circle's views on determinism is that the group developed in the vicinity of, and in contact with, the community of Vienna physicists. As well documented by M. Stöltzner (1999, 2003), around the turn of nineteenth century Mach, Boltzmann, and Exner argued that Laplacean determinism might break at the level of micro-physical processes. Thus, a natural question is whether the philosophers were informed by what Stöltzner aptly calls the "Vienna Indeterminism" of the physicists, and how their views were influenced by the physicists' perspective. Here the work of the theoretical physicist Philipp Frank, a member of the Circle who wrote extensively on (in)determinism, is especially relevant.

The paper is organized as follows. Section 2 provides a preview of the Circle's notion of determinism. Section 3 discusses a line of thoughts that led the group to identify determinism with predictability. Frank's analysis of determinism is the topic of section 4. Section 5 contains a discussion (including a discussion of Schlick's reaction to Łukasiewicz's project) and wraps up the paper.

2. Preview

Let us document by a few quotations the concept of (in)determinism held by the Vienna Circle. To begin with Schlick (1962/1931, 291),[4] we find him declaring:

> The word 'determined' therefore means exactly the same as 'predictable' or 'computable in advance'.

A similar view is voiced by Carnap (1966, 192-193)[5]

---

4 I use here the convention that in the bibliographical data the first year refers to the work's publication in English, whereas the second year—to the original publication. Page numbers refer to pages of the English translation.

5 In this book Carnap identifies determinism and causality, cf. pp. 216-217.

Causal relation means predictability. [...] It does not imply the possibility of someone actually predicting the event, but rather a potential predictability. [...] This prediction is a logical consequence of the facts and the laws. In other words, there is a logical relation between the full description of the previous condition, the relevant laws, and the prediction of the event.

This view is echoed by Feigl (1953, 411):

Determinism [means] ideally complete and precise predictability, given the momentary conditions, the pertinent laws, and the required mathematical techniques.

This consensus of equating determinism with ideal predictability (calculability) seems to be spoiled by Frank's (1941/1907) apparently dissenting view. To explain his terminology, instead of "determinism" Frank writes "strong causality" and he uses "law of causality" for "law of strong causality", which implies that "law of causality" is synonymous with determinism. Determinism (law of causality) is then explained in terms of sequences of states:

The law [of causality] states that if, in the course of time, the state A of the universe is once followed by the state B, then whenever A occurs B will follow it (Frank (1941/1907, 18).

We will further investigate Frank's analysis to find out that, at the end of the day, he explicates determinism in terms of epistemic notions. Frank deserves our special attention, since the task of understanding determinism comes to the fore in his philosophical *oeuvre*. Symptomatically, determinism (law of causality) was a theme of his earliest papers, "Kausalgesetz und Erfahrung" (1907)[6] as well as a dominant topic of his last book, *Philosophy of science: the Link Between Science and philosophy* (1957). Frank was an accomplished theoretical physicist, a successor of Einstein (by his recommendation) at the chair of the theoretical physics at the (German) University of Prague in the years 1912–1938 and later a lecturer on physics and mathematics at the University of Harvard. As we will see, although Frank's views on determinism diverge from those of Schlick, Feigl, and Carnap, they are driven by a similar motivation to escape the trivialization problem.

3. Trivialization Problem

I have used above the term "Laplacean determinism" as if it did not require elucidation, which is clearly not the case. Getting to grips with this notion was an essential part of the Vienna Circle's reflection on determinism. In a well-known passage, Laplace (1951/1820, 4) invokes an omniscient intelligence who, having a knowledge of the positions and motions of all bodies in the universe at a given

6 Though it is hard to grasp today the thrust of this paper, it made quite a stir: it attracted Einstein's attention, which ultimately led to the developments of a personal and scientific contact between the two. It also brought in the fierce criticism of Lenin, in his *Materialism and Empirio-Criticism* (1909).

instant of time, and a grasp of all forces acting in the universe, "has before its eyes" both the future and the past of the universe. It is worth observing that the Laplacean intelligence acts upon two kinds of data: a knowledge of an instantaneous state of a system (which in relativistic physics should be further generalized to a state of an appropriate spacetime region) and a grasp of the mathematical form of physical forces. The picture is epistemological, as it appeals to seeing. The Viennese read "having before its eyes" as implying the existence of a function that assigns, for each instantaneous state of a system, a global time evolution of the system's states, that is, a maximal set of mutually consistent instantaneous states. We will look at Schlick's and Frank's formulations, beginning with the former.

To begin with Schlick's (1961/1931) terminology, he identifies determinism with causality (p. 179), so the question about a concept of determinism translates into that of the content of the concept of causality. He introduces a set of observed magnitudes and values thereof, where these values are capable of being instantiated at a region of space and time. One may thus consider how the mentioned values are ordered in time.[7] Schlick's important decision is then to relate a causal (deterministic) development of values to these values being related by a law of nature. The meaning of causality (determinism) is thus reduced to the question "what characteristic must the spatio-temporally ordered group of values have so that it may be regarded as an expression of a 'law of nature'?" (*ibid.*, p. 180) Obviously, in a temporal development of states, the states are already ordered. So the question emerges ("most puzzling one", notes Schlick, *ibid.*, p. 181) what a distinctive feature of an orderly succession is, as opposed to a chaotic succession of the magnitudes' values?

Schlick tries a few answers to this question, but finds none satisfactory. However, in discussing the answers he brings two ideas to the light. The first is the observation that no matter how seemingly non-orderly a succession of values is, there is a function from time to the set of values instantiated at moments of time. As he emphasizes, there is such a function even for values of spontaneous jumps of electrons between orbits in Bohr's theory of atoms (*ibid.*, p. 183). This brings us to the second thought, that a causal course of events should be described by a function of a simple form. To focus on the "function argument", let us reflect upon this quote:

> Now one is in the habit of saying that the principle of causality is valid if from the state of the system during a very short time and from the boundary conditions all other states of the system may be deduced. Such a deduction is, however, possible *under all circumstances* for, according to what has been said, functions may always be found that represent all observed values with any desired accuracy. (*ibid.*, p. 186)

Consider the totality of functions from (linearly ordered) time to the set of values, capable of being instantiated at moments of time. In this totality, there are functions that represent the actual course of events with any specified degree of

7 Schlick says "spatiotemporal ordering" of values, but in this paper considers temporal ordering only.

accuracy. If we now want to explicate determinism in terms of deduction (which clearly is Schlick's intention), we obtain that, if we set a degree of accuracy, there are functions that describe the world within this limit of accuracy; from such a function the actual values (within the assumed limit of accuracy) are deducible.[8] Accordingly, identifying determinism with the existence of a function from time to the set of magnitudes' values, makes determinism trivially true. Schlick observes, however, that the mentioned functions can be too complicated to be written down, and generally, to be grasped by humans. Accordingly, he suggests linking determinism to the existence of simple functions of the mentioned kind.

An appeal to simplicity or some other pragmatic criterion is independently motivated by the Circle's doctrine of meaning as testability. In this spirit Schlick advises the reader to turn to the question of how scientists test whether a given formula expresses a law of nature, or not. A simple form of the function is a virtue, but more important for Schlick is the function's projective feature:

> We observe whether the formula obtained [from some observational data] also represents correctly those observation that *were not used* in achieving the formula. Only when this is true does [the physicist] consider his formula to be a law. [T]he only important [...] thing is that equations derived from certain data be applicable to other, *new* data. [...] [T]he true criterion of law, the essential sign of causality, is the *success of prediction*." (*ibid.*, p. 187)

A similar topic, that analyzing determinism in terms of a function makes determinism tautological, occurs in Frank's writings. We turn then to his work, beginning with this quote:

> It is clear that for the world there is only one course of events that really happens. If there is an all-embracing intelligence that knows this course in advance it can also predict it. Every other being also knows that there can be only one course of events for the world; only he does not know in details what it will be like. If the proposition 'everything is predetermined' is meant to say that in reality there is only one course of events, it is a tautology (Frank 1997/1932, 35).

Uncontroversially, if determinism means that there is only one (actual) course of events, determinism is a tautological doctrine. Frank's later texts, however, suggest a strengthening of this observation. Assuming global linear time, the intuition of a single actual course of events translates into the existence of a function f from time to the set of instantaneous global states of the universe. Since it is always possible to define a new function f' by adding 'inert' arguments to the old function, $f'(t,s) := f(t)$, we get a function that seemingly underwrites Laplacean determinism: For a moment t of time and the universe's instantaneous state $s(t_0)$

8 There is a problem of characterizing determinism in terms of deducibility, discussed first by R. Montague (1962), in reference to Nagel's (1953) explication of determinism. A first-order language is typically required to have countably many terms; this requirement is violated by permitting the variable to range over continuous time. Montague's observation lies behind his project of defining determinism in model-theoretic terms.

at a given time t_0, the function yields state $s(t)$, i.e., $f'(t, s(t_0)) = s(t)$. The existence of such a function suggests that one can calculate the universe's state at time t from the knowledge of the universe's state at a single time t_0. What is paradoxical about this conclusion, however, is that the state $s(t_0)$ is an inert argument of function f', as the states $s(t)$ depends on time t only. This kind of trivialization argument was put forward by Russell (which Frank acknowledges). Let us follow Frank in considering how the argument applies to a simple system, a mass point of the Newtonian point particle mechanics:

> Whatever the future may hold in store, it will be described by giving the coordinates x, y, z of the mass point as functions of the time (t). In other words, the future is determined by the three equations: $x = f_1(t)$; $y = f_2(t)$; $z = f_3(t)$. If such functions [...] exist, the "future of the world is determined". Since the world exists only once, these functions are determined by the course of events in the world. "It is true" says Russell, "that the formulae involved (f_1, f_2, f_3) may be of almost infinite complexity, and therefore not practically capable of being written down or apprehended." [...] Thus the material universe must be subject to formulae and the future must be determined. (Frank 1957, 262)

Given the danger of trivializing the claim of determinism, what does Frank suggest to resolve the problem? After observing that a development of states of a given system can always be described by a formula, he points out that in some cases determinism (causal law, in his terminology) might be "of factual value in the prediction of the future". In order that determinism allowed for prediction, he continues,

> [...] we must introduce the vague quality of "simplicity" which is, of course, dependent on the psychological and sociological status of the scientists of a certain period. A "simple" formula means, in this context, a "workable" formula. Since "prediction of the future by ordinary human beings" is an activity within the whole range of human activities, it is understandable that the criterion of "simplicity" can be applied to a causal law. (*ibid.*, p. 268)

In Frank's reaction to the trivialization problems, there is thus an appeal to the same concepts of simplicity and predictions that Schlick uses. What comes to the fore in his analysis of determinism, are cases in which determinism is associated with a formula of a particular mathematical form, which allows us, human beings, to draw a prediction. To have a word for such cases, we call the formulas permitting predictions "simple". But, since whether a formula permits a prediction, or not depends to some extent on the theoretical knowledge and technological abilities of mankind, the notion of simplicity is relative to such social factors. Simplicity is then, writes Frank, a criterion for a formula being a deterministic law. But, bearing in mind a distinction between the content of a concept and a criterion for applying the concept, Frank must opt for the simplicity of an involved formula being part of the content of determinism. As we saw, he appears to be persuaded by the "function argument" that determinism might be vacuously true. To block this option, he must then give determinism some non-trivial meaning, which exactly the appeal to simplicity achieves.

We have seen above two reactions to the trivialization argument: explicating determinism in terms of predictability (Schlick) and appealing to simplicity of formulas, with a knowledge-relative notion of simplicity (Frank). A still another approach, suggested by Schlick (1961/1931, 186) and earlier discussed by Russell, was to require that the functions representing the development of a system's states "do not contain the space and time coordinates" as arguments. Schlick does not elaborate on this suggestion, however.

Before closing this section, let us address a related matter: what was the Circle's view on the status of the determinism thesis (provided it is not a tautology)?

The options on the table were: (1) the determinism thesis is a testable empirical claim, (2) it is a Kantian condition of experience, (3) it is a conventional definition, and (4) it is an imperative of research. The group generally opted against (1)–(3).[9] Option (4) was generally accepted by the group; its clear-cut statement is the following passage:

> The principle of causality does not directly express a fact to us, say, about the regularity of the world, but it constitutes an imperative, a precept to seek regularity, to describe events by laws. Such a direction is not true or false but is good or bad, useful or useless. And what quantum physics teaches us is just this: that the principle is bad, useless, impracticable within the limits precisely laid down by the principle of indeterminacy (Schlick 1962/1931, 285).

4. Frank on Laplacean causality

To first explain Frank's terminology, he calls determinism "strict causality", and call a law "causal" if it "allows us to infer from information about one region of space and time some information about another region of space and time" (Frank 1957, 264).[10] In this sense, the laws Laplace appeals to are causal. Frank emphasizes that Laplace's vision is a philosophical reflection upon Newtonian mechanics and gravitation and hence, to get the gist of Laplacean determinism, one needs to investigate these theories and in particular, the mathematical features of their defining equations. Now, Newton's second law can be written as a set of equations of

9 We note, however, that Frank defended the view (3) in his early paper (1907). And Carnap (1966) talked about (in)determinism of a *theory* (quantum mechanics), anticipating the today's orthodoxy.

10 Frank wrote on determinism for almost fifty years, and his views evolved from a Kantian position in his (1941/1907) paper to an empiricist stance of his later works.The present analysis neglects the former paper, and assumes a hypothesis that in his later works, starting with the 1932 book, Frank was developing a single position on determinism.

the form

$$\frac{d\xi_k}{dt} = F_k(\xi_1, \ldots, \xi_{6n}), \quad k = 1, \ldots 6n, \tag{1}$$

where ξ_i's are generalized coordinates representing position coordinates and momentum coordinates of n point masses (Frank calls them "state variables"), and F_k are generalized forces, that is, components of forces acting on point masses as well as momenta of point masses. (In other words, equation (1) encapsulates the second law, $dp/dt = F$, as well as the formula for momentum, $dx/dt = p/m$.) Frank's diagnoses concerning the determinism of Newtonian point particle mechanics is then the following:

The mathematical theory of differential equations teaches us to "integrate" a system of the form

$$\frac{d\zeta_k}{dt} = F_k(\zeta_1, \zeta_2, \ldots), \quad (k = 1, 2, \ldots, n).$$

This means that if the values of the state variables for one instant of time (e.g., $t = 0$) are given, we can find the values of $\xi_1, \ldots, \xi_n$ at any time t if the value for $t = 0$, and the differential equations (in other words, the functions F_k) are given. The differential equations are an instrument that allows us to compute the values of the state variables at all times t if the values for one instant $t = 0$ are given. We can call this a "prediction" because we compute the future values from the present values (Frank 1957, 266).

Frank's verdict that the solutions can be found "at any time t" is not correct, however. His formulation claims the existence of global solutions, whereas what is guaranteed (under a certain condition) is the existence of local solutions, that is, in a neighborhood of every point in the domain of F_k's. Moreover, the condition, known as the Lipschitz condition, is crucial: its failure may result in there being no local solution at some t from the domain of F_k's.[11] The failure of the Lipschitz condition lies behind a currently widely discussed indeterministic mechanical system, known as the Norton dome (Norton 2008). Undoubtedly, being a top mathematical physicist, Frank must have been conversant with the subtleties of differential equations. After all, he edited (with R. von Mises) a handbook on differential and integral calculus (Frank and von Mises 1925). A contribution to this volume, written by L. Bieberbach, contains a discussion of the initial value problem for systems of differential equations, including an analysis of the Lipschitz condition.[12] For some reason, however, Frank does not inform the philosophically oriented reader about when and what kind of solutions to differential equations exist. He appears to think that there is a graver problem with equations (1) than the non-existence of solutions for some particular functions F_i's: Even if solutions to Eq. (1) exist, can anyone get to know them, if we allow for arbitrarily complex function and arbitrarily complex initial conditions? Or, to put it differently, what

11 That is, the Lipschitz condition is a sufficient but not necessary condition for the existence of local solutions of differential equations like Eq. (1).

12 Cf. his "Gewöhnliche Differentialgleichungen", in Frank and von Mises (1925), in particular pp. 282ff.

idealizations does one need to assume to link the (abstract) existence of solutions to its being knowable? We face a dilemma:

If we do not introduce a Superior Intelligence which "knows" the functions F_k or present explicitly a formula for the F_k, the word "determined" can only mean that we assume some property of the F_k that restricts the possibilities. [...] Newton's laws become practically useful only when we add the assumption that the "forces" are "simple" functions of the coordinates. [...] If we admitted an arbitrary complexity of the functions $F_k(\xi_1, \ldots, \xi_n)$, the mere statement of "existence" would be no statement about facts, but a tautological statement that could not be refuted by any experiment (Frank 1957, 298).

In other words, if we do not assume the agent's powers to be equal to those of the omniscient intelligence, linking the existence of solutions to calculability requires imposing a restriction on how complex the formulas for initial conditions and forces are; otherwise the (abstract) existence of solutions is beyond our reach in the sense that it cannot be tested whether a temporal evolution of a system is properly described by the solutions. Presumably, since a claim of the existence of solutions might fail to have experimentally testable consequences, Frank calls it "tautological".

Although it is not immediately clear, Frank appears to prefer the option with non-testable claims to the *ad hoc* restrictions of the form of forces and initial conditions. The former means explaining determinism in terms of the Laplacean omniscient intelligence, which Frank appears to accept. He emphasizes, for instance, that Laplace *really* needed an appeal to the omniscient intelligence, without suggesting that modern adherents of Laplacean determinism can do without it.[13] The option looks acceptable to Frank since even if one allows for forces of arbitrary complexity, and accepts that a solution to the defining equation could be as complex as a piecemeal description of a system's states, the existence of solutions implies a claim about observable phenomena:

There is, however, one conclusion that can be drawn from the principle of causality that is independent of the special form of the functions F_k and which perhaps could be tested by experiment. Whenever these variables [i.e., ξ_k's] assume their original values, the changes in time reassume their values too. In other words, if a state of the system repeats itself, all following states repeat themselves too. If we call a value system of the ξ_k the "state A" of our system, we can say: If a state A of our system is followed by a state B, every time that A returns it is followed by B (Frank 1957, 269).

We thus have come to the re-occurrence view: a system is deterministic means that if in the development of its states a state A followed by a state B occurs, then whenever A re-occurs, it is followed by B. But with this formulation as well,

13 Cf. Frank (1957, 267): "It is interesting to note that Laplace, who did not need the hypothesis of God in his book on astronomy, needed a Superior Intelligence in his formulation of the principle of causality. He believed, of course, that this is only a convenient way of speaking, and that this Superior Intelligence could be eliminated and the principle of causality could be formulated by reference to human ability alone."

a trivialization creeps in, as Frank observes (*ibid.*, p. 282ff.). First, if no state of a system ever reoccurs, the system is qualified as deterministic, which seems counter-intuitive. Secondly, there is a cheap move of assessing any system as deterministic by playing the incompleteness card: if a state A reoccurs, but is not followed by same states as before (which provides *prima facie* evidence that a system in question is indeterministic) one may declare that A does not fully represent the underlying reality. That is, although A occurred on two occasions, the underlying maximal states were different on these occasions. By positing more informative states (including "hidden" states, or as Frank says, "hidden qualities"), determinism is restored—so the argument goes.

To counter this argument, Frank appeals to the practice of physics: one aim in a theory-construction is to come up with a small number of variables representing physical states. To illustrate, in Newtonian point mass particle mechanics, the state variables are position and momentum coordinates of the point masses involved; in standard non-relativistic quantum mechanics, the state variables are closed subspaces of the Hilbert space of an appropriate dimensionality. More precisely, it is not so much the small number of state variables, but rather that they are uniform, and that adding a new type of variables to a given theory is a bad practice. In this spirit, to keep trivialization at a bay, Frank suggests to focus on a small number of state variables, and investigate if "the return of this small number of variables to their original values is followed by the return of the state B that had followed the first time" (Frank 1957, 274).

To take a bird's-eye view of Frank's journey to explicate determinism, we are not getting any closer to a firm grip on determinism. There is a note of despondency in his wrapping up of his study of determinism:

> In order to prevent the principle of causality from becoming tautological, we must introduce vague terms like "small number of state variables" and "simple laws of forces (F_k)". We have a choice between making the principle of causality precise and tautological or vague and factual (*ibid.*, 274).

5. Discussion

In the 1960's, when Carnap, Frank, and Feigl were still active, two logically rigorous frameworks were developed with the aim to define and analyze (in)determinism. Montague (1962) put forward a model-theoretic definition of Laplacean (in)determinism for theories, with "theory" understood in the logical sense. In a nutshell, a theory is said to be Laplacean indeterministic if it has two partial models such that an initial segment of one and an initial segment of the other are identical. In this way Montague removed epistemological overtones from the concept of Laplacean (in)determinism. On his account, whether a theory is (Laplacean) deterministic is an objective matter, decided by relations between abstract set-theoretic structures—models of the theory. Montague's framework forms a core of the current orthodoxy of analyzing indeterminism in the philosophy of science, as shaped by the works of J. Butterfield and J. Earman,[14] although after Montague's work, the concept of history got relaxed, and the strict machinery of partial models gave way to looser notions of possible worlds and laws or theories being true in a possible world.

The second development occurred in the Aristotelian (modal) tradition of thinking about (in)determinism. Arthur Prior worked on a project of introducing modalities to models with linear time, when he received S. Kripke's letter (dated September 3, 1958) with a passing suggestion to represent possibilities (or indeterminism) by branching trees.[15] Prior briefly discussed the model, now known as the branching-time semantics, is his (1967) book; the theory was later fully worked out by Thomason (1970). Today the notion of branching possible histories underlies a variety of theories of branching time and of branching space-times.

What was the Viennese's reactions to the two developments? As far as I know, they have not discussed Montague's or Prior's work, and, most likely they did not know it. One might be nevertheless tempted to ponder how the Circle's members might have responded to accounts of (in)determinism that Montague, Kripke, and Prior initiated.

I am skeptical of this enterprise, however. The Circle's epistemological approach and its worry that the determinism thesis might have no empirical consequences make the circle's investigation hardly connectible to later debates over (in)determinism. Observing the Vienna Circle grappling with the concept of indeterminism is like watching an entirely different kind of sport. Today the truism that there is only one actual course of events does not look related to (in)determinism. We freely construe the latter as a claim that there are many *possible* courses of events, or that a theory has more than one realization of appropriate kind. The orthodox view takes (in)determinism to be an attribute of theories; accordingly it does not declare the determinism thesis "tautological" merely on the ground that it might have no testable consequences. Finally, the question whether a theory's defining equations (typically, differential equations) admit the existence of unique and stable solutions is taken to be of utmost importance: the behavior of the defining equations is believed to provide the main evidence about the theory's (in)determinism.

In relation to indeterminism-friendly logic, Schlick's (1962/1931) reaction to Łukasiewicz's (1930) many-valued logic suggests a similar diagnosis. Łukasiewicz's avowed aim was to accommodate indeterminism: he was persuaded by Aristotle's sea battle argument that bivalent truth is in conflict with indeterministic future, and in particular, with the logician's freedom of creativity. Schlick responds that by equating determinism with predictability, the principle of bivalence is saved, even if indeterminism reigns. Indeterminism, he claims, does not mean that a proposition about a future event "is not unambiguously true or false, but only that the truth or falsity of that proposition cannot be *calculated* on the

14 Cf. Butterfield (2005) and Earman (1986, 2007).

15 This is mentioned by Øhrstrøm and Hasle (1995, p. 173).

basis of propositions about present events" (*ibid.*, p. 292) Assessing Łukasiewicz's project, he writes:

There must of course be an error here, for logical propositions, which are only rules of our symbolism, cannot depend as regards their validity upon the existence of causality in the world; every proposition must have truth or falsity as a timeless characteristic (*ibid.*, p. 291).

Clearly, the argument is a non sequitur: its premise that propositions must have timeless truth-values is rejected in various logics, which relativise truth to moments of time, events, possible worlds or histories. It is again hard to see how Schlick's position can be connected to a later research in the Prior-Kripke tradition.

References

Butterfield, J. N. (2005): "Determinism and indeterminism", In: *Routledge Encyclopedia of Philosophy*, vol. 3, Routledge, London.

Carnap, R. (1966): *Philosophical foundations of physics*. Basic Book, Inc., New York.

Earman, J. (1986): *A primer on determinism*, G. Reidel, Dordrecht.

Earman, J. (2007): "Aspects of determinism in modern physics", In: J. N. Butterfield and J. Earman, editors, *Philosophy of Physics*, pp. 1369-1434, Elsevier, Amsterdam.

Feigl, H. (1953): "Notes on causality", In H. Feigl and M. Brodbeck, editors, *Readings in the philosophy of science*, pp. 408-418. Apleton-Century-Crofts, New York.

Frank, P. (1941/1907): "Law of causality and experience", in his *Between Physics and Philosophy*, Harvard University Press, Cambridge, Mass.; translation of Frank, P. (1907): "Kausalgesetz und Erfahrung", *Ostwald's Annalen der Naturphilosophie*, 6: 443-50.

Frank, P. (1957): *Philosophy of Science: the Link between Science and Philosophy*, Prentice Hall, Engleford Cliff, NJ.

Frank, P. (1997/1932): *The Law of Causality and Its Limits*, Springer, New York; translation of Frank, P. (1932): *Das Kausalgesetz und seine Grenzen*, Julius Springer, Wien.

Frank, P. and von Mises, R. (1925): *Die Differential und Integralgleichungen der Mechanik und Physik*, F. Vieweg, Braunschweig.

Goldblatt, R. (2006): " Mathematical modal logic: a view on its evolution", In D. M. Gabbay and J. Woods, editors, *Hanbook of the History of Logic*, vol. 2, pp. 1-98. Elsevier, Amsterdam.

Laplace, P. (1951/1820): *A Philosophical Essay on Probabilities*, Dover Publications, New York; translation of Laplace, P. (1820): *Théorie analytique des probabilités*, V. Courcier, Paris.

Lenin, V. I. (1909): *Materialism and Empirio-criticism. Critical Comments on a Reactionary Philosophy*, Zveno Publishers, Moscow.

Łukasiewicz, J. (1920): "O logice trójwartościowej", *Ruch Filozoficzny*, 5: 169-171.

Łukasiewicz, J. (1930): "Philosophische Bemerkungen zu mehrwertigen Systemen des Aussagenkalkuls", *Comptes rendus des seances dela Societe des Sciences et des Lettres de Varsovie*, 32: 51-77.

Montague, R. (1962): "Deterministic theories", *Decisions, Values, and Groups*, 2: 325-370; reprinted in Montague, R. (1979): *Selected papers of Richard Montague*, Yale University Press.

Nagel, E. (1953): "The causal character of modern physical theory". In H. Feigl and M. Brodbeck, editors, *Readings in the Philosophy of Science*, Appleton-Century-Crofts.

Norton, J. D. (2008): "The dome: An unexpectedly simple failure of determinism", *Philosophy of Science*, 75: 786-799.

Øhrstrøm, P. and Hasle, P. F. V. (1995): *Temporal logic: from ancient ideas to artificial intelligence*, Kluwer AP, Dordrecht.

Prior, A. N. (1967): *Past, present, and future*, Oxford University Press, Oxford.

Russell, B. (1912): "On the notion of cause", *Proceedings of the Aristotelian Society (New Series)*, 13: 1-26.

Schlick, M. (1961/1931): "Causality in contemporary physics (I)", *British Journal for the Philsophy of Science*, 12(47):177–191; translation of the first part of Schlick, M. (1931): "Die Kausalität in der gegenwärtigen Physik", *Naturwissenschaften*, 19(7): 145-162.

Schlick, M. (1962/1931): "Causality in contemporary physics (II)", *British Journal for the Philsophy of Science*, 12(48):281–298; translation of the second part of Schlick, M. (1931): "Die Kausalität in der gegenwärtigen Physik", *Naturwissenschaften*, 19(7) :145-162.

Stöltzner, M. (1999); "Vienna indeterminism: Mach, Boltzmann, Exner", *Synthese*, 119(1-2): 85-211.

Stöltzner, M. (2003): "Vienna indeterminism II: From Exner to Frank and von Mises". In P. Parrini, W. C. Salmon, and M. C. Salmon, editors, *Logical Empiricism: Historical and Contemporary Perspectives*, pp. 194-231. University of Pittsburgh Press.

Thomason, R. H. (1970): "Indeterminist time and truth-value gaps", *Theoria*, 36: 264-281.

Department of Philosophy, Jagiellonian University
52 Grodzka
31-044 Krakow
Poland
Tomasz.Placek@uj.edu.pl

JOHN D. NORTON[1]

INFINITE IDEALIZATIONS

1. Approximations of arbitrarily large but finite systems are often mistaken for infinite idealizations in statistical and thermal physics. The problem is illustrated by thermodynamically reversible processes. They are approximations of processes requiring arbitrarily long, but finite times to complete, not processes requiring an actual infinity of time.
2. The present debate over whether phase transitions comprise a failure of reduction is confounded by a confusion of two senses of "level": the molecular versus the thermodynamic level and the few component versus the many component level.

1. INTRODUCTION

In my talk at the celebration of the 20th Anniversary of the Vienna Circle Institute, I sketched results of recent work on approximation and idealization (Norton, 2012). A goal of that work was to clarify the widespread use of infinite limits in statistical physics to introduce what are informally described as idealizations. This literature examines the behavior of systems composed of very many – but always finitely many – components. Certain properties of these systems settle down to stable values if the number of components is very large, as is the case with ordinary thermal systems. These stable values can be approximated quite closely by taking a limit of the property. Each of these properties is a function of the number of components and one merely takes the limit of these functions. The key point is that the limit of these functions can be taken without considering a system that consists of an actual infinity of components.

Infinite limits can be treacherous, and especially so, if we take a limit to a system consisting of an actual infinity of components. When things go badly, as they often do, this limit may yield a limit system of infinitely many components with quite unintended properties. Worse, if the limit is sufficiently badly behaved, there may be no limit system at all. What results is confusing talk about limit systems that have properties different from those announced; or, worse, the attribution of properties to systems that cannot exist at all.

1 I thank Jos Uffink for helpful discussion.

M.C. Galavotti et al. (eds.), *European Philosophy of Science – Philosophy of Science in Europe and the Viennese Heritage*, Vienna Circle Institute Yearbook 17,
DOI 10.1007/978-3-319-01899-7_14,

To help clarify these issues, I called for a more disciplined use of the related terms "idealization" and "approximation." It then followed that many of what are called idealizations in the present statistical physics literature are really approximations. Below in Section 3, I will give another example of a misleading infinite limit, the notion of a reversible process in thermodynamics. My goal is to show how the terms "approximation" and "idealization" can be used to separate two quite distinct forms of analysis. The example has been chosen because it is technically much simpler than the infinite component number limits employed in statistical physics. The infinite time limit of reversible processes will display the confusion of approximation and idealization with fewer distractions from the technical difficulty of the infinite component number limit of statistical physics.

Infinite idealizations in statistical physics are often associated with claims of a failure of reduction; and there is presently some debate in the literature over when and whether the reductions succeed. My concern is that much of this literature conflates two distinct senses of reduction in statistical physics. They are a reduction between macro-thermodynamic and micro-molecular systems; and a reduction between system of many components and few components. While I do not think that separating these two senses will resolve all disputes, I do think that attending directly to the two senses may dissolve some of the incomprehension in the disputes. In particular, recent work on phase transitions displays a failure of reduction between many component and few component system; but it does not display a failure of reduction between thermodynamic and molecular systems. These two senses of reduction will be developed in Section 4.

2. A Well-Behaved Idealization

My goal is to demonstrate how easy it is to mistake an idealization for something less. To prepare for these troublesome cases, I will review an unproblematic case in this section. An ideal gas provides an example of an idealization in which everything works as we would expect. Describing it will also enable me to introduce the definitions that will be employed for approximation and idealization.

Dilute gases under normal conditions of temperature and pressure conform quite closely to the ideal gas law

$$PV = nRT \tag{1}$$

Here P is the pressure, V the volume, n the number of moles, T the temperature and R is the ideal gas constant. The law has surprisingly wide scope. It applies also to the osmotic pressure exerted by dilute solutions. While often good, the fit of the law to the behavior of real systems is never perfect. Errors become larger if, for example, we consider regimes of high pressure and high density.

The law expressed in (1) is a proposition that describes the target system inexactly. These I shall approximations:

> An *approximation* is an inexact description of a target system. It is propositional.

So far, the ideal gas law has been introduced as an empirical generalization. There is a second way to introduce it. We consider another fictitious system of very many non-interacting, spatially localized components. The system is fictitious since there are no real systems like that. All molecules occupy some volume of space and interact at least weakly with other molecules. The same is true of solute molecules in solution. While these real, target systems conform only inexactly to the ideal gas law, it turns out that the fictitious system obeys it exactly.[2]

This fictitious system, I will say, is an idealization of the target systems of real dilute gases and dilute solutions. Loosely speaking, it is an idealization because it looks like those systems in the particular aspects that interest us; that is, in the relationship between their pressures, temperatures and other magnitudes in the ideal gas law (1). More carefully, this fictitious system is an idealization of those real systems in that an *exact* description of the fictitious system for the aspects of interest turns out to provide an *inexact* description of the target system. Put more tersely:

> An *idealization* is a real or fictitious system, distinct from the target system, some of whose properties provide an inexact description of some aspects of the target system.

This characterization describes the essential relationship between idealizations and approximation. We take the description that is exactly true of the idealization. It will be propositional, such as the ideal gas law (1). This same proposition then turns out to be an inexact description of the target system.

It follows that we can always demote an idealization into an approximation by discarding the idealizing system and preserving only the relevant propositions that describe it. A corresponding promotion from approximation to idealization will not always be possible. Cases in which the promotion fails are just those that will interest us here.

---

2 It is not generally recognized how few assumptions are needed to generate the ideal gas law. One does not need the familiar assumptions of many point-like particles propagating in straight lines in vacuo and bouncing elastically off the walls confining the gas. One merely needs to assume that there are many, spatially localized components that do not interact with each other. They may interact with other components present in the bulk of the system without compromising the derivation. For a short demonstration employing very few assumptions, see Norton (2006, §2.1).

3. Reversible Processes in Thermodynamics

An idealization is a system, fictitious or real, whose properties, propositionally described, supply inexact descriptions of the target system. The essential starting point of the notion of idealization is that we have a consistently describable system, even if it is fictitious. Without that, there is no idealization. Reversible processes in thermodynamics, I shall now argue, fail this minimal condition. Whatever they may be, they are not idealizations. Recognizing that fact helps us dispel the fog of paradox that surrounds them.

3.1 The Informal Idea of Reversible Process[3]

All real thermodynamic processes are dissipative in the sense that they increase the total thermodynamic entropy of the universe. For example, when a quantity of heat Q passes from a hotter body at temperature T_{hot} to a colder one at T_{cold}, the entropy created is[4]

$$\Delta S = Q/T_{cold} - Q/T_{hot} = Q(T_{hot} - T_{cold})/ (T_{hot}.T_{cold}) = Q.\ \Delta T/(T_{hot}.T_{cold}) \qquad (2)$$

where $\Delta T = T_{hot} - T_{cold}$. This creation of entropy amounts to a loss of the ability of the heat to perform work, for a smaller portion of the degraded heat at the lower temperature can be converted to useful work. It is an inefficiency that the science of thermodynamics was founded to minimize.

The inaugural insight of thermodynamic theory, in Sadi Carnot's original work of 1824, is that the least dissipation and greatest efficiency is approached in the unachieveable limit of reversible processes. No real thermodynamic process is reversible. Heat will only pass from a hotter body to a colder one if there is a temperature difference between them; and, as along as that temperature difference is maintained, the process cannot be reversed. Heat cannot spontaneously pass from the colder to the hotter body.

3 For a detailed examination of the notion of reversibility in thermodynamics, see Uffink (2001).

4 It is assumed that the heat passes by conduction and without appreciable change in the temperature of either body. For that Q may be very small in relation to their internal energies; or their internal energies may be kept constant by work performed reversibly on each system.

We can convert this process into a thermodynamically reversible process by making the temperature difference ΔT between the hot and cold body "infinitesimally small," so that the temperature difference that drives the heat transfer is negligible. To assure that the desired quantity of heat, Q, still passes, we must wait an arbitrarily long time, since we make the temperature difference arbitrarily small. However taking the limit of infinite time returns a process with remarkable properties. Throughout, the two bodies are at the same temperature, that is, they are at thermal equilibrium. As a result, the process now no longer has a favored direction for the passage of heat. It is the same process as would be created had we started with a temperature difference that passed heat in the reverse direction. In this sense, the process produced by this ideal, infinite time limit is reversible; it proceeds equally well in either direction. Finally, since its temperature difference ΔT is zero, equation (2) tells us that no entropy is created by the passage of heat Q between the two bodies. We have achieved minimum dissipation.

3.2 Why it is Nonsense

The account of the last section is familiar from thermodynamics treatises. But is it also nonsense in the precise sense that there is no possible referent for the term "reversible process." For the reversible process just described cannot exist if we want to preserve some of the simplest elements of thermal physics. The process described must satisfy two incompatible conditions:

(a) A non-zero quantity of heat Q passes from one body to the other.

(b) The two bodies are at the same temperature.

Once the bodies are at the same temperature as (b) requires, no heat passes between them and (a) fails. This follows directly from the fact that heat only passes from one body to a second if the first body is hotter than the second. Condition (b) allows no driving temperature difference. Even an "infinitesimal difference" is too much since it will favor passing heat in one direction and not the other, thereby contradicting reversibility.

We may hope that the magic of taking the infinite time limit will somehow preserve the passing of heat. For, in every finite time process considered in approaching the limit, we do succeed in passing Q of heat. When the limit delivers a process consisting of the two bodies at the same temperature in contact over an infinity of time, we imagine that somehow the passage of heat is preserved. It is not.[5]

Our vain hope is that a zero rate of heat transfer over an infinity of time balances out to the passing of the requisite non-zero quantity Q of heat. In effect we are pinning our hopes on the faux equation:

$$\text{“}Q = 0 \times \infty\text{”}$$

5 The infinity of time also makes it a process that never ends, so we will never see its results. Moreover, we will not be able to append a subsequent process. For there is no finite time at which the subsequent process can start.

Alas, we know that this equation is nonsense. Once we take the limit of zero temperature difference and infinite time, we have a static system. It consists of two bodies at the same temperature that pass no heat between them in any finite time. Waiting longer will not change the fact that no heat passes. Even waiting an infinity of time will make no difference. That infinity of time consists of infinitely many periods of finite time in each of which no heat passes. The sum of these infinitely many zeroes of heat passed is still zero.

In short, we have used a limit to infinite time incautiously. We take a process that is well defined for all finite times. We let the time go to infinity and then speak uncritically about the process that results, attributing properties (a) and (b) to it. There is no such process on pain of violating the basic notion that heat passes between bodies only when driven by a temperature difference.

3.3 Reversible Processes as Approximations

The prospects look bleak for reversible processes in thermodynamics. How it is possible that they can figure so centrally in thermodynamic theory? The answer is that they figure as approximations and not as idealizations. The infinite time limit procedure does not generate a new system that is our idealization. Rather, the limit procedure takes the exact propositional descriptions of the finite time processes and converts them into a simpler, inexact description of the finite time process. That is, the limit process merely generates approximations.

To see how this works, recall the rules that govern heat transfer. The total heat transferred between two bodies with temperature difference ΔT over some time is given as:

$$\text{heat} = \text{constant x } \Delta T \text{ x time} \tag{3}$$

and the entropy created in the process is given by (2). Let us tabulate these quantities as the limit of infinite time is approached. In order to maintain passing of a fixed, non-zero quantity of heat, ΔT and time are adjusted in concert so that (ΔT x time) retains the same value. To keep the numbers simple, we can set the constant in (3), so that heat=100 when ΔT = time = 1:

	T_{hot}	T_{cold}	ΔT	time	heat passed Q	entropy created ΔS
	10	9	1	1	100	1.11
	10	9.5	0.5	2	100	0.526
	10	9.9	0.1	10	100	0.101
	10	9.99	0.01	100	100	0.010
	10	9.999	0.001	1000	100	0.001
	10	9.9999	0.0001	10000	100	0.0001
	…	…	…	…	…	…
Limit of columns at time → ∞	10	10	0	∞	100	0
Properties of limit process	10	10	0	∞	0	0

Table 1. The infinite time limit of a heat transfer

The table shows how the various quantities stabilize in their approach to limiting values:

$$\Delta T = 0 \quad Q = 100 \quad \Delta S = 0 \tag{4}$$

What are to make of these limiting values? They are the properties that we had uncritically attributed to a single process that we called "the reversible heat transfer." However, as we have just seen, they cannot be the properties of a single process. Once we set ΔT = 0, then we must have Q = 0 as well, no matter what the time. To underscore the point, the last row of the table displays the corresponding values for the static system that arises when we take the limit. That limit process consists of two bodies at the same temperature passing no heat between them.

The importance of the propositions comprising (4) is that they are inexact descriptions of each of the heat transfer processes that arise when we have a very small ΔT and a correspondingly large time. For, by virtue of their construction, the limiting values reported in (4) must be very close to the actual values possessed by heat transfer operations in the table that require very large, but finite times. That is, the limiting value of ΔS = 0 is a very good approximation of the correct values of ΔS = 0.0001 for the process with time = 10000 and ΔS = 0.001 for the process with time = 1000.

To revert to the terminology introduced above, the function of the infinite time limit is to create propositions (4) that are *inexact descriptions* of finite, long time heat transfers; that is, the function of the infinite time limit is to generate an *approximation* of finite, long time heat transfers. Confusion arises only if we try to present the propositions of (4) as describing just one idealized process exactly, for then we are asked to imagine a fictitious process with impossible properties.

This analysis tells us how we should reform talk of reversible processes in thermodynamics. If there is a need to be precise, we would not talk of a reversible heat transfer in which heat Q = 100 is passed at ΔT = 0 while creating no entropy, ΔS = 0. Rather we would say that real heat transfer processes can be slowed so that they come arbitrarily close to the parameters set in (4).[6] If we manage to slow the heat transfer sufficiently, the differences between the parameters of our real system will become so small that we will not be bothered by the difference. However no amount of slowing will ever allow us to realize a process that is described exactly by the propositions of (4).

In sum, the infinite time limit generates approximations of long, finite time heat transfers in the form of the propositions of (4). The limit does not create an idealized limit process, that is, an idealization, of which these propositions are an exact description.

3.4 Other Examples in Statistical Physics

The example of reversible processes in thermodynamics provides a simple illustration of a problem that arises repeated in statistical physics. In that science, we examine statistically the behavior of large systems of components – molecules, spins, radiation modes – in order to recover their thermal properties. Since the systems have very many components, the taking of an infinite component limit is common. It is also common to treat these infinite limits incautiously as if they were creating idealizations. Yet they less commonly do so. Their role will vary with the particular way the infinite limit is taken and almost every possibility is realized somewhere.

The simplest case arises with the "continuum limit," described by Compagner (1989). In it, the number of components passes to infinity in such a way that both system volume and the fraction occupied by matter remain constant. One finds (Norton, 2012, § 4.2.1) that there is no limit system; some of limits that would return its properties are not convergent.

The Boltzmann-Grad limit (Lanford, 1975, 70-89; 1981) also passes the number of components to infinity while keeping the system volume constant. However it does it such that the fraction of space occupied by matter becomes zero and the matter itself becomes an infinity of dimensionless points. This limit system has unexpected properties. In it, the familiar rules of mechanics are no longer able to resolve collisions uniquely, so that, if there are collisions in the system, they are indeterministic. (See Norton, 2012, § 4.2.2).

6 This arbitrarily close approach is possible only as long as we operate within phenomenological thermodynamics. Once we allow for the atomic constitution of matter, then we operate within statistical mechanics and fluctuations intervene. When the temperature difference becomes small, some creation of entropy is necessary to overwhelm the fluctuations and allow the process to proceed to completion. This last observation, I argue in Norton (2011, manuscript), comprises a fatal no go result for the theory known as the thermodynamics of computation.

The most commonly used limit is the "thermodynamic limit," in which both the number of components and the system volume pass to infinity in such a way that their ratio, the component density, remains fixed. One can find two versions of the limit in the literature. In what I call the "weak" form, such as in Le Bellac et al. (2004, pp. 112), one merely tracks the behavior of the system's properties as the number of components becomes arbitrarily large. In what I call the "strong" form, as in Ruelle (2004, p.2), the limit is used to take us to an actually infinite system whose properties are then examined as a surrogate for large systems of finitely many components. The weak form corresponds to using the limit merely to create approximations of large systems with finitely many components. The strong form corresponds to using the limit to create an idealization that consists of a new system with an actual infinity of components.

These last idealizations are inherently fragile since most infinite systems of components behave differently from their finite counterparts. Most notably, they are generically indeterministic. An infinite system of masses connected by Hooke's law springs, for example, is a natural model of an infinite crystal. It can remain quiescent for an arbitrary time. Then it can spontaneously set itself into motion by excitations propagating in from infinitely far away in the crystal.

If the strong version is to be used, one must contrive additional constraints to block this unwanted behavior. This is the practice of the more mathematically oriented theorists such as Lanford (1975). However it leaves the reader wondering what *other* unwanted behavior unique to infinite systems remains unblocked.

Most other invocations of the thermodynamic limit leave vague just how the limit is used. While they may talk of infinite systems and their properties, it is more often the case that the analysis proves to be not directly of an actually infinite systems. There are, for example, no additional conditions applied to block the unwanted generic indeterminism of infinite systems of components. Rather, the analysis seems to arrive at its results by tracking the limiting behavior of the propositions describing large systems with finitely many components. That is, their results are the analogs of the propositions (4) and provide approximations, not idealizations. I have argued in Norton (2012, § 4.3) that this is case with the analysis of critical phenomena by renormalization group methods. They employ approximations and not infinite idealizations.

4. Reduction and Emergence in Statistical Physics: An Overlooked Distinction

The presence of infinite component limits in the statistical physics literature has ignited a debate in the philosophy of science literature. It is over the question of whether phase transitions are emergent properties, in the sense that they indicate a failure of reduction between thermodynamics and statistical mechanics.[7] This is a frustrating debate, rife with mutual incomprehension. My conjecture is that at least a part of the incomprehension comes from a failure to distinguish two senses of "level." The two senses should be kept separate, since one favors reduction relations and the other does not. Participants in the debate can press toward conclusions favoring reduction or emergence according to which notion of level appears more in their analysis.

4.1 A Familiar Distinction: Ontological and Theoretical Reduction

First a preliminary: the distinction is *not* between ontological and theoretical (or explanatory) reduction, which will be sketched in this section. In this distinction, a thesis of ontological reduction asserts that thermal systems just are systems of many molecules, spins, radiation modes, and so on. A thesis of theoretical reduction asserts that the theories of these microscopic components are able to generate the macroscopic thermodynamic theories.

In the case of ontological reduction, the debate has been settled in favor of the reductionists.[8] While the ontological thesis is quite ambitious, the evidence in its favor is so massive that, now, no one who doubts it is or should be taken seriously. The situation with theoretical reduction is less clear. In some cases it clearly works: the recovery of the ideal gas law from microphysics is a good and popular example. However the case is yet to be made that it will always succeed.

---

7 A few of the contributions to this debate are Batterman (2005), Belot (2005), Butterfield (2011, 2011a), Callender (2001), Jones (2006).

8 Lest one think that this is a trivial outcome, it is well to recall that atomic and similar corpuscular theories of the constitution of matter were little more than imaginative speculation for most of the millennia of their existence. Their correctness was only established within the last century or so. See Chalmers (2009).

The difficulty is that a theoretical reduction requires manipulation within a theory and theories employ human created methods and human selected postulates. That creation and selection must be done sufficiently well to enable computations at the molecular level that will return the requisite results at a thermodynamic level. The reach of derivations within theories can be less than we expect. Consider the simple Euclidean theory of figures built from circles and straight lines. The novice in geometry expects that all constructions are achievable with Euclidean straight edge and compass. Yet there are simple figures in the theory's ontology that outstrip the constructive powers of a straight edge and compass. Merely trisecting an angle is beyond their reach. The more extreme but familiar example is Gödel's celebrated discovery that some truths of arithmetic will escape the theorems of any particular, finite axiomatization. What happens when we pass to more complicated thermal systems? Will their properties always be within reach of the computations with molecules and spins of some nominated theory?

4.2 The Neglected Distinction

The present debate is whether these last limits of theoretical reduction have already been reached with phase transitions. It is often couched in terms of levels. There is a lower level in which we attend to the properties of individual components molecules, spins or radiation modes. There is a higher level in which we attend to the thermodynamic properties of large collections of these components. We ask if a theoretical reduction relation obtains between the two levels.[9] There are two ways of identifying the higher and lower level:

I. Molecular-Thermodynamic.

The lower molecular level is the theory of the components, characterized by the canonical positions and momenta of a phase space. The basic quantities of the theory are functions of the phase space: the Hamiltonian characteristic of the system and a probability distribution over the phase space. Other quantities, including mean energy, free energy and entropy, are derived from them. The higher thermodynamic level is the theory of thermal states, characterized by the state variables: pressure, volume, temperature, density, and so forth. Internal energy, free energy and entropy are functions of these state variables.

9 I use this symmetrical language to retain compatibility with the different usages of the term "reduce" and "reduction" by philosophers and physicists. Following Nagel (1961, Ch. 11), philosophers ask if the higher level theory can be derived from the lower level theory. If so, the higher level theory is said to reduce to the lower level theory. Physicists ask if the lower level theory transforms into the higher level theory when an appropriate limiting operation is applied, such as the taking of a limit of infinitely many components. If that transformation succeeds, then the lower level theory is said to reduce to the higher level theory. This is a reverse of the language of the philosophers in terms of what reduces to what. Both are expressing the same idea, however. Loosely speaking, the higher level theory is some kind of special case of the lower level theory.

This is the familiar distinction that appears explicitly in most of the discussions in the philosophy of science literature. There is a second distinction that delineates the levels differently:

> *II. Few Components-Many Components.*
>
> Both of these levels reside within the molecular level of the first distinction. The lower "few" level consists of the analysis of systems of just a few components. The higher "many" level consists of the analysis of systems of very many components, where the systems of many components are still understood to have molecular constitution.

The ideal gas law provides a success of reduction between these levels. A single molecule can be treated as a gas, albeit one with wildly fluctuating properties.[10] Nonetheless one can determine the mean values of these fluctuating quantities, such as the gas' mean density and the mean pressure it exerts. These mean quantities can be scaled up from the few component level to the level of the very many components of a macroscopically sized samples of a gas of non-interacting molecules. What results is the ideal gas law.

This second delineation of levels is routinely suggested. It is already alluded to in Anderson's (1972) half century old and much celebrated slogan "more is different." However it is rarely distinguished from the first delineation. Rather we pass from a few component, lower level account to a many component and thermodynamic, higher level account without separating the two senses of lower and higher level just described. That obscures the source of any breakdown in the reduction relations. Did the breakdown arise in the passage from few to many components? Or did it arise in the passage from the micro-molecular to the macro-thermodynamic?

For each of these two delineations of levels, we can ask if a theoretical reduction relation obtains between the lower and higher level. In my view, we recover a different answer in each case.

In the molecular-thermodynamic case, the renormalization group method analysis of phase transitions supplies a successful theoretical reduction. Because of the technical brilliance and difficulty of the method and the presence of oddly behaved limits, it is possible to miss this simple fact. The key observation is that the renormalization group analysis is carried out in the lower, molecular level. It is, in the end, a molecular analysis. The basic units of the analysis are large systems of components, understood at the molecular level, with their associated

---

10 The idea that a single component – a molecule or microscopically visible corpuscle – exerts a pressure by virtue of its thermal motions was used to good effect by Einstein in his 1905 analysis of Brownian motion that established the reality of molecules. See Norton (2006, §3). However the same idea in Szilard's 1929 single-molecule engine has engendered another tradition in the thermodynamics of computation that is less successful for its neglect of fluctuations. See Norton (2011).

Hamiltonians. The notable successes of the methods, such as the recovery of critical coefficients, arise within computation in this statistical-molecular context.

In the few-many component case, matters are otherwise. We cannot scale up the properties of a single component molecule or spin and recover phase transition behavior. Something more is needed. Mean field theory was an attempt to provide that something more. It considered the behavior of a few components within the mean field generated by the many more components of the system. Since the analysis focused on just a few components, using the mean field as a surrogate for the rest, it belongs in the few component, lower level. The failure of mean field theory to return correct results quantitatively concerning phase transitions is a failure of reduction. The advance of renormalization group methods was to realize that analyzing the fluctuation behavior of systems of very large numbers of components is essential to representing phase transitions, quantitatively. In this sense, the analysis of phase transitions requires the higher, many component level and is not reducible to the lower, few component level.

The net outcome is that there is a failure of reduction as one passes from the lower level analysis of just a few components to the higher level of thermodynamic analysis. The passage is in two steps: first we pass from the few to the many component level within the molecular analysis; then we pass from the molecular to the thermodynamic level. The failure of reduction arises in the first step and not in the second.

References

Anderson, P. W. (1972) "More is Different," *Science*, 177: 393-96.

Batterman, Robert. 2005. "Critical Phenomena and Breaking Drops: Infinite Idealizations in Physics." *Studies in History and Philosophy of Modern Physics*. 36: 225-244.

Belot, Gordon. 2005. "Whose Devil? Which Details?" *Philosophy of Science*.72: 128-53.

Butterfield, Jeremy. 2011. "Less is Different: Emergence and Reduction Reconciled." *Foundations of Physics*. 41: 1065-1135.

Butterfield, Jeremy. 2011a. "Emergence, Reduction and Supervenience: a Varied Landscape." *Foundations of Physics*. 41: 920-959.

Callender, Craig. 2001. "Taking Thermodynamics Too Seriously." *Studies in History and Philosophy of Modern Physics*. 32: 539-553.

Chalmers, Alan (2009) *The Scientist's Atom and the Philosopher's Stone*. Springer.

Compagner, A. (1989) "Thermodynamics as the Continuum Limit of Statistical Mechanics," *American Journal of Physics*. 57: 106-117.

Jones, Nicholaos. 2006. *Ineliminable Idealizations, Phase Transitions, and Irreversibility.* Dissertation: The Ohio State University.

Lanford, Oscar. (1975) "Time evolution of large classical systems," pp. 1 – 111 in J. Moser, ed., *Dynamical Systems, Theory and Applications: Lecture Notes in Theoretical Physics. Vol. 38.* Heidelberg: Springer.

Lanford, Oscar. (1981) "The Hard Sphere Gas in the Boltzmann-Grad Limit." *Physica*. 106A: 70-76.

Le Bellac, Michel; Mortessagne, Fabrice; Batrouni, G. George. (2004). *Equilibrium and non-equilibrium statistical thermodynamics*. Cambridge: Cambridge Univ. Press.

Nagel, Ernest. (1961) *The Structure of Scientific Theories: Problems in the Logic of Scientific Explanation.* New York: Harcourt, Brace & World, Inc.

Norton, John D. (2006) "Atoms, entropy, quanta: Einstein's miraculous argument of 1905," *Studies in History and Philosophy of Modern Physics*, 37: 71–100.

Norton, John D. (2011) "Waiting for Landauer," *Studies in History and Philosophy of Modern Physics* 42: 184–198.

Norton, John D. (2012) "Approximation and Idealization: Why the Difference Matters," *Philosophy of Science,* 79: 207-32.

Norton, John D. (manuscript) "The End of the Thermodynamics of Computation: a No Go Result."

Ruelle, David. (2004) *Thermodynamic Formalism.* 2nd ed. Cambridge: Cambridge University Press.

Uffink, Jos (2001), "Bluff Your Way in the Second Law of Thermodynamics," *Studies In History and Philosophy of Modern Physics*, 32: 305-394.

Department of History and Philosophy of Science
Center for Philosophy of Science
University of Pittsburgh
Pittsburgh PA 15260
jdnorton@pitt.edu

Günther Sandner

Political Polyphony
Otto Neurath and Politics reconsidered[1]

1. Introduction

Otto Neurath is known as the most political intellectual among the Viennese philosophers of Logical Empiricism (LE). This theme has, however, not received as much attention as other Neurath-related topics and projects such as Unity of Science or ISOTYPE. Nevertheless it is a matter of fact that none of these scientific or educational projects can be strictly separated from politics. Instead, in Neurath's case, science and politics are interlinked in various ways. But what does this mean exactly?

What I will try to do in this essay is the following: First, I will recapitulate some of the arguments that were put forth in recent decades on the relation between political involvement and scientific program in Neurath's work and life. After that I will re-evaluate two particular questions that were (to various degrees) addressed in those contributions: What kind of political approach did Neurath follow (not necessarily *qua* scientist) – how did he, in political terms, become what he was? And how was his political attitude – or, better: set of attitudes – related to his scientific and pedagogical projects?

The case of Neurath, however, is most remarkable not only because he repeatedly brought politics into philosophy of science, social sciences and visual education. Beyond that, he actually intervened in politics – and did so not only on a theoretical level. This may sound a little trivial but, anyway, it constitutes a substantial difference from many other philosophers and philosophers of science – not only from those of a 'scientific-technocratic' and seemingly apolitical neo-positivism or LE but also some of those who were usually considered as to be extremely political.

2. Some interpretations

In the following I will try to summarize the arguments of those approaches that discussed the relation between Neurath's political engagement (and confession) on the one hand and his scientific program on the other in the last decades explicitly.

1 FWF – Austrian Science Fund (P 20595-G14): Otto Neurath (1882–1945) – An intellectual and political biography (director: Günther Sandner).

M.C. Galavotti et al. (eds.), *European Philosophy of Science – Philosophy of Science in Europe and the Viennese Heritage*, Vienna Circle Institute Yearbook 17,
DOI 10.1007/978-3-319-01899-7_15,

Although they may not be the only ones[2] there are – until now – not many others and these examples are quite appropriate for illustrating the status quo of debate.

The first systematic analysis was Rainer Hegselmann's introductory essay to a 1979 edition of selected writings of Neurath in which he discussed the relation between Neurath's socialism and the 'scientific world conception' (SWC).[3] He states that because of the fact that normative statements are meaningless according to LE, the "inner link" between the SWC and certain political movements and developments maintained by the manifesto is a problematic assertion. LE, Hegselmann concludes, is neutral with respect to political systems or ideologies as long as they do not need any metaphysical foundations. It may provide arguments against fascism and totalitarianism – but it also may be consistent not only with socialism but, for instance, also with an enlightened form of bourgeois liberalism and (therefore) capitalism.

About ten years later, Gideon Freudenthal submitted a different interpretation.[4] He claimed that Neurath's empiricist philosophy and epistemology were *dependent* on his political views. For Freudenthal, there is a sort of primacy of politics because, as he put it, "socio-political views involve interests which may affect reasoning while no such interests are directly involved in epistemology".[5] In Neurath's case, however, these socio-political views were anything but Marxist. In contrast, he was an "authoritarian liberal" who followed the conception of a neutral state which installs the rules of distribution according to political objectives formulated by technocratic experts. Central features of Marxism such as the close correlation between human practice and socio-political changes or the labour theory of value in economic theory were simply absent or rejected in Neurath's writings.

In a quite recent essay, Joshua Stuchlik followed in a way Freudenthal's approach whereby an ethico-political program may be the key element in a 'holistic' interpretation of Neurath's work.[6] He assesses this program, however, completely differently than Freudenthal, whose essay, by the way, isn't mentioned in Stuchlik

2 Cf. e.g. Thomas E. Uebel (2005), *Political Philosophy of Science in Logical Empiricism: The Left Vienna Circle*, in: Studies in History and Philosophy of Science 36 (2005), pp. 754-773; Nancy Cartwright / Jordi Cat / Lola Fleck / Thomas E. Uebel, *Otto Neurath. Philosophy between Science and Politics*. Cambridge: Cambridge University Press 1996.

3 Rainer Hegselmann, „Otto Neurath – Empiristischer Aufklärer und Sozialreformer", in: Rainer Hegselmann (Ed.), *Otto Neurath, Wissenschaftliche Weltauffassung, Sozialismus und Logischer Empirismus*. Frankfurt/M.: Suhrkamp 1979, pp. 7-78.

4 Gideon Freudenthal, "Otto Neurath: From Authoritarian Liberalism to Empiricism", in: Marcelo Dasca / Ora Gruengard (Eds., 1989), *Knowledge and Politics*. Boulder, San Francisco, London: Westview Press 1989, pp. 207-240.

5 Freudenthal, Otto Neurath, loc.cit., p. 234.

6 Joshua Stuchlik, "Felicotology: Neurath's Naturalization of Ethics", in: *HOPOS: The Journal of the International Society for the History of Philosophy of Science*, 1 (2010) 2, pp. 183-208.

at all. His thesis: Neurath's oeuvre (which includes philosophy of science, economy, social science, empirical sociology, visual education and a number of other fields) is best understood on the basis of his social and political program of felicitology, the aim of which is nothing less than to improve the happiness of mankind.

I am completely aware that I may have simplified the essays in summarizing their differentiated arguments in just a few phrases. But I will come back to some of their points in the next sections, in which I will first discuss Neurath's political views and then, secondly, raise the question of how far these views determine or are dependent on, or are just in a contingent relation to his epistemological and educational projects.

If Freudenthal is right in claiming that socio-political views involve interests that affect reasoning, then the *early*, juvenile or adolescent formation of such views that have proceeded or accompanied the formation of his epistemology is of particular importance. What do we know about Neurath's political socialization?

3. Political socialization

Some of the early political influences such as those of his father Wilhelm Neurath, who had made him sensitive to the failures and negative social consequences of capitalism and market economy, or Ferdinand Tönnies, whose book 'Community and Society' he had read several times, are obvious and – at least in part – profoundly analyzed.[7] There is, however, one set of documents that has not been introduced into the research on young Neurath: His letters to Swedish popular writer, social reformer and pedagogue Ellen Key (1849–1926). I will briefly summarize why these documents are highly relevant not only as biographical evidence but also to the question of Neurath's politicization.

Ellen Key was a very popular figure in European turn-of-the-century debates on social reform, women's rights and, above all, education.[8] Her books were bestsellers that provoked enthusiasm as well as harsh critique and rejection. Already as an 18-year-old Viennese schoolboy, Otto Neurath probably attended one of her lectures in an overcrowded auditorium in Vienna[9] – and immediately afterwards began writing letters to her. What do we learn from this collection of early letters that already had started at the turn of the century?

---

7 This is at least the case for Wilhelm Neurath's influence on his son. Cf. Thomas E. Uebel, "Otto Neurath's Idealist Inheritance: The Social and Economic Thought of Wilhelm Neurath", in: *Synthèse*, 1995, 103, 87-121.

8 Probably the best intellectual portrait of Ellen Key is Reinhard Däbring, *Der Traum vom ‚Jahrhundert des Kindes'. Geistige Grundlagen, soziale Implikationen und reformpädagogische Relevanz der Erziehungslehre Ellen Keys*. Frankfurt am Main, Bern, New York, Paris: Peter Lang 1990. The Ellen Key estate (including the letters from Otto Neurath) is held by the National Library of Sweden in Stockholm.

9 Cf. *Neue Freie Presse*, June 27, 1901, p. 5.

First of all, young Neurath was highly attracted by the *issues and themes* Key addressed: Matters of education, the changing role of women in contemporary society, sexual relationships and marriage, ethics, and the relation between individuals and the state. The young Neurath was not only an admirer of Key who praised her in poetry as a "northern seer".[10] Moreover, he used her writings for developing his own approach. He even made some attempts at a logical systematization of her (in his view: unsystematic) thought and surprisingly the bestselling author did not reject the suggestions for improvement formulated by a young man who had not even started his university studies.[11]

Before Neurath went to Berlin in 1903, he introduced a young female poet to Key and asked her how to promote this extraordinary talent with reviews in Nordic journals. This female poet was none other than his future wife Anna Schapire. When Neurath met Schapire (probably at the very beginning of his university studies) she was an intellectual and highly politicized woman who had already in 1898 written a Marxist critique of Ellen Key in Karl Kautsky's social democratic newspaper "Die Neue Zeit".[12] From then on, both Neurath and Schapire wrote reviews of Key's books in several journals and newspapers (these texts are hardly known)[13] and it becomes very obvious that they used this intellectual exchange to clarify their own social and political positions.

Neurath was strongly influenced by Schapire who was five years older than he and had already written a number of articles on social policy, women's rights and other social and political issues in journals and periodicals. She was a woman of letters, a translator, a political commentator and a radical feminist whose proposals sound modern even today. Her doctoral thesis that she had submitted at the University of Bern was on "the protection of workers and the political parties in Germany".[14] It is obvious that Neurath was politicized not exclusively but considerably by the woman who would later be his first wife. It was after he met Schapire that his views on women and their role in society changed considerably. Henceforward he defined the question of women's rights as one of the most important issues (in addition to social reform and peace).[15] He already started to attend many

10 Otto Neurath to Ellen Key, December 1901.

11 Cf. Otto Neurath to Ellen Key, August 4, 1901 and October 1, 1901.

12 Anna Schapire, „Mißbrauchte Frauenkraft!“, in: *Die Neue Zeit. Revue des geistigen und öffentlichen Lebens*, 16 (1898), pp. 535-537.

13 Otto Neurath, „Ellen Key“, in: *Neue Bahnen. Halbmonatsschrift für Kunst und öffentliches Leben*, 3 (1903) 1, pp. 341-342; Anna Schapire, „Ein neues Buch von Ellen Key“, in: *Die Wage. Eine Wiener Wochenschrift*. 7. Jg. (1904), pp. 1035-1036; O.(tto) N.(eurath), Rezension zu „Ellen Key: Über Liebe und Ehe“. Berlin: S. Fischer 1904, in: *Neues Frauenleben* (1904) 7, pp.19-20; O.(tto) N.(eurath), „Ellen Key“, in: *Neues Frauenleben* (1905) 4, pp. 16-19.

14 Anna Schapire, *Arbeiterschutz und die politischen Parteien in Deutschland*. Bern 1906.

15 „Drei Fragen treten im alten Griechenland sowohl als heute in den Vordergrund: die soziale Frage, die Frauenfrage und die Friedensfrage“. Otto Neurath, "Sozialwissen-

of the lectures and congresses of early feminism in Vienna and continued to do so particularly in Berlin. Some of his later proposals in the socialization debate of the early 1920s such as those on payment for bringing up children and household work or on the representation of housewives in the soviet system may be seen as a later consequence of this early influence.[16]

The letters in the Key estate in the National Library in Stockholm demonstrate, however, not only the relevance of Schapire. Although he became more and more a critic of Key and accused the famous author of practicing a sort of "pseudo-science"[17], there are at least two points that functioned as stimuli for his later intellectual life. The first is the assessment that educational measures should not only address cultural elites or intellectual circles but large segments of the society. That was something which Key did with her lectures and books, and Neurath admired it enormously. And there was a second topic that he shared with Key from this point on: A secular understanding of human happiness as the principal aim of politics and social reform. And this, by the way, supports Stuchlik's argument for the centrality of this issue in Otto Neurath's work.

A provisional summary is the following: Neurath was already a political character at the very beginning of his intellectual career. Politics was not something that joined in later (around 1918 as is often maintained) – it was there even in the early stages.

4. Marxism

In 2009 Sarah Richardson published a controversial two-part essay in which she aimed to challenge (if not disprove) the Left Vienna Circle (LVC) thesis (as she called it).[18] In his refutation[19] and also in some previous essays, Thomas Uebel demonstrated why there definitely has been a LVC, why its philosophy did not prescribe but allowed for political engagement, and in what respect its philosophy of science was political (for instance in recognising the influence of extra-theoretical values). I will not continue this debate in which many important arguments were

schaftliches von den Ferial-Hochschulkursen" in Salzburg (1903), in: Rudolf Haller / Ulf Höfer (Eds.), *Otto Neurath. Gesammelte ökonomische, soziologische und sozialpolitische Schriften (Band 1)*. Wien: Hölder-Pichler-Tempsky 1998, pp. 1-7, p. 5.

16 Cf. Otto Neurath, *Betriebsräte, Fachräte, Kontrollrat und die Vorbereitung der Vollsozialisierung*, Berlin: Buchverlag Rätebund 1920, p.28.

17 O.(tto) N.(eurath), „Ellen Key", in: *Neues Frauenleben* (1905) 4, pp. 16-19, p. 19.

18 Sarah S. Richardson, "The Left Vienna Circle, Part 1. Carnap, Neurath and the Left Vienna Circle Thesis", in: *Studies in History and Philosophy of Science* 40 (2009), 14-24. Part 2. "The Left Vienna Circle, disciplinary history, and feminist philosophy of science", in: *Studies in History and Philosophy of Science* 40 (2009), pp. 167-174.

19 Thomas E. Uebel, "What's right about Carnap, Neurath and the Left Vienna Circle thesis: a refutation", in: *Studies in History and Philosophy of Science* 41 (2010), pp. 214-221.

brought forward by Thomas Uebel. With respect to Neurath, however, there is one argument in which Richardson is in accordance with other studies, and that is when she points out that the precise character of Neurath's Marxism has remained rather unclear (or controversial) until today.[20] I will take up this open question in sketching first the development of Neurath's reception of Marxism and then try to give a characterization.

I have explained that the political socialization of Otto Neurath began quite early. But also his examination of Marxism started not much later – at the time of his university studies in Berlin. In one of his letters to Ferdinand Tönnies[21] he discussed Franz Oppenheimer's book on Marx in which he rejected the theorem of capitalist accumulation and the emergence of an industrial reserve army which was, as he put it, not in accordance with empirical evidence.[22] But if the theory of accumulation is wrong, Oppenheimer concluded, there will be no collapse of capitalism and no future socialist society either. The young Neurath, however, expressed his sympathy with the book and its author. In 1903 he studied the protocols of the Social Democratic Party of Germany (SPD) convention in Dresden very carefully and sided with the circle around Eduard Bernstein whose revisionism was condemned in an official resolution.[23] Neurath was obviously fascinated by social democracy but he was neither a factionist nor a Marxist. Politically he favored at that time a coalition between social democracy and the liberals around Friedrich Naumann, and he and Tönnies most probably even had a meeting with the legendary and highly contested German politician and political thinker in which they exchanged their thoughts.[24] Neurath claimed that it was Tönnies (who was politically also a non-Marxist socialist) who had introduced him to politics, and as a student he supported his socio-political activities. So it seems to me that Freudenthal is right in stressing Neurath's early sympathy for revisionism. But was he an "authoritarian liberal" or did he ever become one?

Let's turn to the socialization debates shortly after World War I in which – according to Freudenthal – the authoritarian-liberal character becomes most evident. It means that in his radical concept of total socialization in 1919, he followed his own assumptions of 'war economy' and the concepts of economists and social philosophers such as Carl Ballod and Popper-Lynkeus – but not that of Marx. He explicitly rejected the labor theory of value, which is without a doubt an essential part of Marxist economic theory, and many Marxists such as Austro-Marxist theoretician Helene Bauer accused him of not being in accordance with Marxist

20 Sarah S. Richardson (2009), The Left Vienna Circle, loc.cit., p. 21.

21 Otto Neurath to Ferdinand Tönnies, December 9, 1903. (Estate Ferdinand Tönnies, Schleswig-Holsteinische Landesbibliothek, Kiel).

22 Franz Oppenheimer, *Das Grundgesetz der Marx'schen Gesellschaftslehre. Darstellung und Kritik.* Berlin: Reimer 1903.

23 Cf. Carl Schorske, *Die große Spaltung. Die deutsche Sozialdemokratie 1905–1917.* Berlin: Olle und Wolter 1981 [1955], p. 45.

24 Otto Neurath to Ferdinand Tönnies, April 7, 1913.

theory.[25] Nonetheless the socialized economy in Neurath's view was an in-kind economy without money and market, and this has in no way been a liberal concept – even despite the fact that Freudenthal defines liberalism as a theory that understands the state as a "socially neutral institution" and "authoritarian liberalism" as a "liberal social theory which (…) holds that the state can and should install rules of distribution to achieve particular social aims in the interest of society as a whole".[26] And neither was it authoritarian, in my opinion. Neurath did not say that social engineers should make plans that the state would put into practice (even against the people's will). Rather, in his conception, it is the electorate who has to choose democratically between different projects elaborated and proposed by social engineers.[27] There may be a lot of qualified criticism of his socialization plans – but accusing it of being authoritarian is misleading.

What was his political self-definition? Prima facie it seems to be paradoxical. Precisely at the same historical moment (1918–19) in which Neurath declared that he had left behind his life as a scholar and entered the stage of a political 'vita activa', the self-image of the social engineer who writes apolitical expert reports appeared. But what did he want to demonstrate by describing himself as apolitical? Conceptual historians – the successors of Reinhart Koselleck such as the German historian Christian Geulen – have emphasized only recently that there were a number of key terms whose meaning has been transformed considerably in the 20th century[28] – and I think politics is among them. Democratization and participation have influenced our understanding of politics, which is not the same as it was immediately after the Great War. When he spoke of being apolitical, Neurath meant something different than we do today. First of all, Neurath's distance from politics had to do with its negative meanings such as ideological conflicts whose proponents simply lacked the know-how for rational problem solving. There are indications that Neurath followed this understanding even prior to the socialization debates, for instance when he pleaded for civic education – not only for common people but even for politicians![29] Second – and in accordance with this – Neurath

25 Cf. the controversy between Otto Neurath and Helene Bauer: Otto Neurath, „Geld und Sozialismus", in: *Der Kampf* 1923, pp. 145-157; Helene Bauer, „Geld, Sozialismus und Otto Neurath", in: *Der Kampf* 1923, pp. 195-202; Otto Neurath, „Geld, Sozialismus, Marxismus", in: *Der Kampf* 1923, pp. 283-288.

26 Freudenthal, Otto Neurath, loc. cit., p. 210.

27 Cf.e.g. Otto Neurath, Betriebsräte, loc. cit., p. 16. Thomas E. Uebel, "Introduction: Neurath's Economics in Critical Context", in: Thomas E. Uebel/Robert S. Cohen (Eds.), *Otto Neurath. Economic Writings. Selections 1904–1945.* Dordrecht–Boston–London: Kluwer Academic Publishers, 2004, pp. 1-108, p. 44.

28 Although „politics" is not among the examples mentioned by Christian Geulen, Plädoyer für eine Geschichte der Grundbegriffe des 20. Jahrhunderts, in: Zeithistorische Forschungen/Studies in Contemporary History, Online-Ausgabe, 7 (2010), H.1, URL: http://www.zeithistorische-forschungen.de/16126041-Geulen-1-2010.

29 Cf. Otto Neurath, „Die allgemeine Einführung des volkswirtschaftlichen und staatsbürgerlichen Unterrichts" (1908), in: Rudolf Haller/Ulf Höfer (Eds.), *Otto Neurath.*

wanted to make it clear that it is not he but the people who will make the decision on proposed plans. The social engineer was not the counter-image of the politically engaged intellectual, but, instead, the appropriate figure to participate as an expert in political proceedings. Apolitical simply meant not being a politician, a representative of a political party.

Neurath's closest examination of Marxism took place in Vienna (1920–1933). In this time he tried to harmonize the different parts of his social, economic and political thinking with LE and the SCW. For him, Marxism was first of all an empirical and scientific sociology based upon materialism – and not a political ideology. In this context, the rejection of Max Weber's thesis on capitalism and Protestantism[30] and his own interpretation of the base-superstructure problem, the critique of Karl Mannheim's 'Ideology and Utopia',[31] and of Bertrand Russell as a petit bourgeois revolter[32] were formulated. Within the Vienna Circle only a few shared Neurath's interest in Marxism—most likely those who joined the regular sessions on Marxism that took place in the Boltzmanngasse in Vienna as those of the Vienna Circle did. It was a group that consisted of only a few individuals such as Rudolf Carnap and Hans Zeisel but not, for instance, Moritz Schlick.[33]

How could someone who rejected fundamentals of Marxist economic theory, opposed Hegelian dialectics and blamed Marxism for having discredited the useful concept of utopia[34] have been welcomed in the Austro-Marxist scene of Red Vienna? First, Austro-Marxism was not an orthodox and dogmatic movement, and left enough room for controversial positions.[35] There were 'Machians' such as Neurath and Friedrich Adler as well as Neo-Kantians such as Max Adler. How strongly leading Austro-Marxist politicians and diplomats supported Neurath – despite political antagonisms – becomes most evident in the political quarrels about his deportation from Bavaria to Austria after he was sent to prison for high treason following his role in the Munich soviet republic. In fact, there was a diplo-

Gesammelte ökonomische, soziologische und sozialpolitische Schriften (Teil 1), Wien: Hölder-Pichler-Tempsky 1998, pp. 119-132, p. 120.

30 Otto Neurath, "Marxismus eines Jesuiten" (1931), in: Rudolf Haller/Heiner Rutte (Eds.), *Otto Neurath. Gesammelte philosophische und methodologische* Schriften (Band 1), Wien: Hölder-Pichler-Tempsky: 1981, pp. 401-405.

31 Otto Neurath, "Bürgerlicher Marxismus" (1930), in: Haller/Rutte, Otto Neurath, pp. 349-356.

32 Otto Neurath, "Bertrand Russell, der Sozialist" (1929), in: Haller/Rutte, Otto Neurath, pp. 337-343.

33 These protocols were analyzed only recently by the author as part of a collection of Neurath papers and documents held by the special archive Moscow, now Austrian State Archive (AdR 1433-1-5).

34 Otto Neurath, „Ein System der Sozialisierung", in: *Archiv für Sozialwissenschaft und Sozialpolitik* 48 (1920) 1, pp. 44-73, p. 44.

35 Cf. Günther Sandner, „Austro-Marxism", in: William A. Darity, Jr. (Ed.), *International Encyclopedia of the Social Sciences,* 2nd edition, Farmington Hills, MI: Macmillan Reference USA 2007, pp. 212-13.

matic tug-of-war on Neurath's release and deportation between the Bavarian and the Austrian government (Federal Chancellor Karl Renner and Foreign Minister Otto Bauer) that lasted more than half a year.[36] Second, in Austro-Marxism it was education that played a pivotal political role (in this respect, Austro-Marxism was, in a way, exceptional within the socialist movement) and this was an ideal framework for the project in visual education which he had already started to develop.

In sum, there were obviously different periods of Neurath's dealing with Marxism. The revisionist period during his university studies, the explicitly socialist but still non-Marxist period of 1918/19, the longer Austro-Marxist one until 1933/34, and the period of emigration to which we turn finally.

5. Democratisation and Unification

Was there a Neurathian political project in the time of exile – and if so, how did it relate to his scientific and educational programs and activities? In any case, Marxism did not play a role anymore. It almost disappeared from 1934 onwards. There are, however, three interconnected intellectual strands of which the political approach of the émigré Neurath consisted.

The first one is 'democratization of knowledge'[37] which is formulated most clearly but also differentiated in an essay on Visual Education of the 1940s (published posthumously only in 1996)[38] in which he not only continued the empowerment project of Red Vienna – giving the people knowledge, tools and instruments for democratic participation – but also reflected theoretically on democratic ways of decision making for instance in making proposals for non-hierarchical cooperation among experts and citizens. Many of his activities in exile – writings, exhibitions, and not least of all the documentaries for the British Ministry of Information[39] – are part of this endeavor.

The second one includes the debates on 'unity of science', particularly the dispute with Horace Kallen after the end of World War II and the beginnings of the Cold War. In this context, however, Neurath made it clear that for him LE had

36 There are numerous documents in the Bayerisches Hauptstaatsarchiv that support this view. Cf. Telegramm Otto Bauer to the Austrian embassies in Berlin and in Munich, May 10, 1919 and letter Otto Bauer to Otto Neurath, July 3, 1919 (both: BayHstA, MJu 18589).

37 Günther Sandner, „Demokratisierung des Wissens. Otto Neuraths politisches Projekt", in: *Österreichische Zeitschrift für Politikwissenschaft* 38 (2009) 2, pp. 231-248.

38 Otto Neurath, "Visual Education. Humanisation versus Popularisation", in: Elisabeth Nemeth/Friedrich Stadler (Eds.), *Encyclopedia and Utopia. The Life and Work of Otto Neurath (1882–1945)*. Dordrecht, Boston, London: Kluwer Academic Publishers 1996, pp. 245-335.

39 Cf. Michael Burke/Toby Haggith, "Words divide: pictures unite. Otto Neurath and British propaganda films of the Second World War", in: *Imperial Museum Review* 12 (1996), pp. 59-70.

a political meaning. Around 1940 Kallen claimed repeatedly that there would be a general contemporary tendency towards unification (economically, politically) of which the unity of science is a part. As part of that broader unification movement, however, it ran into danger of becoming a totalitarian enterprise that nullifies any alternatives. To be successful, as he put it, "the proponents of the program would have to establish themselves in a position to exact conformity and to control education. They would need, in the twentieth century much the same powers and privileges which certain schoolmen enjoyed in the thirteenth."[40] Not surprisingly, Neurath interpreted LE and unity of science in exactly the opposite way. He insisted that its proponents are followers of an even sceptical pluralism, absolutely anti-totalitarian, and oriented on the cooperation of different approaches in both science and politics. On various occasions he linked the epistemic and the political meanings, when he set pluralism against absolutism, democracy against totalitarianism, the encyclopedia against the system, multiple possibilities against the one and only solution. "The encyclopedism of logical empiricism", he put it, "with the unified science encyclopedia are the children of the tolerant approach of democratic cooperation."[41]

In a letter to educationalist Joseph Lauwerys,[42] he wrote about his (Neurath's) book "Foundations of the Social Sciences" (which Lauwerys planned to review and Neurath obviously wanted him to share his self-interpretation) that LE is particularly pluralistic. The book was meant to be his "SONGS [sic!] OF SONGS OF TOLERANCE, as far as a theoretical attitude may support persecution or tolerance." This is a remarkable formulation because it is one of the few in which he addresses directly the question of the influence between a theoretical approach or program and political engagement. A later paragraph of the same letter, however, is more concrete:

> Through the whole essay goes the argument, that a scientist is not better equipped for making decisions, than other people, because the better a scientist, the more he must be prepared to see MORE THAN ONE PREDICTION, MORE THAN ONE POSSIBILITY. (...) That is important for any scientific analysis of parliamentary government against any kind of "technocratic" government executed by scientific experts who say they KNOW THE OPTIMUM SOLUTION.

One can paraphrase this statement as stating that it is LE that supports participation and democratic decision making against technocracy and expert leadership.

40 Horace Kallen, "A Discussion of the Unity of Science – The Meanings of "Unity" among the Sciences once more", in: *Philosophy and Phenomenological Research. A Quarterly Journal*, 6 (1946) 4, pp. 493-496, p. 494.

41 Otto Neurath, "The Orchestration of the Sciences by the Encyclopedism of Logical Empiricism", in: *Philosophy and Phenomenological Research. A Quarterly Journal*, 6 (1946) 4, pp. 496-508, p. 508.

42 Otto Neurath to Joseph Lauwerys, January 9, 1945 (Isotype Collection, University of Reading).

After all, however, in a third strand Neurath wanted to employ his political interpretation of LE as a democratic, tolerant and pluralistic force in particular for the current political situation of ending World War II. In his theory on the German climate, he explained why a 'human climate' that is traditionally strongly influenced by metaphysics and idealism was more liable for totalitarianism and National Socialism than the Western cultures, especially the English 'climate' in which empiricism and utilitarianism were dominant and human happiness played a more important role than duties or obedience to higher – and necessarily metaphysical – ideals. I have analyzed this approach elsewhere in detail.[43] Now I only want to point out that at the very end of his life he drew a very clear line to connect epistemic and political elements including both their political application and necessary political consequences, for instance in civic educational and re-educational measures in which he participated considerably.

6. Conclusion

I have taken up Freudenthal's assumption that socio-political views affect reasoning more than vice versa and considered this in the context of Neurath's early political socialization in which social reform, education and – probably most importantly – the orientation on human happiness started to play a crucial role. This underlines Stuchlik's argument that 'felicitology' as an approach that aimed to improve the happiness of humans, was a foundation of his lifework that is interlinked with the economic, the sociological, the philosophical and the pictorial pedagogical parts of it. As a socio-political program, however, it needs to be specified and complemented in terms of politics.

I would agree with the thesis in Hegselmann's early essay that the SWC is not exclusively linked to Marxism or socialism. But if we regard – beyond the Vienna Circle's manifesto – Neurath's writings and activities in economics and visual education, it becomes evident that his works provide not only arguments against concepts such as elitism, nationalism, fascism or totalitarianism. His insistence on the compatibility (if not interdependence) of planning and freedom, the orientation on the standard of living and the happiness of even the common people instead of apotheosizing the liberal economic freedom that produces inequality, an egalitarian-oriented pedagogy that empowers the less educated to participate politically instead of elitist scientific communication – all this was without doubt part of a larger left wing political project.

Considering Neurath as proponent of a political concept means, however, not only analyzing its relation to his scientific and pedagogical work but also looking

43 Günther Sandner, "The German Climate and Its Opposite: Otto Neurath in England, 1940–45", in: Anthony Grenville / Andrea Reiter (Eds.), *Politics in British Exile – Exile Politics in Britain after 1933* (Yearbook of the Research Centre for German and Austrian Exile Studies, vol. 12). Amsterdam, New York: Rodopi 2011, pp. 67-85.

closely at both its strengths and weaknesses – and as such it is open to critique. I don't think that there can be a strong determination of a scientific lifework by a political program – there are always other influences and multiple possible developments. In Neurath's case, however, certain socio-political elements formed a coherent project together with his philosophy of science, his economic writings, his sociological works, unity of science and visual education – but these elements, however, were not determined to appear only in this particular combination.

Anyway, Neurath never swam in the mainstream of a political ideology and always formulated his own, peculiar way. Beside Kallen's 'orchestration' he much earlier used another word from the world of music to characterize his approach. In the early 1920s in a letter to his friend, art historian Franz Roh, he argued for a political "polyphony" which – for him – did not mean arbitrariness but providing a political framework in which different groups and cultures could coexist and try to maintain a political consensus and cooperation, and deliberate on common political targets with opposing attitudes – as long as it is possible.[44]

Institute Vienna Circle
Spitalgasse 2-4, Hof 1
1090 Wien
Austria
guenther.sandner@univie.ac.at

44 Otto Neurath to Franz Roh, June 19, 1924 (The Getty Research Institute, Los Angeles, California).

Herlinde Pauer-Studer

Kelsen's Legal Positivism and the Challenge of Nazi Law[1]

1. Introduction

In this paper I am going to examine Kelsen's legal positivism in the light of Nazi legal theory. My claim will be that Kelsen's thesis that law and morality constitute two distinct normative spheres is highly plausible, but that some of his metaethical assumptions are seriously flawed. Kelsen's insistence on the separation of law from morality seems much more persuasive if one takes into account the Nazi legal theorists' program of a 'unification of law and morality'. The critique that legal positivism was responsible for the compliance of the German judiciary with the Nazi's interpretation and use of law is not viable. I will, however, argue that Kelsen's relativist account of morality made his position vulnerable to the post-war objections that legal positivism provided no safeguard against the Nazi perversion of law.

The paper is structured in the following way: Section II explains briefly why Nazi law is thought to pose a special problem for legal positivism, Section III outlines the main assumptions of Kelsen's legal positivism and his relativist understanding of morality. In section IV, I depict the anti-positivist view of some prominent Nazi legal theorists. Section V reflects on the lessons we should draw from the arguments of the Nazi legal theorists. I conclude by arguing that the weak point of Kelsen's position is not the separation between law and morality, but his skepticism towards an intersubjectively valid morality.

2. The challenge for legal positivism

Legal positivism is famous for denying any internal connection between law and morality. Positivists consider law and morality to be distinct normative spheres. That a legal system fails to meet standards of morality does not justify denying it legal validity: a bad legal system can still be law.

In the continental tradition of legal theory, Hans Kelsen certainly was the most prominent advocate of such a position. Kelsen's succinct statement reads:

1 Research for this paper has been funded by the ERC-Advanced-Research Grant „Distortions of Normativity". I would like to thank the European Research Council for its generous funding of the project. Thanks also to J.D. Velleman and Julian Fink for helpful comments on an earlier version of this paper.

M.C. Galavotti et al. (eds.), *European Philosophy of Science – Philosophy of Science in Europe and the Viennese Heritage*, Vienna Circle Institute Yearbook 17,
DOI 10.1007/978-3-319-01899-7_16,

"The validity of a positive legal order does not depend on its conformity with some moral system."[2] In the Anglo-American tradition legal positivism finds most vivid expression in Hart's *separability thesis*, which maintains that it "is in no sense a necessary truth that laws reproduce or satisfy certain demands of morality, though in fact they have often done so".[3]

Natural law theorists, on the other hand, argue that legal systems which violate basic demands of morality and justice cannot qualify to be genuine law. They are, as John Finnis puts it, "peripheral" and "watered-down versions" of the central cases of law.[4]

Legal positivism has been repeatedly accused of failing to provide the resources for rejecting bad and evil law. The separability thesis, so the critique goes, rules out a substantial critical assessment and rejection of immoral legal orders. If we accept the framework of legal positivism, we are stuck with the judgment that wicked legal systems, notwithstanding their failure to meet basic moral standards, still amount to valid law.

A prominent reaction after World War II, when the crimes, atrocities and horrors of the NS-regime became apparent, was to blame the legal profession for submitting to the Nazi regime because of its blind acceptance of the principle 'law is law'. The responsibility for this failure was attributed to legal positivism and its dominating influence on German jurisprudence in the pre-Nazi area, which, so the critics say, continued to shape the jurists' attitude towards the Nazi-regime.

In his ground-breaking article *Gesetzliches Unrecht und übergesetzliches Recht*, published in 1946, Gustav Radbruch called for a revival of natural law thinking in post-war German jurisprudence, writing:

> Positivism, with its thesis that ‚law is law', has indeed rendered the German judiciary defenseless against laws with arbitrary or criminal content.[5]

How should we assess this critique of legal positivism which was addressed specifically against Kelsen's legal theory? Let us turn to Kelsen's characterization of legal positivism.

2 Hans Kelsen, *Pure Theory of Law*. Translation from the Second (Revised and Enlarged) German Edition by Max Knight, Berkeley and Los Angeles: University of California Press, 1967, p. 67.

3 H. L. A. Hart, *The Concept of* Law. With a Postscript edited by Penelope A. Bulloch and Joseph Raz (2nd edn Oxford: Oxford University Press, 1994), pp. 185, 186.

4 John Finnis, *Natural Law and Natural Rights.* Oxford: Oxford University Press, 2011, p. 11

5 See Gustav Radbruch, Gesetzliches Unrecht und übergesetzliches Recht, in: *Süddeutsche Juristen-Zeitung* 1946; reprinted in: Radbruch*: Gesetzliches Unrecht und übergesetzliches Recht*. Baden-Baden: Nomos Verlagsgesellschaft, 2002, p. 10. (Translation H. Pauer-Studer. All German citations in this text are translated by H. P. St.)

3. Kelsen's legal positivism

Kelsen's aim in developing a 'pure theory of law' is to provide a descriptive-scientific account of law. A science of law should depict law as it is, not law as it ought to be from the point of view of legal politics or morality. "The aim of the pure theory of law", Kelsen writes, "is to explore what law is, not what it ought to be."[6] Therefore law should be seen in its normative purity, free from all "alien elements" like statements of psychology, ethics, and politics.[7]

In conformity with his project of developing a pure theory of law, Kelsen states that the definition of legal validity must be internal to the law. Whether certain norms are legally valid or whether a certain order of norms amounts to a legally valid system cannot depend on external standards like criteria of political or moral correctness. Legal validity has to be defined and assessed on merely legal grounds.

Kelsen develops a procedural account of legal validity. A certain norm declaring certain acts to be obligatory or forbidden is a legally valid norm if the norm is part of a legal system. Crucial is the grounds on which the norm is considered to be part of a legal system: a particular norm belongs to a particular legal system and is legally valid, according to Kelsen, if and only if the norm has been generated exactly in the manner which the particular legal system defines as its norm-creating procedure. "The norms of a legal order", Kelsen writes, "must be created by a specific process."[8]

The procedure of norm-creation is defined by the constitution. But regress threatens if we consider the act by which the constitution has been created as itself an act which has to be authorized by a higher legal authority, and so on. The process of law-generation must have a beginning; but in a legal framework which is not based on a God-given natural order the human designers of the legal order have to set this beginning point. The constitution must be simply declared to be the source of law.

The creation of positive legal norms thus depends on the organs and institutions to which the constitution confers the norm-creating authority. The authority of the constitution itself, however, can only rest on the constitution ascribing to itself the normative power of establishing its own authority. In "the logical sense", this requires according to Kelsen, that the constitution be read as a theoretical presupposition, as a sort of transcendental condition of making possible the process of generating law. This theoretical presupposition Kelsen calls the "basic norm" (*Grundnorm*). Accordingly he writes: "The basic norm is the presupposed starting point of a procedure: the procedure of positive law creation."[9]

---

6 Kelsen, The Pure Theory of Law, *loc. cit.*, p.1.

7 *Ibid.*, p. 1.

8 *Ibid.*, p. 198.

9 *Ibid.*, p. 199.

What Kelsen presents is a thoroughgoing legal constructivism: the norms of a legal system do not reflect a natural order, but are man-made. At the basis of a legal order is a decision about the constitution and the manner of generating legal norms via a settled procedure.

An accurate scientific account of law must, Kelsen maintains, depict the normative authority of law. Law requires that certain acts *ought* to be done and that the omission of acts which ought to be done and the performance of acts which *ought not to be done* will be sanctioned. If the law, for example, requires one to pay taxes, then omitting payment of taxes, especially by acting on a skilfully planned scheme of tax evasion, will have legal consequences in the form of punishment. Sanctions are for Kelsen a constitutive element of law. Legal orders – and this distinguishes law from morality – are backed by coercive power. As Kelsen notes:

(T)he social orders designated as 'law' are coercive orders of human behavior. They command a certain human behavior by attaching a coercive act to the opposite behavior. This coercive act is directed against the individual who behaves in this way (or against individuals who are in some social relation to him).[10]

How does Kelsen set law apart from notorious coercive schemes, such as the coercive system set up by a gang of robbers or mafia bosses? If coercion is a constitutive feature of law, what rules out coercive acts which we perceive to be mere threats or even crimes?

Kelsen concedes that the command of a robber – 'your money or your life' – addresses an 'ought' to another individual. However, we ascribe the objective meaning of a norm only to the commands of a legal organ, defined as such by the constitution. The command of the legal organ but not that of the robber amounts to an objectively valid norm. The request of the robber contains a merely subjective 'ought' reflecting an individual wish, not an objectively valid norm.[11] Kelsen maintains that the subjective meaning of the command of the gangster and the command of an income-tax official are the same. Both correspond to a wish that a certain action be performed. But the command of the tax-official has, in addition to the subjective meaning, the objective meaning of a valid norm. In Kelsen's own words:

Only the one order, not the other, is a norm-positing act, because the official's act is authorized by a tax law, whereas the gangster's act is not based on such an authorizing norm. The legislative act, which subjectively has the meaning of *ought*, also has the objective meaning – that is, the meaning of a valid norm – because the constitution has conferred this objective meaning upon the legislative act.[12]

---

10 *Ibid.*, p. 33.

11 *Ibid.*, pp. 44, 45.

12 *Ibid.*, p. 8.

Note that 'objective validity' is here defined in terms of the internal legal validity of the legal order. The objective validity of a norm depends on what the higher legal norms and finally the constitution define to be law.

There is an alternative way to draw the line between the tax-official's request and the gangster's command. Many legal theorists consider the appeal to norms of justice essential for distinguishing between the valid commands of a legal organ and those of a robber gang. The command of the legal organ is considered to be a form of coercion aiming at justice; the command of the gangster amounts to coercion by a purely arbitrary will, acting outside all moral and legal restrictions.

Kelsen rejects this. Justice, he claims, cannot be the demarcation line between law and other coercive systems. Justice, Kelsen maintains, "cannot be made an element of law".[13] The reason is that Kelsen regards justice to be a relative value, understood and defined differently in different social communities.

The example on which Kelsen bases his argument is telling. He refers to St. Augustine's claim in the *Civitas Dei* that there cannot be law if justice is not in place. Augustine's conception of justice, Kelsen objects, is defined with reference to God. Acting justly means giving to God what is owed to him. Therefore any coercive order which is not based on the belief in the God of the Christian religion has for Augustine merely the status of a robber gang. Given that standpoint, Kelsen concludes, the Roman Law would not be law. Equally, he continues, "the capitalistic coercive order of the West is not law from the point of view of the Communist ideal of justice, nor the Communist coercive order of the Soviet Union from the point of view of the capitalist ideal of justice."[14] The upshot is: Legal orders can be assessed to be just or unjust from the standpoint of a certain conception of justice. But standards of justice are community-relative and world-view-dependent. To judge a legal order to be unjust is therefore no reason to deny that it is a legal order.

It is crucial for an appropriate assessment of Kelsen's legal positivism to keep in mind that Kelsen was a moral relativist. He recognizes only two forms of values: absolute values and relative values. Absolute values are based on the belief in God or in another form of trans-human authority. The notion of objective "good" accordingly only makes sense in the context of assuming that there is a normatively binding trans-human will or authority, foremost God. All values which rest on human foundations are merely relative values, worse, they are arbitrary. To quote Kelsen:

> Inasmuch as the norms that are the basis of the value judgments are enacted by human, not superhuman, will, the values constituted by them are arbitrary … Therefore the norms enacted by men and not by divine authority, can only constitute relative values.[15]

13 *Ibid.*, p. 48.

14 *Ibid.*, p. 49.

15 *Ibid.*, p. 18. Kelsen's relativism is also apparent in: Hans Kelsen, Foundations of Democracy, in: *Ethics* 66, 1955, pp. 1-102, esp. 16, 17 and pp. 42-48.

Kelsen distinguishes between a subjective and an objective meaning of acts that address another person's behavior by an 'ought'. The subjective meaning of an 'ought' is just the individual's personal and subjective wish that the persons to whom the 'ought' is addressed comply with the norm. The objective meaning of an 'ought' is a valid norm binding upon the addressee of the 'ought'-statement. Kelsen illustrates this with the following example: If X who is in need asks another man, Y, for help, the subjective meaning of this request is the wish of X that Y ought to help him. But in an objective sense Y ought to help, that is to say, he is morally obliged to help, only if a general norm – established, e. g., by the founder of a religion – is valid, such as the norm 'love your neighbour'. And this latter norm is objectively valid only in case one ought to behave as the religious founder has commanded.[16]

In a nutshell, Kelsen's relativism amounts to the following argument:

a. To assess a norm and the value stated by the norm as objectively valid and therefore as good, the norm must emanate from a superhuman authority, i.e. from God or from a natural order created by God.
b. The object of a scientific theory of value are only norms (and the values constituted by these norms) enacted by a human will.
c. Therefore: It is not possible to assess the norms and values enacted by a human will to be objectively valid (if we operate in the framework of a scientific theory of value).

Kelsen, however, offers no argument for the first premise. The dichotomy – that values are absolute and based on the will of God or they are man-made, based on human willing and therefore relative – is just presupposed. Kelsen rules out by definition that there might be norms and values that are the result of human willing and construction and that can be justified by rational deliberation and argument as objectively, i.e., intersubjectively valid. Though Kelsen was an antagonist of natural law, especially religion-based natural law, he was obviously still in the grip of religious thinking when it came to explaining the objectivity of values.

Kelsen's relativism is thoroughgoing. We are of course familiar with conflicting value judgments: What A judges to be good, B can claim to be bad. A and B can also hold conflicting moral norms to be valid: A maintains that norm x is valid; B claims that norm not-x is valid. Kelsen's relativism, however, is not restricted to admitting that individuals might entertain different value judgments and moral norms, and that conflicts of value among individuals might therefore occur. Kelsen thinks it impossible to rationally resolve the question which of two conflicting norms is the correct one, as the following passage shows:

> For example, a norm could be valid that forbids to commit suicide or to lie under all circumstances, and another norm could be valid permitting or even commanding suicide or lies

16 *Ibid.*, p. 8, 9.

under certain circumstances, yet it would be impossible to prove rationally that only one of these two norms, but not the other, is the truly valid one.[17]

The norm-conflict and the dispute between contradictory conceptions of what is valid could be decided only if one of the two conflicting norms could be proved to constitute an absolute value. This means that the norm must be backed by a superhuman authority, namely, God or a natural order created by God. But for Kelsen such assumptions remain outside a scientific theory of value, which can take into account only norms and values that are the result of human willing.

Kelsen's way of talking of a 'science of value' reminds one of course of the 'Scientific World Conception' propagated by the Vienna Circle.[18] Most of the members of the Vienna Circle acknowledged only two categories of meaningful sentences, namely analytic or empirical sentences, thereby excluding normative sentences. The most radical version of non-cognitivism was advocated by Rudolf Carnap who denied that value judgments have any cognitive content and compared them with mere emotional exclamations expressing approval or disapproval.[19]

Kelsen, however, was not an emotivist. He did not share Carnap's views. One of the passages in the *Pure Theory of Law* reads like a direct response to Carnap:

If somebody says that something is good or bad, but if this statement is merely the immediate expression of his emotional attitude toward a certain object, if it expresses that he wishes something or does not wish it but its contrary, then the statement is no value 'judgment', because it is no function of cognition, but a function of the emotional component of consciousness; and if this emotional reaction refers to the behavior of another individual, then it is the expression of an emotional approval or disapproval, akin to the exclamation 'bravo' or 'phooey'![20]

---

17 *Ibid.*, p. 18.

18 In a personal letter from 1963 Hans Kelsen states that though he was not formally a member of the Vienna Circle, he had contact with several of its members like Moritz Schlick, Otto Neurath, Philipp Frank, and Viktor Kraft. Kelsen explains in that letter that it was the Vienna Circle's „antimetaphysical thrust" which connected him to its philosophy. The letter is cited in Clemens Jabloner, Kelsen and his Circle: The Viennese Years, in: *European Journal of International Law* 9, 1998, pp. 368-385, here: 378, 379; cf. also Clemens Jabloner and Friedrich Stadler (Eds.), *Logischer Empirismus und Reine Rechtslehre*. Beziehungen zwischen dem Wiener Kreis und der Hans Kelsen-Schule, Wien: Springer 2001; cf. also Friedrich Stadler, *The Vienna Circle. Studies in the Origins, Development and Influence of Logical Empiricism*, Vienna-New York: Springer, 2001.

19 Carnap, R., 1937, *Philosophy and Logical Syntax*, London: Kegan Paul, Trench, Trubner & Co., pp. 23-29.

20 Kelsen, *The Pure Theory of Law*, *loc. cit.*, p. 20.

Kelsen argues that expressions of one's emotional attitudes are not value judgments, but merely "a function of the emotional component of consciousness"; to be a value judgment, a judgment must be "a function of cognition".[21]

Kelsen draws here a twofold distinction. One way to distinguish between value judgments is to take into account the value they state, namely whether they state a subjective or an objective value. Another way to look at the question is to ask whether an utterance is a function of emotion or of cognition.

With respect to the value they state, we can distinguish between subjective and objective value judgments, Kelsen maintains. A subjective value defines the relation of an object to the wishes of an individual. Subjective value judgments are judgments that refer to a subjective value, i.e. a value which consists in the relation of an object, in particular someone's behavior to the wish or will of an individual. A value judgment is merely subjective if the judgment that x is good means that x is wished by the individual who makes the judgment; the judgment that x is bad would then just mean that the individual wishes the opposite behavior of x. This way the values 'good' and 'bad', Kelsen claims, are relative to the individual who wishes x or non-x.[22]

An objective value, according to Kelsen defines the relation of an object x, e.g. a certain behavior, to an objectively valid norm. An objective value judgment has no relation to the wishes of an individual; it merely states whether x (a certain behavior) corresponds to an objectively valid norm.

In regard to the emotional or cognitive function of an utterance things are different, Kelsen holds: the category of subjective value judgments drops out of the picture because a statement that would be merely a function of one's emotional attitudes, is, according to Kelsen, not a judgment in the strict sense.

The emotive/cognitive distinction thus determines what is a *judgment:* judgments cannot be expressions of emotion. The objective/subjective view thus depends on whether a judgment is about someone's wishes. There can be subjective judgments: they would be cognitive attitudes about someone's wishes.

As a function of cognition any judgment must be objective, in the sense that it must be given and hold independently of the wishes of the judging individual. So, in regard to the function of cognition all value judgments are objective.

Note, however, that Kelsen has a merely descriptive reading of 'objective value judgments'. To utter an objectively valid judgment means for him that one compares the behavior of x with the norms of an existing normative order and then assesses whether the behavior conforms to the norms of that order. This purely descriptive exercise can be done without any emotional involvement, without any way of approving or disapproving of the normative order or the particular norm at stake. One can, to cite Kelsen's own example, state that according to Christian morality it is appropriate and good to love one's enemies, but one can make that

21 *Ibid.*, p. 20.
22 *Ibid.*, p. 20.

value judgment without taking any stance on whether one approves or disapproves of the norm that one ought to love one's enemy.[23] Or take the following explanation by Kelsen:

> The answer to the question whether according to a valid law the death penalty ought to be imposed upon a murderer and consequently whether – legally – the death penalty in case of murder is valuable, can and must be given without regard to whether the one who has to give the answer approves or disapproves of the death penalty. Only then is a value judgment objective.[24]

Kelsen thereby rules out that it is meaningful to assess the validity of value judgments from a normative perspective. In Kelsen's framework, the question whether x is objectively valuable or not, or whether y ought to be done or not, is not decided on normative grounds. To decide whether x is objectively valid or whether y ought to be done depends solely on which norms are held to be valid by the normative system against which one makes the judgment, either a particular legal system or a particular group morality.

Kelsen simply does not take into account that we might interpret value judgments to make a normative claim which we are then going to assess on normative grounds, namely asking whether the normative claim amounts to a mere partial wish or interest of a particular individual or whether the claim is viable and can be normatively endorsed from the perspective of (all) others. For Kelsen issues of morality are not something which we can deliberate about from a universal perspective. There are in his normative world no universally valid principles of morality, i.e. a morality whose central principles could be shared normative ground across various religious, social, ethnic and political communities. This is so since Kelsen recognizes only two kinds of moral norms: all moral norms or principles either rest on the belief in God and depend therefore on the acceptance of a certain religious system, or they are the result of human willing and part of a particular moral system which, however, expresses merely the moral beliefs of a specific group or community of people. In the first case the moral norms are absolute, in the second case relative. Given this categorization of moral norms into absolute and religious-based ones on the one hand, and man-made and therefore relative on the other hand, for which Kelsen provides no reasons and justification, all non-religious moral principles or norms are by definition relative.

This radical form of relativism, not his separation of law and morality, is, I think, the weak point of Kelsen's position and his legal positivism. Before elucidating this point more clearly, let us take a look at the Nazi conception of law.

---

23 *Ibid.*, p. 21.

24 *Ibid.*, p. 21.

3. The Nazi legal philosophy: The unification of law and morality

The program of the Nazi lawyers was to call for the unification of law and morality (*Sittlichkeit*)[25]. Leading Nazi legal theorists like Reinhard Höhn, Otto Koellreutter, Karl Larenz, Ernst Forsthoff, and the notorious Rudolf Freisler thought that eliminating the distinction between law and morality was crucial for achieving a National Socialist form of law.[26] The idea was that an appropriate conception of law should embrace not only positive legal norms but a'higher law idea' (*höhere Rechtsidee*) reflecting 'the right and the just'. Ethical principles should be embedded in law.

Höhn demands that considerations of justice „find immediate expression in the legal norm"[27]; they are not merely relevant in the case of discretionary decisions of the judge. "Justice", Höhn claims, "does not stand outside law".[28] Whereas positivism implies that „the idea of justice cannot be incorporated into positive law",[29] the dichotomy between the idea of justice and statutory law should be overcome in the National Socialist state, because *"(l)aw is lived morality."*[30] Law amounts, in Höhn's words, not merely to a simple technique for applying independent ethical principles; "law can only mean the lived ethics of a people".[31] Thereby "*the sharp separation of law and morality loses its meaning*".[32]

The *volks*-community (*Volksgemeinschaft*), Höhn states, "*is according to the new German conception of law not only a social fact, but also a principle of law (justice)*".[33] The ethical values which are mandatory for the National Socialist conception of community, such as honor, loyalty, and race, must be part of law.[34] Only from the perspective of an individualist system of law would such concepts seem

---

25 The Nazi legal theorists usually use the term 'Sittlichkeit' for 'morality' which was common at the time.

26 Reinhard Höhn, Volk, Staat und Recht, in: Reinhard Höhn, Theodor Maunz und Ernst Swoboda, *Grundfragen der Rechtsauffassung*. München: Duncker & Humblot, 1938, pp. 1-27; Reinhard Höhn, *Rechtsgemeinschaft und Volksgemeinschaft*. Hamburg: Hanseatische Verlagsanstalt, 1935, pp. 49-83; Otto Koellreutter, *Deutsches Verfassungsrecht*. 3rd edn., Berlin: Junker und Dünnhaupt, 1938; Karl Larenz, *Deutsche Rechtserneuerung und Rechtsphilosophie*. Tübingen: Mohr, 1934; Karl Larenz, *Rechts- und Staatsphilosophie der Gegenwart*, 2nd edn., Berlin: Junker und Dünnhaupt Verlag 1935; Ernst Forsthoff, *Der totale Staat*, Hamburg: Hanseatische Verlagsanstalt 1933.

27 Höhn, Volk, Staat und Recht, *loc. cit.*, p. 13.

28 *Ibid*, p. 13 (italics in original).

29 *Ibid.*

30 *Ibid.*

31 *Ibid.*

32 *Ibid.*

33 *Ibid.*, p. 9 (italics in original).

34 *Ibid.*

to be „foreign in their essence and 'non-juridical'" Höhn maintains.[35] Because of the reassessment of the *volks*-community as a legal notion, the principles of racial sameness and species sameness should also be constitutive of law, according to the National Socialist conception of a legal system.[36]

Otto Koellreutter equally emphasizes the relatedness of ethics, law, and community. It is crucial that members of the *volks*-community recognize the paramount political and legal status of the *volks*-community as well as the personhood and honor of every other member. [37] This sense of community also enhances the capacity for moral and legal judgment:

> If the corporate feeling is awakened and nourished and sustained in each *volks*-comrade (*Volksgenosse)*, then there develops the "right" legal feeling which matches the volks-conception of the right and just. The *volks*-conception of justice thus has the same roots as the *volks*-conception of the political.[38]

The standards of justice include, in Koellreutter's words, those norms and principles that have the "task to protect and develop the national life order in its existence".[39] The community sentiments of the people define what is just.

Höhn also ascribes to the members of the aryan *volks*-community the decisive epistemic competence in legal and evaluative questions: their judgment determines what counts as just and unjust. It becomes especially clear how misguided and ideologically entrenched this conception is when Höhn cites approvingly Alfred Rosenberg's statement: "Law (justice) is that which aryan persons consider to be right, and injustice is what they dismiss."[40]

Though the standards of justice and correctness are according to that definition clearly dependent on the acceptance of the Nazi worldview, Koellreutter, for example, has no reservations about the presumptuous claim: "Therefore the National Socialist rule of law state is a just state as well as an order-based state."[41]

Karl Larenz, too, endorses the unity of law and morality. Consider the following passage:

> Law is not a matter of arbitrary discretion, it is also not a matter of external expediency and utility, but a live order closely connected with the moral and religious life of the community, which confronts the individual with its own claim to validity and internally binds the individual.[42]

35 *Ibid.,* p. 9.

36 *Ibid.*, p. 8.

37 Koellreutter, *Deutsches Verfassungsrecht*, *loc. cit.,* p.12.

38 *Ibid.*, p. 11.

39 *Ibid.* According to Koellreutter the Nazi racial laws belong to the constitution of the Third Reich.

40 Höhn, Volk, Staat und Recht, *loc. cit.,* p. 8 (Italics in the original).

41 *Ibid.*, p. 56.

42 Larenz, *Deutsche Rechtserneuerung und Rechtsphilosophie*, *loc. cit.,* p. 5.

Larenz claims that law must be connected to a specific understanding of morality: morality must "not (be) a matter of the individual person, but a life form of the community, in Hegel's words, 'the true spirit of a family and a people'".[43] The meaning and end of law is according to Larenz located in the „substance and promotion" of the *volks*-community which is of unconditional value. Positive legal norms must be in line with the "ethical consciousness, the mores of the *volk*".[44] Larenz claims that the "common will", not the particular will of the individual, must be the basis of the "spiritual", but also "real validity" of the legal norms.[45]

The Nazi program of a unification of law and morality entails that in National Socialism, as the Nazi jurist Walter Hamel writes, „a distinction between moral duty and legal duty … cannot be made any more."[46] The same requirement is voiced by Roland Freisler:

> There can be no gap between a legal imperative and an ethical imperative. This is so because imperatives of law are imperatives of decency; however, what is decent is determined by the conscience of the *volk* and the individual member of the *volk*.[47]

Because the moralization of law eliminates the distinction between legal and ethical obligations, honorable, decent and loyal behavior is legally required. Because law and morality form a unity, the law binds the members of the community also ethically.

The National Socialist reading of law as expressing the ethics of the *volks*-community extends the authority of the state and the Nazi regime. In a liberal framework, the fulfilment of ethical and moral duties is entrusted to the person's own conscience. The Nazi legal theorists' elimination of the distinction between legal and ethical requirements entails transforming ethical duties into legal duties, enforceable by state coercion.

Crucial for the new understanding of law is, as the criminal law theorist Georg Dahm expresses it, that "law, ethics, criminal law and the worldview of the *volk*

43 *Ibid.*, p. 6.

44 *Ibid.*, p. 9.

45 *Ibid.*, p. 30.

46 Walter Hamel, Wesen und Rechtsgrundlagen der Polizei im nationalsozialistischen Staate, in: Hans Frank (Ed.), *Deutsches Verwaltungsrecht.* München: Zentralverlag der NSDAP Eher, 1937, pp. 381-398, esp. p. 384. Koellreutter equally emphasizes the connection between ethics, law, and community. Koellreutter, *Deutsches Verfassungsrecht, loc. cit.,* p. 12. Compare also Ulrich Scheuner, Die Rechtstellung der Persönlichkeit in der Gemeinschaft, in: Hans Frank (Ed), *Deutsches Verwaltungsrecht, loc. cit.*, pp. 82-98.

47 Roland Freisler, *Nationalsozialistisches Recht und Rechtsdenken*, Schriften des Reichsverbandes Deutscher Verwaltungs-Akademien. (Ed.) Dr. H.H. Lammers unter Mitwirkung von D. Fritz Müßigbrodt, Heft 4, Berlin: Industrieverlag Spaeth & Linde 1938, p. 56.

grow together."[48] The values of National Socialism and the legal and ethical convictions and sentiments of the *volks*-community form the new foundations of law.

In many respects the required moralization of law remains on the level of a programmatic rhetoric and reduces to mere vague invocations of the ethical values of the *volks*-community. Some Nazi legal theorists, however, look for more precise formulations, appealing thereby to virtue ethics. Wilhelm Sauer, for example, considers the "original German (gothic) duties", which are part of the „German ethical and cultural history", to form the content of the "*volks*-ethics".[49] Central for Sauer are the four „gothic virtues": truthfulness, reliability, goal-directedness and self-control, where goal-directedness finds expression in a community-sense and a spirit of sacrifice; the virtue of self-control includes "restrained self-assertion". The opposite of those virtues are the following "typical criminal vices" which also count as the strongest violations of the "community ethos": obliqueness, betrayal, ravenousness (egoism), and license and intemperance (envy).[50]

The practical consequences of the Nazi moralization of law also have an effect on the function and official self-understanding of the judge. Since ethical norms are of crucial importance, the judge has to follow his conscience and display "duty-bound discretion".[51] The judge must also respect the "juridical basic law, the law of the well-being of the *volk* as the conscience of the nation".[52] It is crucial that the judge does not merely adjudicate in a formal mode, but that he also avenges the material wrong connected with an offense. The subject of adjudication is not merely the formal violation of law, but also the infringement of the higher trans-legal ethical order.

Identifying law with morality entails distinguishing it from statute. The separation between statute and law (between *Gesetz* and *Recht*) which is repeatedly emphasized by the Nazi legal theorists holds equally for criminal law. Freisler, for example, writes:

> (W)e believe that throughout criminal law a material conception of law and of legal wrongdoing must be decisive since we do not put on a par law and legal statute, but develop it out of the ethical order of the *volk*. From our point of view one should therefore not deduce criminal law solely from the legal statute; we rather should equally recognize an extra-legal source of recognition of the law.[53]

48 Georg Dahm, Die Erneuerung der Ehrenstrafe, in: *Deutsche Juristenzeitung*, 39. Jahrgang, Heft 13, pp. 822-832, here: p. 826.

49 Wilhelm Sauer, Die Ethisierung des Strafrechts. Über die Prinzipien der Strafrechtserneuerung und ihre praktische Auswirkung, in: Roland Freisler (Hg.), *Deutsches Strafrecht. Strafrecht, Strafrechtspolitik, Strafprozeß*, Zeitschrift der Akademie für deutsches Recht, Neue Folge, 1. Band, Berlin: R.v. Decker's Verlag, G. Schenck 1934, pp. 177-190, here: p.188.

50 *Ibid.*, p. 188.

51 *Ibid.*, p. 184.

52 *Ibid.*, p.184.

53 Freisler, *Nationalsozialistisches Recht und Rechtsdenken*, *loc. cit.*, p. 57.

The material conception of lawfulness and of legal right and legal wrong should first and foremost be secured by the skilled sentence of the judge. This means that the judge must, besides matters of facts and the descriptive features of an offense, also take into account specific normative elements which make possible a value-based adjudication. The list of "genuine normative elements" cited by Freisler includes: well-being of the Reich; well-being of the *volk*; social evaluations like esteem, depreciation and exploitation; good mores (*Sitten*), a concept of of decency, and sanitary *volks*-feelings. These „genuine normative elements" are according to Freisler decisive for establishing a material conception of legal right and wrong as a second source of legal cognition besides and equal to the positive legal statute.[54]

The fusion of law and morality serves the NS-jurists as a welcome means to extend the authority and power of the Nazi regime. Larenz's writings are an illuminating example. On the one hand, Larenz claims to follow Kant's philosophy of law, especially Kant's way of defining the authority and dignity of law independently from considerations of utility. He argues, for example, that Kant "rightfully removes not only from the sphere of ethics, but also from the philosophy of law" a "form of utilitarianism which claims that law exists for the interests of the individual human beings and is created by them".[55] The *volks*-community as a value for its own sake rests on the interest-independent ethical self-obligation of its members.

On the other hand, Larenz clearly gives up the basic principle of Kant's political philosophy, namely that the force of the state merely extends to regulations in the sphere of external freedom.[56] The state is, Kant argued, not permitted to interfere into the sphere of internal freedom. The authority and power of the state must not extend to the control of our personal convictions, including our ethical attitudes. In regard to law the consequences of this Kantian restriction are: The state is entitled to enforce the subjects' compliance with legal norms. But the state is not authorized to require that compliance to rest on specific moral or ethical motives.[57]

54 Freisler, Aufbau und Aufgabe des Besonderen Teils, Gestaltung seiner Tatbestände. Einbau des Schutzes der Bewegung in das Gesetz, in: Franz Gürtner (Hg.) *Das kommende deutsche Strafrecht. Besonderer Teil.* Bericht über die Arbeit der amtlichen Strafrechtskommission, p. 61.

55 Larenz, *Deutsche Rechtserneuerung und Rechtsphilosophie, loc. cit.*, pp. 5, 6. By ‚utilitarianism' Larenz means a self-interest-based theory of morality.

56 Immanuel Kant, Metaphysical First Principles of the Doctrine of Right, in Kant, *The Metaphysics of Morals.* Ed. by Mary Gregor, Cambridge: Cambridge University Press, 1991, p. 31, 6:239.

57 It is of ethical value if citizens recognize the norms of the state and feel committed to *ethical juridical duties*, but that is not something the state can require citizens to do. This of course does not mean that Kant excludes all intentional and internal elements of morality from the sphere of law. A good discussion of the point is Allen D. Rosen: *Kant's Theory of Justice*. Ithaca and London: Cornell University Press, 1993, p. 85f.

According to Larenz, however, the subjects' internal ethical commitment to law and legal norms is crucial for the National Socialist conception of law.[58]

The elimination of the separation between state, law, and morality blurs the line between the spheres of external and internal freedom. In a liberal-democratic framework the state and therefore law regulate the relations of external freedom. Ethics, on the other hand, is a standard of internal freedom whose recognition is due to the autonomy of the person. The Nazi state rescinds the neutrality of the state towards the citizens' personal ethical convictions and world-view principles. The personal sphere of freedom is now subject to the state's comprehensive normative control.

Dahm's remarks clearly display that new orientation when he writes:

> We recognize today the independence and neutralisation of law, the separation and opposition of state and law, law and politics, law and world view, law and ethics to be the core of evil. Overcoming those oppositions and the realization of unity within law amounts to a renewal of our legal life.[59]

The elimination of the distinction between law and morality thus strengthens the dictatorial power of the Nazi regime. An attitude of inner loyalty and commitment towards law is much more conducive to the ends of the Nazi state than a mere compliance with orders and legal regulations due to brute state force. The 'ethical recognition' of Führer orders by officials and judges was, as historians have pointed out, crucial for the relatively high inner stability of the Nazi state in the pre-war years.[60] Morality thus turns into one of the most effective tools of the Nazi regime to realize its political goals.

4. Conclusions

What conclusions should we draw from the Nazi philosophy of law? What follows in regard to the assessment of Kelsen's legal positivism? Is the claim that legal positivism contributed to the jurists' compliance with the Nazi regime tenable?

The Nazi legal theorists aggressively attacked legal positivism. Ernst Forsthoff, for example, holds the strict separation of law and morality and the "formalistic conception of law" defended by legal positivism to be responsible for „the ethical disorientation“ of Germany during the Weimar republic. Forsthoff claims that the liberal state is unable to distinguish between "true – false, just – unjust, good – bad, moral – immoral, since it lacks the substance which is the basis for

58 Larenz, *Rechts- und Staatsphilosophie der Gegenwart*, *loc. cit.*, pp. 157, 158.

59 Georg Dahm, Die Erneuerung der Ehrenstrafe, p. 826.

60 See e.g. Hans Mommsen, *Beamtentum im Dritten Reich.* Mit ausgewählten Quellen zur nationalsozialistischen Beamtenpolitik, Stuttgart: Deutsche Verlagsanstalt, 1966.

drawing such distinctions."[61] The Weimar republic was for him nothing other than a mere "apparatus-state"("*Apparaturstaat*"), guided by functional expediency and utility.[62] Instead of the formalism of the Weimar constitution, the National Socialist State is according to Forsthoff in need of a material understanding of law, requiring the rejection of positivism and individualism.

These sharp objections are addressed specifically against Kelsen, who was the most prominent academic representative of legal positivism in the German-speaking world at the time. Karl Larenz, for example, describes Kelsen's *Pure Theory of Law* as „a phenomenon of intellectual foreign infiltration ("*eine Erscheinungsform der geistigen Überfremdung*").[63] And Larenz emphasizes that it was the achievement of Carl Schmitt to bring again to life the "essence of the political" and "the simple reality of political life against the artificial abstractions of the Vienna School (of legal thought)."[64] The special target of critique is Kelsen's demand to treat law and morality as two distinct normative spheres. Legal positivism, Larenz argues, strips law "of its natural connection with national mores and morality".[65]

Instead of endorsing legal positivism, the Nazi lawyers approve the unification of law and morality in a way which at least formally corresponds to a normative demand the post-war anti-positivists articulate as well.

The Nazi conception of morality is of course deeply distorted and permeated by ideology. Nevertheless, the moralization of law by Nazi legal thinkers should be a clear warning that the call for bringing law and morality together will not be the way to remedy the perversions in the Nazi legal system.

What follows in regard to our understanding of law? Law should be an instrument for solving moral problems. That means that law itself should not turn into a moral problem. Sometimes – and totalitarian regimes are a paradigmatic case – law itself becomes so deeply moralized that it cannot function any more as a reliable normative guideline.

In order for law to avoid getting stuck in ideological moralized territory and losing its capacity for being an impartial forum for resolving disputes, requires expressing the deepest insights of morality in normative requirements applying to legal systems. Requirements like publicity, reliability, predictability, prospectivity, consistency, understandability, avoidance of arbitrariness and retroactive legislation[66] support and work towards a legal system which resists deterioration. All

61 Forsthoff, *Der totale Staat*, *loc. cit*., pp. 13, 14.

62 *Ibid*., pp. 13, 34.

63 Larenz, *Deutsche Rechtserneuerung und Rechtsphilosophie*, , *loc. cit*., p. 11.

64 *Ibid*., p.18.

65 *Ibid*., p.12 (italics in the original).

66 The mentioned criteria are of course the principles which Lon Fuller considers to constitute the „inner morality of law" and they formed the basis of his claim that law without a core of inner morality does not deserve to be called law any more. See Lon L. Fuller, *The Morality of Law*, revised edn., New Haven and London: Yale University,

these criteria the Nazi legal system was far from meeting. However, to claim that a legal system should fulfill those standards by no means amounts to demanding that morality and law should form a unity.

To sum up: The post-war critique that legal positivism is responsible for the submitting of the German judiciary to Nazi law is not viable. Yet there is a deep deficiency in Kelsen's theoretical position.

Kelsen defends the separation of law and morality as two distinct normative spheres with the following arguments:

a. Law is a coercive order, sanctions are enacted by force. Morality is not a coercive order, the sanctions are milder.
b. A justification of positive law through morality is only possible if there exists a contrast between moral and legal norms and between morally good and morally bad law.
c. Law cannot form a unity with justice and morality; otherwise the demand *that law ought to be just* does not make sense. In Kelsen's own words: "(I)f law is identified with justice, and if that which *is* is identified with that which *ought* to be, then the concept of justice as well as the concept of the Good have lost their meanings."[67]

The first assumption (a.) is certainly uncontroversial. The second and third argument (b. and c.), claiming that we need to draw a distinction between moral and legal norms in order to be able to criticize certain legal norms as morally bad, maybe even obnoxious, are equally plausible. They are especially so in the light of the Nazi program of a moralization of law.

However, in the second and third argument Kelsen talks as if there were principles of morality and justice which do have the resources to allow us to criticize bad and evil law. He talks as if there were principles of justice functioning as normative guidelines directing us in the process of deliberating what sort of law and legal norms *we ought to create.* In reminding us that we need to distinguish between law and morality in order to get a clear notion of what law *ought* to be, Kelsen seems to make use of 'ought' in a normative sense. Recall, however, that Kelsen considers moral values to be relative; morality is for him only a form of a partial personal or group morality.

Let us look briefly at what he had to say about the Nazi legal system even many years after the war. In the English edition of the *Pure Theory of Law* which appeared 1967 Kelsen writes:

> The legal order of totalitarian states authorizes their governments to confine in concentration camps persons whose opinions, religion, or race they do not like; to force them to perform any kind of labor; even to kill them. Such measures may morally be violently

Press, 1969, pp. 39-41. But Fuller's assumption of an inner morality of law does not draw the line between law and morality in a clear way. It is, I think, too close to the problematic idea of uniting law and morality.

67 Kelsen, *Pure Theory of Law*, p. 67.

condemned; but they cannot be considered as taking place outside the legal order of these states.[68]

From a descriptive point of view Kelsen's statement which applies also to the Nazi legal system is accurate. But of course we would also like to know what he had to say about the Nazi legal system from *a normative point of view*. In this respect the passage tells us that one might "violently condemn" such a system on moral grounds.

But in terms of Kelsen's relativistic account of morality, such a moral condemnation would be relative to the personal moral standpoint of an individual or relative to a specific group-morality. Given Kelsen's background, personal history, and his deep engagement for democracy there is of course no question that he personally despised the Nazi regime.[69] Yet his theoretical views do not allow the condemnation of the Nazi regime from an agent-neutral and impartial point of view. The problem of Kelsen's position thus is not the separation between law and morality, but his relativism in regard to morality.

In regard to Kelsen's legal positivism we can state:

- Kelsen's insistence on the separation of law and morality is correct.
- Kelsen's moral relativism is implausible and his conception of morality needs to be revised.
- Kelsen's conception of the science of law as a mere descriptive analysis of law has to be corrected. Law remains a normative enterprise, the question what law *ought* to be is normatively crucial.

But given the dramatic political developments and circumstances facing Kelsen – namely a politically influential religious absolutism in Austria resulting eventually in an authoritarian Christian-catholic *Ständestaat* eliminating parliamentary democracy from 1934 to 1938 on the one hand, and the rise of Nazi power in Germany on the other hand – we can understand Kelsen's deep wish that at least law should be free from ideological and moralizing elements, though we might not share his moral relativism.

Institut für Philosophie
Universität Wien
Universitätsstraße 7
A-1090 Wien
herlinde.pauer-studer@univie.ac.at

68 *Ibid.*, p. 40)

69 Kelsen was one of the drafters of the Austrian constitution in 1919. He left Austria in 1930 to take up a professorship at the University of Cologne. In April 1933, due to the „Law for the Restoration of the Civil Service" Kelsen was dismissed as a Jew from his position at the University of Cologne. After positions in Geneva and Prag he emigrated to the US in 1940. Cf. Oliver Lepsius, Hans Kelsen und der Nationalsozialismus, in: Robert Walter, Werner Ogris and Thomas Olechowski (Eds.), *Hans Kelsen: Leben – Werk – Wirksamkeit*, Wien: Manzsche Verlags- und Universitätsbuchhandlung 2009 (=Schriftenreihe des Hans Kelsen Instituts Vol. 32), 271-288.

JEANNE PEIJNENBURG AND DAVID ATKINSON

BIASED COINS
A MODEL FOR HIGHER-ORDER PROBABILITIES

ABSTRACT

Is it coherent to speak of the probability of a probability, and the probability of a probability of a probability, and so on? We show that it is, in the sense that a regress of higher-order probabilities can lead to convergent sequences that determine all these probabilities. By constructing an implementable model which is based on coin-making machines, we demonstrate the consistency of our regress.
Keywords: Higher-order probability, infinite regress, consistency.

1. INTRODUCTION

If it makes sense to express the probability that a proposition is true as a specific number, it also makes sense to doubt whether that number itself is accurate. One can further consider the probability that the probability of the proposition is equal to the number in question. Hence we are led ineluctably into an infinite regress, for then we must consider the probability of the probability of the probability that the proposition is true – and so on *ad infinitum*. It has indeed been argued that this sounds the death knell of the idea of higher-order probabilities (Hume 1738/1961, Rescher 2010). However, in this paper we shall show that the regress in question is generally benign. Far from producing a *reductio ad absurdum*, as we will see, the regress usually engenders a convergent sequence that leads to a well-defined probability of the truth of the original proposition.

Some have seriously entertained the notion of a second-order probability (Uchii 1973, Skyrms 1980, Domotor 1981, Kyburg 1987, Gaifmann 1988). Skyrms in particular discusses and demolishes a number of attempts aimed at showing that the concept is inconsistent. One of those attempts has to do with Miller's paradox, to which Skyrms gives short shrift, dismissing it as "simply a fallacy of equivocation".[1] Skyrms continues by presenting a short proof in the form of a model, which makes it clear that second-order probabilities are formally consistent. The model uses relative frequencies, but Skyrms, tongue-in-cheek, argues

1 Skyrms 1980, p. 111. For the original statement of this paradox, if it deserves the name, see Miller 1966.

M.C. Galavotti et al. (eds.), *European Philosophy of Science – Philosophy of Science in Europe and the Viennese Heritage*, Vienna Circle Institute Yearbook 17,
DOI 10.1007/978-3-319-01899-7_17,

that it is implicit in the work of that champion of subjective probabilities, Bruno de Finetti:

> I would say nothing more about formal inconsistency were it not that some reputable philosophers continue to have suspicions (if not arguments) ... Though it may be a case of bringing out a cannon to swat a fly, I therefore feel obliged to point out that there is implicit in de Finetti's work a proof of formal consistency for a theory of second order probabilities: simply interpret $[Pr]$ as relative frequency probability. ... This is not the intended interpretation, but it suffices to settle the question of consistency.[2]

Here Skyrms is interpreting a relation like $PR[Pr(E) = x]$, where Pr is a first-order probability, and PR a second-order probability, by making the former an objective, relative-frequency chance, and the latter a subjective probability about a proposition concerning that chance.

Skyrms' model of a second-order probability is assuredly successful. However he remained silent on the question whether the same goes for an infinite regress of higher-order probabilities: does the concept of a regress of probabilities of probabilities, and so on, make sense? In a recent paper we have explained how to set up such a regress, and we also showed how to calculate the probability of the original proposition by summing a convergent series[3]. What we did not do, however, was to consider the formal question of the *consistency* of our equations. In the present paper we fill this lacuna, i.e. we construct a model of an infinite regress of probabilities of probabilities, where 'model' is used in the logical sense of a structure that makes all the sentences of a theory true. Our model is based on coin-making machines and can in principle be implemented. We show that when the coin-making structure is inserted into the abstract equations defining a regress of higher-order probabilities, the resulting statements are indeed all true.

We start in Section 2 by sketching the relevant results of our earlier paper.[4] In Section 3 we work out a numerical example of a regress of higher-order probabilities. Finally in Section 4 we offer our model, which involves coin-making machines. We show that, in this model, all the formulas of the abstract theory developed in Section 2 are true; thus the regress is consistent.

---

2 Skyrms 1980, p. 112.

3 Peijnenburg and Atkinson 2012.

4 As in that paper, we assign a particular number (rather than an interval) to the probability that a given proposition is true. This is meaningful only if the probabilities in question are discrete. The generalization to a continuous probability distribution will be given elsewhere – Atkinson and Peijnenburg, to appear.

2. PROBABILITIES OF PROBABILITIES

In this section we set up a regress of higher-order probabilities in general terms. For convenience we will talk about the probabilities as if they were subjective; but in fact some could be objective and some subjective: it is only the abstract system that is of importance here.

Let q_0 stand for some proposition of which we doubt the truth. Perhaps we think the probability that q_0 is true is merely v_0, this being some number between 0 and 1. Define q_1 to be the proposition $P(q_0) = v_0$. If we were quite sure that q_1 itself is true, we would effectively be asserting that the probability of q_0 is simply equal to the conditional probability of q_0, given q_1. Let us designate this, our first estimate of the probability of q_0, as $P^{(1)}(q_0)$, where

$$P^{(1)}(q_0) = P(q_0|q_1) = P(q_0|P(q_0) = v_0) = v_0 \,. \tag{1}$$

The last equality in (1), namely $P(q_0|P(q_0) = v_0) = v_0$, is warranted by what Skyrms has called Miller's principle. In words, it says that the probability of q_0, given that the probability of q_0 is v_0, is v_0.[5]

However it may be that we are not at all sure that q_1 is true, but only think that its probability is v_1. This thought can be expressed by the proposition $P(q_1) = v_1$, which we will dub q_2. If were sure that q_2 is true, then we would be asserting that the probability of q_1 is v_1. In that case, our next estimate of the probability of q_0 would be $P^{(2)}(q_0)$, which, with use of the rule of total probability, is given by

$$\begin{aligned} P^{(2)}(q_0) &= P(q_0|q_1)P(q_1) + P(q_0|\neg q_1)P(\neg q_1) \\ &= v_0 v_1 + w_0(1 - v_1)\,. \end{aligned} \tag{2}$$

Here we have used Miller's principle as above, and we have abbreviated the conditional probability $P(q_0|\neg q_1)$ by the symbol w_0. Clearly this new estimate of the value of the probability of q_0 is *not* equal to v_0; it will be somewhere between v_0 and w_0. The new estimate of the probability of q_0 can be thought of as an update on the previous estimate.

The update followed from the provisional assumption that q_2 is true; but suppose next that q_2 is not known to be true, and let q_3 be the proposition $P(q_2) = v_2$. On the assumption that q_3 is true, we would set $P(q_2)$ equal to v_2, so we could

---

5 As Skyrms has it: "Those who have followed the development of modal logic will already know that we invite no additional difficulty by universally generalising Miller's principle to

$$\text{for any } x,\, x = PR[E \text{ given that } Pr(E) = x]\text{"}$$

(Skyrms 1980, p. 112). As we have noted, Skyrms uses the symbol Pr for the first-order probability and PR for the second-order probability. Miller's principle has the same form as David Lewis's Principal Principle, but Lewis would interpret Pr as an objective chance, PR as a subjective credence (Lewis 1980).

then write, at this level of estimation for the probability of q_1,

$$\begin{aligned} P(q_1) &= P(q_1|q_2)P(q_2) + P(q_1|\neg q_2)P(\neg q_2) \\ &= v_1 v_2 + w_1(1 - v_2)\,, \end{aligned} \tag{3}$$

where $P(q_1|\neg q_2)$ has been designated by the symbol w_1. The next update for the probability of q_0 is $P^{(3)}(q_0)$, which has the form

$$\begin{aligned} P^{(3)}(q_0) &= P(q_0|q_1)P(q_1) + P(q_0|\neg q_1)P(\neg q_1) \\ &= v_0 P(q_1) + w_0[1 - P(q_1)]\,, \end{aligned} \tag{4}$$

where for $P(q_1)$ we are to understand the expression on the right of (3), namely $v_1v_2 + w_1(1 - v_2)$. When this is inserted into (4), $P^{(3)}(q_0)$ is thereby expressed as a function of v_0, w_0, v_1, w_1, and v_2.

It should be clear now how to continue the sequence of updates. The nth step leads to $P^{(n)}(q_0)$, which can be evaluated in an analogous way. By way of illustration, a numerical example is worked out in the next section. The sequence of updates $P^{(n)}(q_0)$ converges to a limit, except in extreme cases.[6]

3. A Numerical Example

Consider the uniform situation in which $P(q_0|q_1) = P(q_1|q_2) = P(q_2|q_3) = \ldots = 0.9$ and $P(q_0|\neg q_1) = P(q_1|\neg q_2) = P(q_2|\neg q_3) = \ldots = 0.5$. This assumption of uniformity is introduced purely to facilitate the calculation. The method of Section 2 also works when the conditional probabilities do differ from step to step in the regress, on condition that the constraint (5) in footnote 6 is respected.

We begin with

$$P^{(1)}(q_0) = P(q_0|q_1) = 0.9\,,$$

The next update is calculated from Eq.(2), where in this uniform example $P(q_1)$ is the same as $P^{(1)}(q_0)$, so it is equal to 0.9. Thus

$$\begin{aligned} P^{(2)}(q_0) &= 0.9 \times 0.9 + 0.5 \times (1 - 0.9) \\ &= 0.86\,. \end{aligned}$$

6 Convergence is guaranteed if the difference, $v_n - w_n$, does not approach one too rapidly as n increases. To be precise, the sequence converges so long as the difference between $v_n - w_n$ and 1 does not tend to zero faster than does $1/n$. In mathematical notation

$$1 - v_n + w_n = O\,(1/n) \quad \text{as } n \text{ tends to } \infty\,, \tag{5}$$

where $O\,(1/n)$ means "of order $1/n$". For a proof of this convergence, see Atkinson and Peijnenburg (2010), Appendices A and B.

To calculate $P^{(3)}(q_0)$ we first need $P(q_1)$ at order 2, and because of uniformity this is the same as $P^{(2)}(q_0)$, namely 0.86. Hence

$$\begin{aligned} P^{(3)}(q_0) &= 0.9 \times 0.86 + 0.5 \times (1 - 0.86) \\ &= 0.844\,. \end{aligned}$$

For update number 4 we need $P(q_1)$ at order 3, which, again thanks to uniformity, is the same as $P^{(3)}(q_0)$, which we have just calculated to be 0.844, so

$$\begin{aligned} P^{(4)}(q_0) &= 0.9 \times 0.844 + 0.5 \times (1 - 0.844) \\ &= 0.8376\,. \end{aligned}$$

and so on.

Here is an overview of the values of $P^{(n)}(q_0)$ after an increasing number of updates:

n	1	2	3	4	5	6	10	∞
$P^{(n)}(q_0)$	0.9	0.86	0.844	0.8376	0.8351	0.8340	0.83334	$\frac{5}{6}$

Probability of q_0 after n updates

Note that $P^{(\infty)}(q_0)$ is precisely equal to $\frac{5}{6} = 0.83333...$, and that $P^{(10)}(q_0)$ is already very close indeed to this limiting value.

4. Model of Higher-Order Probabilities

In this section we set up a model of the abstract system of equations in Section 2. Although we talked about the system in terms of subjective probabilities of subjective probabilities, it is indeed abstract in its form. The only requirement is that the probabilities satisfy Kolmogorov's axioms. The probabilities in the model we are about to describe are however all objective chances, which we will couch in the language of relative frequencies. The purpose is twofold: (a) to show that the equations governing the model are precisely those of Section 2, and (b) to show that all the implications of the abstract equations are true in the model. This demonstrates that the abstract system of higher-order probabilities is consistent.

Suppose there are two machines, each of which produces trick coins. Machine V_0 makes coins each of which has bias v_0, by which we will mean that each has probability v_0 of falling heads when tossed; whereas machine W_0 makes coins each of which has bias w_0. An experimenter tosses a coin from machine V_0. We identify the propositions q_0 and q_1 as follows:

q_0 is the proposition "this coin will fall heads"

q_1 is the proposition "this coin comes from machine V_0".

The probability that q_0 is true is based on the datum that the coin comes from machine V_0, in other words that proposition q_1 is true. Clearly

$$P(q_0|q_1) = P(\text{“this coin will land heads”} \,|\, \text{“this coin comes from machine } V_0\text{”})$$

is the same thing as $P\big(q_0|P(q_0) = v_0\big)$, for if the coin has come from machine V_0, the probability of a head is v_0; and conversely, if the probability is v_0, the coin must have come from machine V_0. As in Section 2, we have $P(q_0|q_1) = v_0$.

The experimenter is now instructed to take many coins from *both* machines, and to mix them thoroughly in a large pile. Moreover, the numbers of coins that have been added to the pile from machines V_0 and W_0 are not entirely left to the whim of our experimenter, for their relative number is determined by a second experiment, which is performed by a supervisor. This second experiment is much like the first one, but it involves two new machines, V_1, which produces trick coins with bias v_1, and W_1, which produces trick coins with bias w_1. The supervisor extracts a coin from machine V_1; and he instructs the experimenter to make sure that the relative number of coins that she takes from her machine V_0 is equal to the probability that his coin falls heads when tossed. That is to say, the number of coins that she must add to the pile from machine V_0 is equal to v_1 multiplied by the total number of coins removed from machines V_0 and W_0.

The experimenter takes one coin at random from her pile and, understanding q_0 now to refer to this coin, we can deduce the probability of q_0 in the new situation. Indeed, if

$$q_2 \quad \text{is the proposition} \quad \text{“the supervisor's coin comes from machine } V_1\text{”}\,,$$

and q_2 is true, then $P(q_1) = v_1$. We conclude that, if the experimenter were to repeat the procedure of tossing a coin from her pile many times (with replacement), the resulting relative frequency of heads would be approximately equal to $P^{(2)}(q_0)$, as given by (2) (and the approximation would get better and better as the number of tosses increases – more carefully: the probability that the relative number of heads will differ by less than any assigned $\varepsilon > 0$ from v_0 will tend to unity as the number of tosses tends to infinity). This concludes the description of the model of the first iteration of the regress, constrained by the condition that the supervisor's coin comes from machine V_1, that is by the veridicality of q_2.

In the next iteration, the supervisor receives instructions from an AI (artificial intelligence) that simulates the working of yet another duo of machines, V_2 and W_2, which produce simulated coins with biases v_2 and w_2, respectively. The supervisor makes a large pile of coins from his machines V_1 and W_1; and he adjusts the relative number of coins that he takes from V_1 to be equal to the probability that a simulated coin from V_2 would fall heads when tossed. That is to say, the number of coins in the instructor's pile that have been taken from V_1, divided by the total number that have been taken from V_1 and W_1, is equal to v_2.

If the supervisor were to repeat the procedure of tossing a coin from his pile many times (with replacement), then the resulting relative frequency of heads would be approximately equal to $P(q_1)$, as given by (3) (with the usual probabilistic proviso). This value of $P(q_1)$ is handed down to the experimenter, and she runs her experiment as above, but with the updated value of $P(q_1)$. The relative frequency of heads that she would observe will be approximately equal to $P^{(3)}(q_0)$, as given by (4). The above constitutes a model of the second iteration of the regress, constrained by the condition that the AI's simulated coin comes from the simulated machine V_2, that is by the veridicality of q_3, where

$$q_3 \quad \text{is the proposition} \quad \text{"this simulated coin comes from simulated machine } V_2\text{"} .$$

Of course this procedure must be repeated *ad infinitum*. A subprogram must simulate the working of yet another duo of machines, V_3 and W_3, which program the production of coins with biases v_3 and w_3, and so on. In this way an implementable model has been produced for an arbitrary number of iterations of the abstract system of Section 2, thereby showing that the whole regress is consistent.

Acknowledgements
We would like to thank the audience of the conference "European Philosophy of Science and the Viennese Heritage" for their helpful remarks, in particular Dennis Dieks, John Norton, Wlodek Rabinowicz and Miklós Rédei.

References

Atkinson, David, and Jeanne Peijnenburg. 2010. "The Solvability of Probabilistic Regresses. A Reply to Frederik Herzberg", *Studia Logica*, vol. 94, no. 3, 347-353.

Domotor, Zoltan. 1981. "Higher Order Probabilities", *Philosophical Studies*, vol. 40, no. 1, 31-46.

Gaifmann, Haim. 1988. "A Theory of Higher Order Probabilities", in B. Skyrms and W.L. Harper (eds.), *Causation, Chance and Credence*. London: Kluwer, pp. 191-219.

Hume, David. 1738/1961. *A Treatise of Human Nature, Volume 1*. London: J.M. Dent.

Kyburg, Henry E. 1987. *Higher Order Probabilities and Intervals*. Volume 36 of Technical Report Department of Computer Science, University of Rochester.

Lewis, David. 1980. "A Subjectivist's Guide to Objective Chance", in R. C. Jeffrey (ed.), *Studies in Inductive Logic and Probability. Vol II*. Berkeley: University of California Press, 263-293.

Miller, David. 1966. "A Paradox of Information", *British Journal for the Philosophy of Science*, vol. 17, 59-61.

Peijnenburg, Jeanne, and David Atkinson. 2012. "An Endless Hierarchy of Probabilities", *American Philosophical Quarterly*, vol. 49, no. 3, 267-276.

Rescher, Nicholas. 2010. *Infinite Regress. The Theory and History of Varieties of Change*. New Brunswick, NJ: Trabsaction.

Skyrms, Brian. 1980. "Higher Order Degrees of Belief", in D.H. Mellor (ed.), *Prospects for Pragmatism: Essays in Memory of F.P. Ramsay*. Cambridge: Cambridge University Press, 109-137.

Uchii, Soshichi. 1973. "Higher Order Probabilities and Coherence", *Philosophy of Science*, vol. 40, no. 3, 373-381.

Faculty of Philosophy
University of Groningen
Oude Boteringestraat 52
9712 GL Groningen
The Netherlands
d.atkinson@rug.nl
jeanne.peijnenburg@rug.nl

Matthias Neuber

Is Logical Empiricism Compatible With Scientific Realism?

Scientific realism is the view that the theoretical entities of science exist. Atoms, forces, electromagnetic fields, and so on, are not merely instruments for organizing observational data but are real and causally effective. This view seems to be hardly compatible with the logical empiricist agenda: As common wisdom has it, logical empiricism is mainly characterized by a strong verification criterion of meaning, i.e., by the project of defining the meaning of theoretical terms by virtue of the meaning of purely observational terms. However, it has been largely ignored by the historians of logical empiricism that there indeed existed a *realist faction* within the logical empiricist movement. Among the few authors who have recognized both the historical and the programmatic relevance of this realist faction is Stathis Psillos who, in two recent papers, attempts to emphasize the important role played in this connection by Herbert Feigl[1] and by Hans Reichenbach[2]. According to Psillos, it was these two thinkers who documented in their writings the *compatibility* of logical empiricism and scientific realism.

Like Psillos I am of the opinion that the realist faction within the logical empiricist movement deserves more attention than it has received so far. However, I will come to a different result than Psillos. According to the view I wish to defend, Feigl and Reichenbach (and with them Psillos) are still too optimistic about the ontological impact of language. In order to establish the intended realist account of logical empiricism, more metaphysics is needed than Feigl and Reichenbach (and with them Psillos) would allow. As will be shown, among the logical empiricists themselves it was Eino Kaila who came closest to this—less linguistic and more metaphysical—kind of approach.

I.

To begin with, the historical starting point of the discussion of the realism issue within the logical empiricists' Vienna Circle was no doubt Rudolf Carnap's *Scheinprobleme in der Philosophie* (1928). As is well known, Carnap attempts there to clearly state that any talk of the 'reality' or 'non-reality' of the external

1 See Stathis Psillos, "Choosing the Realist Framework", Synthese 180, 2011, pp. 301-316.

2 See Stathis Psillos, "On Reichenbach's Argument for Scientific Realism", Synthese 181, 2011, pp. 23-40.

M.C. Galavotti et al. (eds.), *European Philosophy of Science – Philosophy of Science in Europe and the Viennese Heritage*, Vienna Circle Institute Yearbook 17,
DOI 10.1007/978-3-319-01899-7_18, © Springer International Publishing Switzerland 2014

world is completely devoid of meaning. More exactly, the realism issue as such is, for Carnap, incapable of being solved by rational means. It is a 'pseudo-problem' and therefore to be excluded from science. Or, as Carnap himself puts it: "In the realism controversy, science can take neither an affirmative nor a negative position since the question is devoid of meaning."[3]

However, in *Der logische Aufbau der Welt* (1928), Carnap explicitly distinguishes between 'metaphysical' realism, which he rejects, and what he calls 'linguistic' realism, which he appreciates as "very useful in the empirical sciences"[4]. A similar distinction was drawn by Moritz Schlick who, in his programmatic paper "Positivismus und Realismus" (1932), demarcated metaphysical from *empirical* realism and categorically claimed that "[l]ogical positivism and realism are [...] not opposed"[5]. Schlick's main point was that—although metaphysical realism had to be banished from science because "the 'problem of the reality of the external world' is a meaningless pseudo-problem"[6]—empirical realism should be regarded as a viable position compatible with the central tenets of logical empiricism. This in turn led to some sort of schism among Schlick's students and interpreters. While, for instance, Ludovico Geymonat held that Schlick never altered his position in a significant way,[7] Herbert Feigl argued (or, rather, complained) that Schlick definitely converted from a realist to a phenomenalistic point of view[8] (thereby implying that 'empirical' realism was no realism at all).[9]

3 Rudolf Carnap, Pseudoproblems in Philosophy, translated by R. A. George. London: Routledge 1968, p. 333. For an extended discussion of Carnap's rejection of metaphysics, see Michael Friedman, "The Aufbau and the rejection of metaphysics", in: Michael Friedman/Richard Creath (eds.), The Cambridge Companion to Carnap. Cambridge: Cambridge University Press 2007, pp. 129-152.

4 Rudolf Carnap, The Logical Structure of the World, translated by R. A. George. London: Routledge 1968, § 52.

5 Moritz Schlick, "Positivism and Realism", in: Henk L. Mulder/Barbara F. B. van de Velde-Schlick (eds.), Moritz Schlick: Philosophical Papers, volume II (1925-1936). Dordrecht: Reidel 1979, p. 283.

6 Ibid., p. 263.

7 See Ludovico Geymonat, "Entwicklung und Kontinuität im Denken Schlicks", in: Brian McGuinness (ed.), Zurück zu Schlick. Eine Neubewertung von Werk und Wirkung. Wien: Verlag Hölder-Pichler-Temsky 1985, p. 25.

8 See Herbert Feigl, "The Power of Positivistic Thinking", in: Herbert Feigl, Inquiries and Provocations. Selected Writings, 1929-1974, edited by Robert S. Cohen. Dordrecht: Reidel 1981, p. 39.

9 It must be seen that the early, pre-Viennese, Schlick advocated what might be labeled 'critical realism.' The essential point of this position was the view that (ontologically interpreted) Kantian things-in-themselves are knowable. This assumption was clearly given up by the Schlick of "Positivismus und Realismus". For the early Schlick's critical realism, see Matthias Neuber, Die Grenzen des Revisionismus. Schlick, Cassirer und das "Raumproblem". Wien/New York: Springer 2012, ch. 2. A more recent position very similar to the 'pseudo-problem' account of Carnap and (the Viennese) Schlick is defended by Arthur Fine. See Arthur Fine, "Unnatural Attitudes: Realist and

II.

It is important to note that both Carnap and Schlick argued in terms of a systematic reflection on the structure of *language*. They both can be viewed as initiators of the famous 'linguistic turn.' Yet, the one who initially tried to *systematically exploit* (in particular) Carnap's ideas was, without any doubt, Hans Reichenbach. As Reichenbach put it in his seminal *Experience and Prediction* (1938), "the difference of the positivistic and the realistic conception of the world" can be formulated as "the difference of two languages".[10] Accordingly, it was Carnap who, in Reichenbach's view, had provided the appropriate means of approaching the whole realism debate.[11] On the other hand, it must be seen that Reichenbach dissociated himself from Carnap's position in one essential respect: For Reichenbach, it was the concept of *probability*, the importance of which Carnap failed to recognize, but which still formed the crucial element in the justification of the choice of the realistic 'language form.'

Indeed, it is the very concept of probability, with the help of which Reichenbach succeeds in shaping the realism issue in his own particular way. According to him, "[t]he key to a theory of meaning corresponding to the intentions of physics lies in the probability problem"[12]. What is more, for Reichenbach, "probability meaning leads to the realistic language of actual science; we start from a rather small domain of our own observations and construct the whole world upon it"[13]. By making use of the notion of *construction* Reichenbach is obviously moving in a Carnapian direction. However, it is at this very point that Reichenbach sees the essential contrast between himself and Carnap. For, in his view, Carnap understands by 'construction' the *reduction* of all theoretical talk to observational predicates, whereas his own, probabilistic, conception amounts to the *projection* of an autonomous realm of unobservable, but causally effective entities, such as atoms, electromagnetic fields, and so on. In order to illustrate his own position, Reichenbach brings the well-known story of the 'cubical world.'[14]

Instrumentalist Attachments to Science", in: Mind 95, 1986, pp. 149-179.

10 Hans Reichenbach, Experience and Prediction. An Analysis of the Foundation and the Structure of Knowledge. Chicago: The University of Chicago Press 1938, p. 145.

11 In § 29 of Experience and Prediction, Reichenbach declares that it is an "important result of Carnap's investigations" (ibid., p. 269) that a sharp distinction between fact and proposition (viz. between "object basis" and "sentence basis") cannot be drawn. The idea that such relations as implication are essentially "relations between sentences" had motivated Carnap's conception of philosophy as the "analysis of scientific language" (ibid., p. 270). Whereupon Reichenbach comments: "We ourselves made use of this conception when we reduced the question of the existence of external things to a question of the meaning of sentences." (ibid.)

12 Ibid., p. 75.

13 Ibid., p. 153.

14 Ibid., § 14.

Without going into the details of the cubical world analogy,[15] it should be mentioned that the peculiarity of Reichenbach's point of view has not gone unremarked. Thus, in the introduction to a *Synthese* special issue comprising the results of a conference on the occasion of the seventieth anniversary of the publication of *Experience and Prediction* one can read the following:

> Reichenbach is now often lumped together with the logical positivists of the Vienna Circle, but his ideas, especially those in *Experience and Prediction,* were often developed in opposition to positivism. Reichenbach frequently sought to save concepts and problems from the many positivists who sought to discredit them. For example, he thought that causality has a place in science and that the problem of the external world is not a pseudo-problem. Indeed, Reichenbach always defended realism against positivism, and his defense was truly ingenious. It has been standard for decades to refer complacently to 'the demise of logical positivism,' but the so-called positivists were more various than the standard stereotype would suggest.[16]

The authors are entirely correct in their assessment: For Reichenbach, logical *positivism,* as it was defended by certain members of the Vienna Circle in its early phase, was essentially at odds with his own, realistically motivated brand of logical *empiricism.*[17] Thus the view of the whole movement as a 'monolithic block' is severely flawed – a diagnosis, by the way, that has already been given by Thomas Uebel in the early 1990s.[18]

---

15 For a compressed discussion of Reichenbach's cubical world analogy, see Matthias Neuber, "Realism as a Problem of Language – From Carnap to Reichenbach and Kaila", in: Richard Creath (ed.), Rudolf Carnap and the Legacy of Logical Empiricism. Dordrecht: Springer 2012, pp. 37-56 (esp. sect. 3). For a more extended and detailed discussion, see Wesley Salmon, "Ornithology in a Cubical World: Reichenbach on Scientific Realism", in: W. Reiter/A. Zeilinger (eds.), Epistemological and Experimental Perspectives on quantum physics. Dordrecht: Kluwer 1999, pp. 303-315. See further Elliot Sober, "Reichenbach's Cubical Universe and the Problem of the External World", in: Synthese 181, 2011, pp. 3-21.

16 Gürol Irzik & Elliott Sober, "Introduction to the Synthese special issue on Hans Reichenbach, Istanbul, and Experience and Prediction", in: Synthese 181, p. 2.

17 See, in connection to this, also his attempt at a demarcation in: Hans Reichenbach, "Logistic Empiricism in Germany and the Present State of Its Problems", in: The Journal of Philosophy 33, pp.141-160. See further the illuminating reconstruction in: Friedrich Stadler, "The Road to Experience and Prediction From Within: Hans Reichenbach's Scientific Correspondence from Berlin to Istanbul", in: Synthese 181, 2011, pp. 137-155.

18 See Thomas Uebel, Overcoming Logical Empiricism From Within. The Emergence of Neurath's Naturalism in the Vienna Circle's Protocol Sentence Debate. Amsterdam: Rodopi 1992.

III.

The systematic merits of Reichenbach's account notwithstanding, it must be noted that his 'probabilistic defense' of the realistic language form did not remain uncriticized. Among the logical empiricists it was in particular Herbert Feigl who objected that a 'linguistic framework,' such as the realistic language form, could not be justified by means of making use of the probability concept. In Feigl's view "the legitimacy of applying the probability concept to the whole realistic frame, instead of merely to inferences within it, remains painfully questionable"[19]. Instead, Feigl maintains, it must be conceded that the choice of the realistic language form as such is a matter of *practical decision.*

This overtly pragmatist train of thought comes very close to Carnap's argumentation in his famous 1950 paper "Empiricism, Semantics, and Ontology." As is well known, Carnap argues there that a distinction must be drawn between "internal" questions, on the one hand, and "external" questions, on the other.[20] The crucial point in this connection is that internal questions are particular questions of existence *within* a given framework, whereas external questions are supposed to concern the existence of the respective framework, viz. system of entities, *as a whole.* While the former kind of questions are, according to Carnap, 'genuine' ontological questions and thus answerable along rational—theoretical—lines, the latter kind of questions are essentially *practical* questions and thus dependent on convention-based decisions. Feigl, in his critique of Reichenbach's attempt at an inductive justification of the realistic language form, argues in fairly straightforward Carnapian fashion when he writes:

> The customary probabilistic realism in trying to justify 'transcendent' hypotheses on the basis of experimental findings has put the cart before the horse. Only after the introduction of the realistic frame can we legitimately argue inductively either from the theory to the outcome of as yet unperformed experiments; or vice versa from the results of experiments to *specific* postulates of the theory. But the presupposed introduction of the linguistic frame, i.e. the semantic-realistic interpretation of the theory, is a step that can be justified only instrumentally: It furnishes the very possibility of a theory that is inductively fruitful.[21]

In short: Scientific ("semantic") realism is, according to Feigl, dependent on a foregoing decision concerning the choice of a language form. It is, in other words, justifiable only in terms of pragmatics.[22] It is for that reason that Feigl, in his 1954

19 Herbert Feigl, "Existential Hypotheses: Realistic versus Phenomenalistic Interpretations", in: Philosophy of Science 17, 1950, pp. 35-62.

20 See Rudolf Carnap, "Empiricism, Semantics, and Ontology", in: Revue Intérnational de Philosophie 4, p. 22.

21 Herbert Feigl, "Logical Reconstruction, Realism and Pure Semiotics", in: Philosophy of Science 17, 1950, p. 195.

22 Interestingly enough, Feigl's relation to pragmatism has not been scrutinized so far. Furthermore, it would be important to clarify Feigl's (frequently declared) indebted-

paper "Scientific Method without Metaphysical Presuppositions", speaks explicitly of the "*relative or pragmatic a priori*"[23] and refers the reader to the positions of C. I. Lewis, Victor Lenzen, Arthur Pap, and Wilfrid Sellars.[24]

IV.

I do not think that Feigl's attempt at a pragmatist justification of the realistic language form was successful. In a nutshell, the decision to be drawn is that between a strong version of pragmatism—then the realist position would certainly be sacrificed; or a weak version of pragmatism—then the whole issue would be trivialized.[25] Surely, it can be conceded that Feigl's criticism of Reichenbach's probabilistic defense of the realist framework was appropriate. But the problem of Reichenbach's approach lies not in the probability concept as such, but rather in the embracing Carnapian conception of the realism issue as a matter of language (which Feigl himself entirely approved of). Furthermore, by a careful reading of *Experience and Prediction* one will recognize that Reichenbach, at least in places, comes remarkably close to making the scientific realist stance *fully dependent* on pragmatist presuppositions. Thus, in § 17, he frankly points out:

> We have said that the choice of a language depends on our free decision but that we are bound to the decisions entailed by our choice: we find here that the decision for the strictly positivistic language would entail the renunciation of any reasonable justification of a great many human actions. The pragmatic idea that the definition of meaning is to be chosen in adaption to the system of human actions, that it is to be determined by the postulate of utilizability, decides, therefore, against the strictly positivistic language.[26]

The decisive distinction here is between "free" and "entailed" decisions. The idea is that the choice of the realistic language form is not dictated by material reality, but that it is rather the product of what Reichenbach calls "volitional bifurcation" concerning the aim of science.[27] As he sees it, the realistic and the positivistic

---

ness to Wilfrid Sellars's conception of "pure pragmatics."

23 Herbert Feigl, "Scientific Method without Metaphysical Presuppositions", in: Herbert Feigl, Inquiries and Provocations. Selected Writings, 1929–1974, edited by Robert S. Cohen. Dordrecht: Reidel 1981, p. 97.

24 The most articulated version of the 'pragmatic a priori' is the one defended by Lewis. See C.I. Lewis, Mind and the World Order: Outline of a Theory of Knowledge. New York: Dover 1929, esp. ch. VIII. See further C.I. Lewis, "The Pragmatic Conception of the a Priori", in: Herbert Feigl/Wilfrid Sellars (eds.), Readings in Philosophical Analysis. New York: Appleton 1949, pp. 286-294.

25 For the details of this evaluation, see Matthias Neuber, "Feigl's 'Scientific Realism'", in: Philosophy of Science 78, 2011, pp. 165-183.

26 Experience and Prediction, p. 150.

27 Ibid., p. 10 and p. 147.

("egocentric") language are not equivalent, so that the free (volitional) decision for one of these two languages makes an essential difference with respect to the decisions *entailed* by it. Reichenbach writes:

The man who speaks the egocentric language cannot express certain ideas which the man with the realistic language may formulate; the decision for the egocentric language, therefore, entails the renunciation of certain ideas, and may, consequently, become highly relevant. We do not thereby say that the egocentric language is "false"; such a criticism would be a misunderstanding of the character of a volitional decision. It is rather the method of entailed decisions which we have to apply here; we can show that the decision for the egocentric language leads to a scientific system of a restricted character which does not correspond to the system constructed by the realistic language in its full extension.[28]

Moreover, as the penultimate quotation makes clear, the crucial criterion for evaluating the entailed decisions themselves is their "utilizability", that is, their adaptability to the "system of human *actions*." Consequently, the motive force in Reichenbach's whole conception is not realism, but (a sometimes disguised form of) pragmatism.

By thus making scientific realism dependent on pragmatist presuppositions, the scientific realist stance is (at least as an autonomous position) eventually sacrificed. It is therefore rather astonishing that Stathis Psillos thinks of the whole framework conception as a promising strategy for *strengthening* the realist point of view. Psillos, who, in his influential 1999 book *Scientific Realism: How Science Tracks Truth*, had argued for a certain form of naturalistic-realist ontology, is now obviously willing to concede that there exists no "ultimate argument" for the scientific realist position.[29] In good old Carnapian fashion, Psillos explicitly declares: "the ontic framework must already be in place before questions about the reality of specific entities are raised."[30] Accordingly, Psillos holds that "there is no framework-free standpoint from which what there is [...] can be viewed"[31]. In my opinion, this amounts to putting the cart before the horse. Psillos obviously assumes that *as soon as* the realist framework is set in place, we can straightforwardly talk of the theoretical entities posited by science. It is, as he argues by invoking Reichenbachian terminology, the "entailed decisions" by which the realist framework gets legitimized. However, why should we bother about our linguistic framework and its entailed decisions and not talk *directly* about material reality? Is it not the case that reality is—completely independently of our 'concepts', 'minds', or 'frameworks'—how it is and that we *therefore* can (and should) attempt to describe it in a truth-conductive way? In order to answer this question it is

28 Ibid., p. 147.
29 See Psillos, "Choosing the Realist Framework", p. 303.
30 Ibid.
31 Ibid., p. 309.

instructive to review Psillos's original account of scientific realism (the substantiation of which he now thinks "is flawed"[32]).

So, in his *Scientific Realism*, Psillos informs us that, according to what he calls the "metaphysical thesis" of scientific realism, "the world has a definite and mind-independent natural-kind structure"[33]. Psillos further points out that "this metaphysical thesis is prerequisite to any meaningful defence of scientific realism"[34], and he motivates this claim by the conviction that "[i]nstead of projecting a structure onto the world, scientific theories, and scientific theorizing in general, discover and map out an already structured and mind-independent world"[35]. Closely connected to this metaphysical point of view is a certain form of "epistemic optimism"[36], according to which it is reasonable to believe that our best scientific theories or, better stated, theoretical assertions are (at least approximately) *true*. The main argument for this sort of epistemic optimism is, as Psillos states, the famous (or rather infamous?) 'no-miracles argument.'[37]

Iimportantly enough, in Psillos's current—revised—account of scientific realism, the status of the no-miracles argument has fundamentally changed. In Psillos's original account that he wrote in 1999, scientific realism was conceived of as being a *theory* (or hypothesis), and the no-miracles argument had the function to provide plausibility to the presupposition that this theory receives support from the undeniable success of science.[38] In his more recent approach, however, the no-miracles argument is 'valid' only *within* the realist framework; that is to say, the realist framework must first be installed so that the no-miracles argument can work at all. Therefore, Psillos concludes, the no-miracles argument "presupposes rather than establishes the realist frame"[39]. The success of science supports the assumption of the (approximate) truth of *particular* scientific theories. However, scientific realism as such is definitely *not* a theory (or hypothesis). Rather, it is—according to Psillos's more recent account—a framework, and this means that it "makes possible certain ways of viewing the world."[40]

But why is the realist framework a *realist* framework? Do we not need some sort of 'meta-framework' in order to secure the most fundamental assumption of scientific realism, namely the metaphysical thesis of an already structured and mind-independent world?[41] This, one might think, is what would be needed in

32 Ibid., p. 312.

33 Stathis Psillos, Scientific Realism: How Science Tracks Truth. London: Routledge 1999, p. xix.

34 Ibid., p. xx.

35 Ibid., pp. xixi-xx.

36 Ibid., p. xxi.

37 See ibid., p. xxii.

38 See ibid., esp. ch. 4.

39 Psillos, "Choosing the Realist Framework", p. 312.

40 Ibid.

41 I owe this point to Alexander Ehmann (Tübingen).

addition to such pragmatic (or methodological) criteria like coherence, simplicity, and so on. Without the assumption of mind-independence at the meta-level, our assumptions within the 'first-level-framework' would remain ungrounded. However, even worse, by establishing the meta-framework as a precondition of our first-level-framework we would still not be able to decide what our position actually is. The question would still remain, though, whether the meta-framework itself is *already* a realist framework. If it were, then one might wonder how choosing the meta-framework can be justified without buying into an infinite regress of meta-meta- and meta-meta-meta-frameworks (and so on). If it were not to be so, then the question would be how we might justify the choice of the realist first-level-framework in another way than by employing such criteria as coherence, simplicity, and so on. We could do that, but then our position would hardly be distinguishable from a pragmatist, conventionalist or even positivist (Machian) point of view. Atoms, forces and electromagnetic fields could then be conceptualized as useful 'heuristic fictions,' and the ontological aspect of mind-independence could be completely neglected.

The only alternative to this scenario (which is unwelcome from the realist standpoint) is obviously the adoption of a 'naturalistic' methodology, such as the one that was defended in Psillos's 1999 original account. On that account, scientific realism is a theory about the causal structure of the world.[42] As such, it is motivated not by the choice of a certain form of language but rather by the *insight* that *there is* a mind-independent causal structure of the world. This is not to say that we possess the 'ultimate truth' about the causal structure of the world. But it still allows us to draw the 'inference to the best explanation' and, accordingly, to avoid the assumption (which is destructive under the aspect of scientific realism) of a linguistic meta-framework.[43]

V.

All of this leads to the following situation: Either (1) the realist framework is viewed as being indispensable as a means of talking about (or referring to) reality; then its assumption is rather trivial and the term 'framework' can (without any change of meaning) be substituted by such terms as 'theory', 'hypothesis,' etc. Or (2) the realist framework is characterized as the constitutive—and *therefore* indispensible—linguistic presupposition of the (putative) realist *ontology*; in that case, we will hardly have any realism but some form of constructivist epistemology.

42 See, in connection to this, especially Psillos, Scientific Realism, ch. 12.

43 For the role of the (abductive) 'inference to the best explanation' within the defense of the scientific realist position, see, for example, Psillos, Scientific Realism, pp. 211-227. See further Howard Sankey, Scientific Realism and the Rationality of Science. Aldershot: Ashgate 2008, pp. 130-133.

In the case of Psillos, it is pretty unclear what exactly his position is. But, as I have tried to show, his revised account of the status of the no-miracles argument amounts to a position very close to the 'pitfall' of constructivist epistemology. This, however, contrasts significantly with Psillos's contention that what—from the logical empiricist perspective—Feigl understood by the 'realist framework' was not intended as a "transcendental argument"[44]. Indeed, this might have been Feigl's *intention*. But in fact, his argument was transcendental in structure and, consequently, not compatible with the scientific realist's naturalistic epistemology.

It must be noted in this connection that the programmatic turn from naturalistic scientific realism to the meta-scientific view of *transcendentalism* is nothing new in the history of philosophy. Hilary Putnam is perhaps the most well-known representative of this kind of turn. What Putnam called "internal realism"[45] was nothing but a species of transcendental (Kantian) idealism.[46] Putnam's former adherence to *scientific* realism was clearly abandoned by this transcendental turn.[47] In the case of Feigl, there is some interesting evidence that he, perhaps, *never* was a scientific realist. In the April 1950 issue of *Philosophy of Science*, one can find an extended discussion of Feigl's position, including contributions by Feigl's logical empiricist colleagues Philipp Frank, Carl Gustav Hempel and Ernest Nagel. It is instructive to see that all these critical contributions agree that what Feigl understands by 'scientific realism' in fact reduces to non-realist positions like positivism, pragmatism or operationalism.[48] Furthermore, Feigl himself, in his reply to these criticisms, concedes that his own point of view—which he says he ought to have rather termed 'pragmatic' realism—simply amounts to the claim that "when we fully and justly explicate the way in which we use the language of science [...] we cannot do without a set of designates that are in principle beyond the

44 Psillos, "Choosing the Realist Framework", p. 309.

45 See Hilary Putnam, Reason, Truth and History. Cambridge: Cambridge University Press 1981.

46 Here, I strongly agree with the diagnosis in: Ian Hacking, Representing and Intervening: Introductory Topics in the Philosophy of Natural Science. Cambridge: Cambridge University Press 1983, ch. 7.

47 This is not to say that the scientific realist is necessarily a metaphysical realist in Putnam's (and in Psillos's) sense. But it is important to keep in mind that the scientific realist is, as it were, obliged by program to categorically reject the later Putnam's central assumption that "[o]bjects do not exist independently of conceptual schemes" and that "[w]e cut up the world into objects when we introduce one or another scheme of description" (Putnam, Reason, Truth and History, p. 52). For the scientific realist, there must be no doubt that this sort of "cookie cutter metaphor" (Ilkka Niiniluoto, Critical Scientific Realism. Oxford: Oxford University Press 1996, p. 217) is untenable, because it is the world itself which determines the structure of the language corresponding to it, and not vice versa.

48 See, especially, Carl Gustav Hempel, "A Note on Semantic Realism", in: Philosophy of Science 17, 1950, p. 173; Philipp Frank, "Comments on Realistic versus Phenomenalistic Interpretations", in: Philosophy of Science 17, 1950, pp. 168-169; Ernest Nagel, "Science and Semantic Realism", in: Philosophy of Science 17, 1950, p. 179.

reach of direct experience"[49]. And he immediately adds: "I maintain that a good many statements concerning theoretical constructs and hypotheses made by Frank, Hempel and Nagel are *de facto* statements in the pure pragmatics of science."[50] As a scientific realist, Feigl should have replied to these criticisms by stressing the mind-independent status of theoretical *entities* instead of invoking the "pure pragmatics of science." What is essential in the debate over scientific realism is that, from the realist's standpoint, theoretical terms like 'atom' or 'electromagnetic field' not only be accepted as indispensable elements of our language system, but that the referred-to entities themselves be characterized as *causally effective*. Feigl never made this claim, because he obviously feared that this would lead him to some sort of obscurantist metaphysics. Or in other words: For Feigl, the problem of realism was *nothing but* a problem of language.[51]

On the whole, Reichenbach and Feigl (and with them Psillos) argue in terms of a, as it were, 'transcendalized' form of pragmatism. In the same way that for Kant the presupposition of "empirical realism" implied "transcendental idealism,"[52] for Reichenbach and Feigl (and, consequently, for Psillos) pragmatism is presupposed as the enabling ground of 'scientific realism.' In both cases, the strategy is *transcendental* insofar as realism is brought into dependence on something mind-related. In the case of Kant, realism is dependent on the faculties of human cognition (or, more exactly, on the "pure forms" of sensibility and understanding); and in the case of Reichenbach and Feigl (and, consequently, of Psillos) realism is dependent on the pragmatic dimension of adopting certain language forms (or linguistic frameworks). Neither approach, however, can be reconciled with the 'metaphysics of mind-independence' that underlies the *naturalistic* methodology of scientific realism.

VI.

Thus a more or less fundamental change of methodological perspective is in order. Among the logical empiricists it is especially Eino Kaila who deserves mention

49 Feigl, "Logical Reconstruction, Realism and Pure Semiotics", p. 193.

50 Ibid.

51 The following passage is fairly representative in this connection: "Semantic realism as I should like to see it understood, is free from the dangers of metaphysics precisely because it does not prescribe anything at all about the nature of the designate of our theoretical constructs. [...] Things are and will always be—as far as we can meaningfully talk about them—what they are confirmably knowable as; and it is up to the advance of science, not to logical or semiotic analysis, to tell us what things are 'really' like. But it is the task of logical analysis to tell us by means of what rules of our language we describe the objects of our knowledge, and [...] what we mean by the surplus of the knowable over the known." (Ibid., p. 192)

52 For the details of this view, see, for example, Henry A. Allison, Kant's Transcendental Idealism: An Interpretation and Defense. New Haven: Yale University Press 1983.

here. The methodologically most remarkable point in Kaila's thinking is that the reflection on the structure of language is *subordinated* to the reflection on the structure of the world. Kaila—himself professor of theoretical philosophy at the University of Turku and (after 1930) at the University of Helsinki—stood in contact with the members of the Circle since 1927 and visited Vienna in 1929 in order to discuss the issue personally with Carnap. Although he was to a large extent inspired by the "exact philosophical method of the Vienna Circle"[53], he refused to take the 'linguistic turn.' Instead, he focused on the explanatory constituents *of science itself.* By taking this broadly naturalistic perspective, Kaila was in a position to circumvent the conception of the realism issue as a pseudo-problem. Thus, in his 1930 *Der logistische Neupositivismus. Eine kritische Studie*, he fundamentally criticizes Carnap's *Aufbau* program and points out:

> Such results—presented in a tone of superior calmness and brilliant logical clarity—have profound effects. If they are correct, they are apt to deprive even empirical research of its *élan*; for the 'realist language' of science is actually more than a mere manner of speaking: it is the expression of the living *soul* of science.[54]

What Kaila has to offer instead is a realistically inspired theory of science and nature which essentially is grounded on the concepts of *invariance* and *probability*.

First, in regard to probability, Kaila completely agreed with Reichenbach that probabilistic reasoning played a fundamental role in defending the scientific realist position:

> The presupposition of empirical science that our perceptions are "sample-like segments" means that probability inferences can be drawn from the given to something not given, that the truth frequencies present in the given can be generalized to the not-given.[55]

Accordingly, Kaila maintained, "[a] certain *realism of all science*"[56] is able to be postulated. That is, for Kaila, the inference to Reichenbachian 'projective complexes' was both epistemologically legitimate and scientifically fruitful.

However, as already indicated, Kaila dissociated himself from Reichenbach's (as well as from Carnap's and from Feigl's) approach toward the realism issue by his refusal to take the 'linguistic turn.'[57] As he saw it, the concept of *invariance*

---

53 Ilkka Niiniluoto, "Eino Kaila and Scientific Realism", in: Ilkka Niiniluoto / Matti Sintonen / G.H. von Wright (eds.), Eino Kaila and Logical Empiricism, Acta Philosophica Fennica 52. Helsinki: Societas Philosophica Fennica 1992, p. 103.

54 Eino Kaila, "Logistic Neopositivism: A Critical Study", in: Robert S. Cohen (ed.), Eino Kaila: Reality and Experience. Four Philosophical Essays. Dordrecht: Reidel 1979, p. 4.

55 Reality and Experience, p. 48.

56 Ibid., p. 49.

57 For the otherwise close connections between Kaila's and Reichenbach's points of view, see Arti Siitonen, "Kaila and Reichenbach as Protagonists of 'Naturphilosophie'", in:

was strong enough to provide us with a convincing *defense* of the probabilistically inspired scientific realist position. The *prima facie* irritating point in that, though, was that Kaila attempted to integrate the invariance concept in a thoroughly Carnapian framework. This thoroughly Carnapian framework entailed the assumption that invariance be conceptualized in terms of a certain form of 'constitutional system.' This is at least what Kaila attempted in his two "contributions to logical empiricism." The first of these two contributions appeared in 1936 and was titled *Über das System der Wirklichkeitsbegriffe. Ein Beitrag zum logischen Empirismus*[58]; the second contribution appeared in 1941 and was titled *Über den physikalischen Realitätsbegriff. Zweiter Beitrag zum logischen Empirismus*[59]. In both essays, Kaila argued from an 'invariantist' point of view: For him, invariances formed the proper subject of scientific inquiry. What is more, according to Kaila, invariances were *ontologically basic*. In his own words: "The 'essence' of a thing consists of the invariances of this thing."[60]

This is not the place to go into the details of Kaila's invariantist ontology.[61] However, by way of conclusion, I would like to point to the fact that by taking invariance seriously, Kaila was in a position to furnish Carnap's constitutional system of the *Aufbau* with an *ontological foundation*. He thereby was able to install invariance as an explanatory principle and, in consequence of this, to account for the causal effectiveness of theoretically postulated entities. His (thoroughly anticonventionalist) *theory of measurement* was intended to drive this point home.[62] Possible problems of Kaila's invariantist ontology notwithstanding, the quest to make logical empiricism compatible with scientific realism thus unavoidably necessitates an affirmative approach toward the role of (non-speculative) metaphysics in the philosophy of science. Rather than reflecting on concerns of internal 'language engineering', the principle aim should be to figure out how and why the language of science *fits with the causal structure of the objective world.* Kaila's invariantist ontology might invest us with the basic means for achieving this goal in a logical empiricist *and at the same time* scientific realist account of science and nature.

Juha Manninen/Friedrich Stadler (eds.), The Vienna Circle in the Nordic Countries: Networks and Transformations of Logical Empiricism, Vienna Circle Institute Yearbook 14. Dordrecht: Springer 2010, pp. 135-152.

58 English translation, "On the System of the Concepts of Reality. A Contribution to logical empiricism," in: Reality and Experience, pp. 59-125.

59 English translation, "On the Concept of Reality in Physical Science. Second Contribution to logical empiricism," in: Reality and Experience, pp. 126-258.

60 Reality and Experience, p. 228. For a similar, more recent, conception of invariance as an ontological category, see Robert Nozick, Invariances: The Structure of the Objective World. Cambridge, Mass: Harvard University Press 2001, esp. ch. 2.

61 For a detailed reconstruction, see Matthias Neuber, "Invariance, Structure, Measurement – Eino Kaila and the History of Logical Empiricism", in: Theoria 78, 2012. See further Neuber, "Realism as a Problem of Language," sect. 5.

62 See in this connection especially Neuber, "Invariance, Structure, Measurement," sect. 3.

University of Tübingen
Department of Philosophy
Bursagasse 1
D-72070 Tübingen
Germany
matthias.neuber@uni-tuebingen.de

Jan Faye

Does the Unity of Science have a Future?

The program of logical positivism gave inspiration to the unity of science movement. The movement carried the belief that all sciences, including the social sciences and the humanities, ought to share some common language if these disciplines were to be considered genuine sciences (*Wissenschaften*). It has been argued that the logical empiricists were very aware of the differences between the sciences, but that they imagined that "the unity of science unites all the sciences together so that they can yield predictions and support action in the world of everyday life. The institution of science and of scientifically minded people has fostered an activist international community of people united linguistically and in their open and critical habits of mind."[1] This might very well be the case. But the reception had it otherwise. According to it, the logical empiricists became devotees of reductionism in spite of the fact that they didn't hold that all disciplines can be traced back to one and the same science, for example, physics. On the contrary, their basic claim was that every science concerned with reality must be based on empirical observations and must expose its findings to empirical control.

Thus, the outlook of the unity of science was focused on epistemology and methodology rather than metaphysics. In 1920s and 1930s the logical empiricists opposed the view formulated by philosophers like Wilhelm Dilthey, Wilhelm Windelband, and Heinrich Rickert that the cultural sciences require a completely different methodology because their objects of study were ontologically different. The logical empiricists did not deny that psychological phenomena are distinct from behaviour but maintained that some physicalistic language was needed if we were to be able to uphold intersubjective communication. Therefore the language of psychological phenomena had to be translated to the language of behaviour or neuroscience in order to be verifiable. In the end, however, according to the standard reception the unity of science program failed because it entailed unrealistic commitments to anti-metaphysics, behaviourism, reductionism, nomothetic descriptions, objectivity, universalism, value-freeness, and verificationism. Today, no philosopher of science would even think of these commitments to hold unrestrictedly for the natural sciences. But neither seems Rudolf Carnap nor Otto Neurath, nor any other leading members of the unified science movement, to have identified the unity of science with reductionism.

---

1 Allan W. Richardson. "The Many Unities of Science: Politics, Semantics, and Ontology", p. 15. In Kellert, Stephen H., Helena E. Longino, and C. Kenneth Waters (eds.) (2006). *Scientific Pluralism*. Minnesota Studies in the Philosophy of Science, Vol. XIX. Minneapolis: University of Minnesota Press, 2006, 1-25.

M.C. Galavotti et al. (eds.), *European Philosophy of Science – Philosophy of Science in Europe and the Viennese Heritage*, Vienna Circle Institute Yearbook 17,
DOI 10.1007/978-3-319-01899-7_19,

The criticism levelled against the logical empiricist's view of the unity of science program was that neither the social sciences nor the humanities were able to meet most of these requirements which were given a strong positivistic interpretation. The opponents pointed out that both the social sciences and the humanities shared an orientation towards idiographic descriptions. As a result Carl Hempel's covering law model of explanation was useless as an account of explanation in history, psychology, sociology or anthropology. In fact, humanistic disciplines were more interested in understanding than explanation. Also the critics emphasized that research within the social and humanistic fields was not value-free, and the humanities at least were interested in subjectivity as much as objectivity. Moreover, they stressed that our understanding of social and humanistic phenomena depends on the cultural and historical context of our inquiry.

In more recent years philosophers of science have turned away from the unity of science program and argued for some sort of scientific pluralism. The first to do this was Paul Feyerabend who was more radical in his view than anybody else. Today we find John Dupré, Ronald Giere, Nancy Cartwright, Hellen Longino and many others embracing forms of disunity. Increasingly more and more philosophers turn their back to any type of scientific monism.

Here I shall nevertheless argue for a methodological unity of all the sciences along the lines of the logical empiricism. I want to point to some mistakes made in the reception but claim that these mistakes do not falsify the main idea.

The ontological issue

To be or not to be a pluralist – that's the ontological question. In 1958 Putnam and Oppenheim put forward a very captivating argument for being a monist.[2] The cosmological argument, as we may call it, builds on two principles which seem to be confirmed by what we know about the universe today. We know that the visible Universe in its present form began 13.7 billion years ago with a Big Bang. What caused the Big Bang, nobody knows. But we know from the physical sciences that the visible Universe began as a very tiny and a very hot place and has expanded ever since. The physical sciences also tell us that there was no ordinary matter at the beginning due to the high temperature of the Universe, but eventually the expansion caused the temperature to drop, and some important symmetry breaks took place creating the known forces necessary for the creation of elementary particles and their interaction. As time goes by we get a long chain of new entities: first, within a couple of seconds after the Big Bang, there are quakes, glyons, leptons, neutrinos, vector bosons, etc. coming into existence. Then elementary

2 Hilary Putnam and Paul Oppenheim, "Unity of Science as a Working Hypothesis." In H. Feigl, M. Scriven and G. Maxwell (eds.) Minnesota Studies in the Philosophy of Science, Vol. II. Minneapolis: University of Minnesota Press, 1958, 3-36.

particles form a new level of atoms, and later atoms form the level of molecules. The first stars and galaxies appear, and after some billion years first cells and then organisms come into existence. Putnam and Oppenheim didn't know as much. In fact they knew very little about it because in 1958 the Big Bang model was not a well-established view of the origin of the Universe. Around this time the steady state model was just as viable as the Big Bang scenario.

The two metaphysical principles, which Putnam and Oppenheim set up, were called *the principle of evolution* and *the principle of ontogenesis*. The first one states that for every level of organization there exists at a certain time of the evolution only levels up to this particular level of organization and no higher levels. The second one says that for each level of organization there is a time of the evolution where this level did not exist but where all its components existed. From those two principles they conclude that everything at a certain level is ontologically dependent on the level below. Therefore every physical object and every property of this object consist of the object's parts and the properties of these parts including the relation between these parts. This claim may be called *ontological physicalism*.

I take these two principles to be unproblematic. There is only one world, and it has developed as described by them according to modern cosmology. In this metaphysical scenario there is also room for seeing human culture as part of nature. Culture is just one of nature's many forms of manifestations. Human beings are products of the biological evolution and the brain-mind relationship has developed as a reaction and an adaption to climatic changes in our forebears' environment. The rise of the human brain with cognitive elements like beliefs, feelings, perceptions and intentions opened up for representing, planning and acting towards a goal. The appearance of language was a step towards the formation of institutional intentions which is a condition for the formation of complex societies and civilized cultures. Therefore human societies and cultures are ontologically dependent on human brains and minds. But this ontological dependency does not provide us with grounds for ontological reduction. The two principles together neither imply ontological reduction nor non-reduction. To act as a reductive argument Putnam and Oppenheim had to operate with a tacit third principle which they never argued for.

This third metaphysical principle might be called *the principle of reduction*: Each level of organization is completely determined by the organization of the level below unless it is the lowest level of organization. The principle comes in two variants: i) every level of organization is identical to simple composition or aggregation of its parts (identity reduction); or ii) every level of organization is completely caused by/correlated with its components and the relations between these components (bridge law reduction).[3] It is only by adding the principle of

3 The two variants of reduction mentioned here correspond with two versions of emergence. One goes back to C.D. Broad who argued that emergence is a synchronic, non-causal covariation of a novel property and its lower-level basis; the other goes even further back to J.S. Mill who distinguished between homopathetic and heteropathic

reduction to the two other principles that one can derive a conclusion in support of *reductive physicalism*.

I do not have space to enter into a detailed discussion of the principle of reduction, but want to point to the fact that the existence of this general principle has not been demonstrated in any satisfactory way. There is strong evidence that the principle of evolution and the principle of evolution hold, but the same cannot be said about the principle of reduction.[4] Even if we look at the various sciences, they yield only little evidence confirming its existence as a general principle. On the contrary, we conceptualize different levels rather differently; and the same level may be described from different perspectives. Consider the following example: A Chinese vase can be described in various ways depending on our cognitive interests. It can be seen as a physical object with a certain volume, mass and position in space and time; it can be understood as a chemical object which consists of a certain type of ware and certain types of glazing; it can be regarded as an economical object which has a certain price if traded; it may be taken as an art object with aesthetic and symbolic values; it may be deem as a historical object with a provenance and an ethnological history; and finally we may look at it as a crafted object to fulfil a certain function, namely to be a container of water for bouquets. From an ontological point of view there is no essential description of such an object to which we may refer all other descriptions. A vase is indeed the product of an intentional action made with a function in mind. Such characterization cannot replace the physical one, but neither can a physical characterization substitute an intentional description. This holds in spite of the fact that human intentions, in my opinion, are identical to certain patterns of brain states. Even though I have little room to argue for it, I maintain that mental states are ultimately reducible to physical-biological-behavioural states but that cultural phenomena like language, social rules, norms, and values supervene on these states.

Our cognitive capacity of having intentions has developed through a biological selection and adaptation of our ancestors' brain. These intentions correspond to certain structures of our brain states. But intentions are directed towards something, and they distinguish themselves from each other in virtue of what they are directed towards. So they are ontologically dependent on something outside the brain. Also their function cannot be completely specified merely as a structure of a brain state. One needs to know, even if each type of intentions has its own particular brain structure, the kind of subject that is the object for a certain type of intentions. One needs to know whether this or that brain pattern corresponds to the intention of drinking or eating or mating, etc. if neuroscientists are to be able to

---

laws of causation while arguing that emergence was an effect of heteropathic causation.

4 Some may accuse me for not making a distinction between ontological reduction and epistemological reduction but this is not my intent. I only want to claim that without any empirical evidence supporting a metaphysical principle it should be discarded as a fiction of the mind.

recognize the meaning of different patterns and define their functional role. As is well known, the reductionist thereby faces an explanatory as well as a predictive gap.[5]

Many intentions have a functional role that can only be understood in terms of social norms and rules. Some intentions are individual and as such they are biologically selected. My intentions of finding something to drink, eat, and mate with have a biological origin whereas many, if not most, intentions are *institutional* intentions, and if you like culturally selected. Those intentions are socially constructed in the sense that they have been formed by our interaction with the culture and the society in which we live. Institutional intentions are indeed also identical to certain brain patterns, but the existence of these intentions is as much ontologically dependent on culture and society as ontologically dependent on the individual brain. Take this example: me stopping at a red light. This is a shared, institutional intention. When I see the red light I get an intention of stopping at a red light; the rise of this intention is determined by my particular perception of red light and so is the corresponding brain pattern. But that I respond to the red light by wanting to halt is a type of intention which is socially and culturally induced and reinforced in me to be the right intention to have in this particular situation. In this case, it simply depends on whether I understand intentions as particulars or as types since I am not able to tell that a particular intention is the right type of intention unless I put it into the social and cultural context. If this is true, we are unable to say about the rise of a particular intention that it is entirely determined by a certain type of brain pattern because this brain state pattern does not tell us that it is the right type of intention.

As a consequence, metaphysical notions like 'emergence' and 'supervenience' have been brought in to replace the notion of reduction. Hence *nonreductive physicalism*. We may put this in terms of a principle. *The principle of emergence* says: At each level of organization there are genuinely novel entities and novel properties that are not identical to their constituents and the relations between these constituents (the strong version); or at each level of organization there are genuinely novel entities and properties which are not causally determined by entities and properties at a lower level (the weaker version). So each level of organization constitutes a different domain of existence. An anti-reductivist view of this kind is usually associated with the idea that the various sciences are autonomous, each having their own distinctive methodologies and descriptive categories.

The principle of emergence is consistent with the two other metaphysical principles, and in combination they express a view saying that at a higher level of organization new entities or new properties may come into existence not only because of what exists at a lower level. This gives rise to what we may call *domain plurality*. It means that the one and same entity may belong to different ontological

5 See, for instance, Jaegwon Kim, The Philosophy of Mind, Cambridge MA: Westview Press, p. 2006, p. 295.

domains because it acquires new features at each level of organization. An entity that belongs to more than one ontological domain demands disparate concepts and theories to be understood and explained with respect to the different domains.

Now the pressing question is: do we need to be reductive physicalists to be able to defend the unity of science? The answer is that in fact plurality does not ruin the unity of the empirical practise of physics. So why should domain plurality violate the unity of science? As we shall see, one can uphold nonreductive physicalism and still argue for a unity of science.

The epistemological issue

The logical empiricists were no friends of metaphysics, but their physicalistic view had only obvious metaphysical implications together with a reductionist stand which they did not embrace. Nonetheless, they were strong advocates of the unity of science in the sense that scientific claims should be intersubjectively evaluable. Hence, if physicalism is required to be the descriptive basis of history, psychology, and sociology, it meant that some statements within any of these fields should be expressible in terms of the physical sciences including anatomy, physiology, and neurology. However, the logical empiricists did not hold that every statement of these disciplines is translatable into a purely physicalistic language. Had they thought so, the implication would then be that in the world of science every expression about social and human phenomena, like 'social mobility' or 'human dignity', would not be a meaningful expression unless we could make such a translation. The logical empiricists abstained from arguing that each statement of science could be reconstructed by means of the same set of basic concepts. But if terms like 'social mobility' and 'human dignity' should be intersubjectively understandable, they must, as Carnap said, "designate kinds of state with some behavioural symptom."[6] Indeed, the logical positivists were deeply aware of the situation that the unity of science program had not been carried through in any of these fields but they were confident that it would be possible to develop the human sciences successfully in this direction.

The motivation behind the logical positivists' defence of the unity of science was first and foremost epistemological: some scientific statements should be expressed by sentences that were empirically verifiable, that is sentences that referred to states of affairs that were observationally accessible. Hence, theoretical terms should somehow be translatable by correspondence rules into observational terms, and observational terms were equivalent to a physicalistic vocabulary. Only by using this language was it possible to express what was intersubjectively accessible and publicly testable by experience. It was this language, and only this

6 Rudolf Carnap. "Logical Foundations of the Unity of Science", 1938, p. 58. In Otto Neurath et al. (1938): 42-62.

language, which gave us cognitively meaningful expressions. So the requirement that the physical language was the language of science implied that a theory of a particular domain of phenomena had to be constructed so its terms were linguistically connectable to observational, physicalistic terms. There might be differences in the way these theories were formulated. But for the positivists it was invariable that the terms of any scientific theory could be construable in physicalistic terms in order for the theory to be verifiable.

So the logical empiricists opposed *semantic* and *epistemic* pluralism: There is one language of science which has a privileged status and this is the physicalistic language which we very easily can determine whether is true or not. The method by which this is done is the use of inductive inference. I think that this insight of the positivists is a lasting one as long as we do not overemphasize the demand of translation of a theoretical vocabulary into an observational one by claiming that the semantic content of theoretical terms can be fully specified by the meaning of observational terms. I think it makes sense to hold that any vocabulary of science must include terms which as part of their meaning contain empirical criteria for their use. This is how the empirical sciences can make predictions and explanations which can be empirically tested. But besides that I don't think that the unity of science is defendable based on the existence of a single language of science.

Domain plurality is the ontological version of scientific pluralism whereas most adherents of scientific pluralism subscribe to an epistemic version thereof by claiming that some natural phenomena at a given level cannot be fully explained by a single model or fully investigated by a single approach. This we may call *account plurality*. It means that some entities can be represented differently within the same domain by incompatible models and theories presumably because they do not have essential properties. A classic example of account plurality is the competition between the wave theory and the corpuscular theory of light. The basic linguistic rules of wave theory are $\lambda = vT$, $f = 1/T$, $\omega = 2\pi f = 2\pi/T$, $k = 2\pi/\lambda$, where λ is the wavelength, v is the phase velocity, T *is* the period of oscillation, f is the frequency, ω is the angular frequency, and k the wave number. Using these definitions it is possible to formulate more complex expressions like the wave equation of motion. The linguistic rules of the corpuscle theory are Newton's so-called laws of motions. Since the days of Newton and Huygens these two theories have struggled to give the most accurate description of light when applied on models of transmission, absorption, reflection, refraction, diffraction, interference, dispersion, and polarization. Each theory can be used to describe and explain certain observations, and neither theory can be said to be a more accurate representation than the other. What one chooses depends on the problem one wants to solve.

Alas, quantum mechanics incorporates both pictures in the wave-particle duality by Einstein's 'particle' expression $K_{max} = h(f - f_0)$, where K_{max} is the maximal kinetic energy of the ejected electron and f and f_0 are the frequency of the incoming photon and the frequency need to knock out an electron, and de Broglie's 'wave' expression $\lambda = h/p$. But the fact is that even though these two forms of thinking of physical phenomena are formally united in quantum mechanics they constitute a duality in the sense that particle-properties and wave-properties cannot be experimentally determined at the same time. They are closely related to different experimental settings that exclude one another. Schrödinger's wave function does not help us here because it does not represent any material wave but a probability amplitude. Rather we do not seem to have anything but perspectival grasp of the wave-particle duality.

The view that scientific explanations would have the same logical structure also became part of the tenets of the unified science program. Explanations of science should be shaped in a logical form in which they were independent of contextual and pragmatic aspects. To meet these conditions Hempel developed his well-known covering law model in which the explanandum should be logically derived from the explanans containing some law statements and statements about initial conditions. Hempel's model is consistent with causal explanations as long as a particular causal relationship is regarded as an instance of a universal law, but the model rules out singular causal explanations which may not be supported by such laws. However, more to the point, it excludes particular types of explanation like functional and intentional explanations to be genuine scientific types of explanation since these forms of explanation are contextual and not universally valid. Hence the aim of the unified science program was to attain genuine scientific explanations by finding generalizations within history, psychology, and sociology that were empirically testable.

Little of this program has come through. I think there are many reasons for this, but I shall only mention a few. First we know that the concept of law is a particularly ambiguous notion. Most laws may be *ceteris paribus* laws. Likewise we have the question concerning how many qualifying clauses can be added within a universal statement (that is within the scope of a universal quantifier) before the expression comes too specific to be considered a law. Can we set up as many strictures as we want and thereby narrowing down the extension of the statement to contain only one instance? These are important problems which Hempel and his supporters never solved. Second, we cannot understand human beings, their thoughts, languages and actions as meaningful unless we attribute intentions to them. 'Intention' is, I believe, the basic naturalistic concept of the human sciences. This means that we cannot get rid of intentional explanations due to the fact that they cannot be completely rephrased in terms of causal explanations. One simply cannot gerrymander with intentional explanations so that they become causal explanations.

Let us look at a brief example. Typically, aesthetic disciplines like art, music, literature, and dance have a biological origin. Music, for instance, comes in great many variations and different forms such that one might think that all music is cognitively determined by personal, cultural or social taste. But if one could insolate some features common for all kinds of music, these would not build on people's history and culture but on biology. They can then be considered impersonal, non-cultural, and non-social. Scientists have found such features. Harmony is one. An octave is an interval between a musical pitch and another with half or the double of its frequency and most music operates with scales with five or seven notes in an octave in spite of the fact that the human ear is able to distinguish 240 frequencies within an octave. So why does most music build on musical scales containing only five or seven notes? The biological explanation of this feature is that the selected scale in an octave corresponds to the same natural overtone components as one find in the human voice. It is our acquaintance with the natural overtones of the human voice that reappears in the selection of a scale.

Granting an evolutionary explanation of some basic phenomena in music, it still leaves us with many contingent features of every musical composition or every performance in virtue of the usage of a particular scale, melody, rhythm, repetition, tempus, etc. These elements are results of cultural impact, historical tradition, and individual intention and are at least as important for our understanding and appreciation of any particular piece of music as any evolutionary explanation of music as such. Among other things it is such historical and contextual features which are the objects of humanistic studies, and it is because of the contingency of contextual phenomena that Hempel's covering law model or similar models were doomed to fail within the humanities. In order to understand the particularities of human beings and their makings we make use of functional, intentional, and interpretive explanations. Scientists and scholars choose between different kinds of explanation according to whether or not they achieve by it the right kind of understanding. Again what counts as the right kind of understanding depends on the scientist or scholar's cognitive interests. Explanations in the humanities, but also in the other empirical sciences, are deeply contextual.

But in spite of these contrasts there is a certain epistemological feature which one might say is universal for all sciences. They have an empirical foundation. It is true for the physical sciences, the social sciences, as well as the humanistic sciences. Whatever is claimed to be knowledge within these disciplines, it has to be justifiable by empirical evidence in one form or another. Any scientific belief or any scientific explanation has to be measured against sensual inspections and be revised if necessary according to our observations. The requirement is, however, that such inspections are done systematically and methodically.

The Methodological Issue

Does the possible failure of reductive physicalism mark the end of the positivists' dream of the unity of science? I don't think so. But the dream will not materialize in one single overarching approach as some scientists and philosophers may believe. In *Rethinking Science. A Philosophical Introduction to the Unity of Science* I argued that what makes physics, biology, history, and literature alike can be found at the level of methodology.[7] Much like the logical empiricists I argued that the human and social sciences should be considered as part of the empirical sciences by adopting some of the same empirical methods as the natural sciences. Whatever discipline we may look at, I think one can extract from the practice of this discipline some basic methodological principles for getting empirical knowledge. You may say that ordinary science is a set of practices aimed at building models and testing hypotheses. I take methods to be those physical or mental procedures that help scientists to produce data, set up hypotheses, and justify these hypotheses.

First, the goal of science is to give us a theoretical and systematic understanding of nature, society, and human beings. For this purpose each and every discipline has developed a particular language in terms of which such a theoretical and systematic approach can be expressed. Usually when we talk about scientific theories, we first and foremost have in mind a set of linguistic rules which explicitly define certain basic properties in terms of some other basic properties. Usually linguistic rules in the natural sciences are called laws of nature. But as linguistic expressions they do not have any descriptive content as such. It is only when they are applied to scientific models that scientists are able to describe something and produce scientific explanations. This holds, I believe, regardless of whether the focus is on the natural sciences, the social sciences, or the humanistic sciences.

Second, theoretical understanding builds on systematic interpretation of data, evidence, or other observational phenomena regardless of the field of investigation. Interpretation takes place when scientists are facing a representational problem such as how a certain unknown phenomenon should be represented or how a given representation should be understood. Every new and undefined phenomenon poses a representational problem of some kind, and it must be interpreted before the scientist can say anything further about it. However, interpretation is not just guesswork as Popper would have us believe; the act of interpretation is guided by analogies and comparisons to what is already known. An interpretation in the sense of constructing meaning consists of creating a conceptual grasp of the phenomenon by placing it in the taxonomy of the proper field (or even by changing the entire taxonomical system). Such a conceptual construction is an important element in any abductive inference where the terms of the hypothesis language are modified with respect to the data language along the dimension of extension and/or intension of selected terms. But sometimes modification is not enough and completely new concepts and definitions have to be introduced.

When the representational problems have been solved, the scientific practise enters another stage where the scientists attempt to formulate explanatory hypotheses. It is here, I think, that the unified understanding of science makes sense. Due to human beings' biological evolution there are common ways by which we process information that we receive through our senses. The particular beliefs we

7 Jan Faye. *Rethinking Science. A philosophical introduction to the unity of science*. Aldershot: Ashgate Publisher, 2002.

form from sensorial information about a certain phenomenon are generalized to experientially similar types of phenomena by the use of ampliative inferences such as simple induction. Afterwards these beliefs must be justified by further observation and actions (experiments). During the human evolution we have always had a need of checking our beliefs against nature because we are often fooled and form false beliefs, which can be very dangerous when these beliefs concern our survival. Hence, human beings cognitive capacity has developed by the methods of trial and error, i.e. the conceptual correlate to biological selection and adaption, and modern methodology of science is originally an abstraction and idealization of our ancient cognitive practise.

If this brief analysis is correct, it means that the natural sciences, the social sciences, and the humanistic sciences have something in common, namely the use of naturally inherited methods to formulate hypotheses and to check whether they are true or false. Apparently, the first philosopher who abstracted modern methodological rules from the rising scientific practice was Ibn al-Haytham, called Alhazen. This Arabic philosopher and scientist lived between 965–1039/40 and became famous for his treatise on optics, which was translated into Latin at the end of the twelfth and the beginning of the thirteenth century. Herein he indicated some methodological prescriptions of science which scientists should follow; procedures by which science could control and examine theoretical hypotheses and inductive based assumptions. Those were 1) observation, 2) formulation of a problem, 3) formulation of a hypothesis as an explanation of the problem, 4) testing the hypothesis with the help of experiments, 5) analysis of experimental data, 6) interpreting the results and making a conclusion, and finally 7) making these results public to secure criticism and discussion. These prescriptions have by now become some of the basic norms for any empirical science. In sciences where direct experimentation is not feasible, other methods have been developed to replace instrumental manipulation. In medicine doctors use placebo in controlled studies of the effects of a certain drug, and in history, for instance, scholars have for centuries developed the method of source criticism.

Some may therefore argue that this proves that different sciences use different methods. Certainly they do. But these different methods are less general than those that are common to all scientific research, and I take them to be more specific practico-instrumental methods. They guide our action as much as our reasoning. The general methods, which are common to every science, are certain ordinary rules of reasoning, namely the methods of induction, abduction, inference to the best explanation, as well as deduction and falsification, all grounded in our cognitive capacity of reasoning which has been essential for our survival. Hence we might distinguish between methods of *data collection* and the methods of *data treatment*. The methods of data collection vary from one science to another, and each of them is designed according to the topic on which the particular science focuses; while the methods of data treatment are abstracted from our general forms of reasoning regardless of the topics.

What counts as a topic is determined by how each discipline defines its research objects. The same object may constitute different topics for different disciplines, and the way it is conceived as a topic may either enforce certain actions towards it or have us abstain from making certain action. How scientists approach a particular topic is caused by which actions they believe give the best and most relevant information about the topic. A chimpanzee, for instance, can be considered as an animal having a particular metabolism under certain research circumstances and be considered as a social being with certain group behaviour under others. These distinct ways of conceptualizing a chimpanzee make us see the same object as giving rise to different research topics that require distinct methods of collecting data. In the first case it may be physiological data reached by experimental methods; in the second case it may be data gained by mere observation.

The general methods of data treatment are what I want to call topic independent; they can be used to reason about any topic that occupies our mind for the purpose of forming justified true beliefs. The data collection methods are more specialized in relation to the topic to be studied since they are tailored for the purpose of directing our actions of investigation. The practico-instrumental methods of data collection are designed according to the topic we want to deal with. So these specialized methods are strongly topic-dependent unlike the general methods.

Conclusion

In general, the logical empiricists were not reductionists; they merely required that scientific claims could be tested against empirical findings. For the logical empiricists "the unity of science is the unity of language of science."[8] Accordingly, the unity of science movement was not preoccupied with ontological questions (there were no such questions) but with questions concerning the logic of scientific language. Today, however, we are not suspicious about ontological issues since scientific research partly provides us with ontological answers. But so does metaphysics. Both science and metaphysics give us different solutions to different ontological problems. The various sciences determine the ontological constituents of their object of study, whereas the metaphysics of science is realism or antirealism in either a reductive or a non-reductive form.

Nonetheless, the logical empiricists had some enduring intuitions when they argued in favour of the unity of science. They believed that every cognitive activity worth calling a science must share some of the same cognitive features. These intuitions are carried over into the naturalistic approach to human knowledge. There is not a principled epistemic division between the natural sciences, the social sciences, and the humanistic sciences. The positivists argued that our language of science had to be physicalistic because this was the only language in

8 Allan W. Richardson, *Op.cit.*

terms of which we could intersubjectively decide whether scientific statements are in accordance with observational facts. However, there is no reason to suggest that the unity of science requires one particular language as long as the languages of the various sciences can be used to express empirical accessible phenomena.

So instead of saying that the unity of science is the unity of the language of science, I will say that the unity of science is the unity of methodology. This claim can be strengthened by adding that human beings as biologically evolved creatures have acquired certain methods of dealing with sensory inputs according to our cognitive needs. These are methods of induction, abduction, inference to the best explanation, etc. which are used in all sciences. Acknowledging this, I believe, promises the unity of science a bright future.

Faculty of Humanities
University of Copenhagen
Njalsgade 80
DK-2300 Copenhagen S
faye@hum.ku.dk

Gereon Wolters

Is There a European Philosophy Science? A Wake-up Call[1]

I. Introduction

The short answer to this question is a firm and unambiguous "yes and no". The long answer will take the whole talk. Indeed, it could easily take an entire book. It is therefore unavoidable to take recourse here and there to simplifying shortcuts and polemical exaggerations, in order to get the message clear.

Let us go back to the short answer. *Yes*, there *does exist* a European philosophy of science. The Viennese meeting and generally our networking program (see fn 1), and all those participating European philosophers of science are ample proof of it. – *No*, there does not seem to be such a thing as European philosophy of science, because there are no genuinely European contributions to philosophy of science, or they are below the international perception threshold.

II. "Globalization": its Impact on the *History* of Philosophy of Science

"Globalization" is a very complex phenomenon that relates to the global flow of goods, services, ideas and people. In almost all European countries we can observe how former economic strongholds have almost completely dissolved in the course of a few decades, following the capital maximizing logic of the markets. At the same time new forms of production or services develop. There are winners and losers of economic globalization. It seems almost certain that in the near future Europe will be among the losers.

What holds for goods and services holds also for ideas. The winners on the market of philosophical ideas receive global professional recognition and sometimes even fame, but also get more mundane goods as invitations to congresses, talks at foreign universities and good positioning on the job market. – I dare say that being a loser on the market of ideas for Europe is not any more the writing

1 The paper is based on a talk given in Vienna, in December 2011, at a joint conference of the "Vienna Circle Institute" and the "European Science Foundation Research Networking Programme *The Philosophy of Science in a European Perspective* (PSE)". The topic of the conference was: *Philosophy of Science in Europe – European Philosophy of Science and the Viennese Heritage*. – The author would like to thank all people who have commented on the paper.

M.C. Galavotti et al. (eds.), *European Philosophy of Science – Philosophy of Science in Europe and the Viennese Heritage*, Vienna Circle Institute Yearbook 17,
DOI 10.1007/978-3-319-01899-7_20, © Springer International Publishing Switzerland 2014

on the wall that may become reality in some distant future; rather, we are losers already. In order to substantiate this thesis that to some might seem a bit exaggerated and even bold, we have above all to consider that ideas do not flow around the globe in a quasi Platonic, disembodied form. They come dressed up in languages. We all know that among the around 6500 languages of the world there is a chosen one. To have it, i.e. to have a *lingua franca*, is a good thing I hasten to add. English as the universal means of communication is of irreplaceable help and enormous importance in international exchange in all fields and on all levels. It can hardly be replaced by any other language or means of communication.

As everything, English as *lingua franca* comes at a price, which is, however, almost exclusively paid by non-native English speakers (NoNES[2]).[3] In the rest of my talk that deals only with *European* non-native English speakers of English the acronym NoNES is restricted to Europeans.[4] "NES" is an acronym for native English speakers and all those, who work at universities in Anglophone countries, whereas "RoW" refers to the rest of the world outside Europe where English is not the first language.

The price that NoNES have to pay for the wrapping of their ideas in English is manifold: *First* of all they do not *naturally* have the wrapping paper, i.e. the English language at their disposal. This means that they have to spend time and money in order to learn English, and then spend more money, in order to have their work translated or corrected by a native speaker.[5] One of the many scarcities of non-Anglophone European Universities is that they do not offer linguistic services. A laudable exception is the Language Services of the University of Helsinki. There might be others, hopefully. I do not know, for example, of any German University that offers anything comparable with "Helsinki" to its faculty, although the average command of English among German academics seems to me clearly inferior to their Finnish colleagues. *Second*, even if NoNES are in a position to somehow wrap their ideas into English, such packing looks in most cases rather poor compared to the original. *Third*, this makes it almost certain that NoNES play a rather secondary role on the market of ideas. *Fourth*, as it sometimes also happens with material goods, the wrapping has an impact on the content. We know that some plastic containers are not suitable for food and drinks, because some containers

2 I take this abbreviation from Clavero 2010, 552. I like pronouncing it "nones".

3 On further reflection, also monolingual Anglophones pay an – intellectual – price (perhaps without realizing it), as long as one finds correct Wittgenstein's connection of understanding languages and understanding forms of life.

4 Note that for reasons of simplicity also native speakers of English working at European universities outside Britain and Ireland are counted as NoNES.

5 A NES reading this paper may have noticed well before that I have saved the money for having it edited by a native speaker. – I gratefully acknowledge, however, that a RoW with excellent knowledge of English has done very much to render an earlier version of the paper more readable. Since then it has been spoiled by many changes and additions.

emit particles that affect the taste of the food and drink in question. Applied to ideas this means that English, or any other *lingua franca*, for that matter, might prevent some ideas from being optimally presented.

These four negative linguistic wrapping effects – I will later add three more and more important ones – are the smaller, the more formal, and the less culturally embedded a discipline is. It is certainly almost negligible in mathematics and formal logic. A bit less so it might be in physics and related "technical" disciplines.[6]

As we all know, modern philosophy of science has multiple origins. Its roots go as far back as into the 19th and early 20th centuries to people like Mach, Helmholtz, Poincaré, Duhem, Russell, Vailati, Twardowski[7] and others. The most important sources seem to be, however, the Berlin and Vienna Circles in the 1920s and 30s with a clear preponderance of the Viennese one. In Vienna Logical Empiricism originated, which has to be regarded as the nucleus of modern philosophy of science. In Vienna we find philosophical giants such as Rudolf Carnap, Carl Gustav Hempel (for a short time), Otto Neurath and Moritz Schlick. And in Vienna we are happy to celebrate these days the twentieth anniversary of the *Institute Vienna Circle* (IVC), ingeniously founded and most successfully guided ever since by Friedrich Stadler. The IVC`s work is dedicated to the preserving, propagating and advancing the precious intellectual heritage of Logical Empiricism.

What seems to us, or at least most of us, as a self-evident truth, namely that Logical Empiricism above all originated in Vienna, is completely ignored by others. Take e.g. the entry "Logical Positivism" in Robert Audi's much used *Cambridge Dictionary of Philosophy* of 1995, that entered unchanged into the second edition of 1999. The author is Richard A. Fumerton, F. Wendell Miller professor of philosophy at the University of Iowa. Here a quote from the first section (p. 514):

> Logical positivism, also called positivism, a philosophical movement inspired by empiricism and verificationism; it began in the 1920s and flourished for about twenty or thirty years. […] In some ways logical positivism can be seen as a natural outgrowth of radical or British empiricism and logical atomism. The driving force of positivism may well have been adherence to the verifiability criterion for the meaningfulness of cognitive statements.

---

6 Mahoney. 2000. identifies respective problems in astronomy. – On the website of the British Parliament (www.parliament.uk) one finds a paper on peer review in form of "written evidence submitted by the Academy of Social Sciences (PR 26)". Among the "weaknesses of peer review" (p. 3) one looks in vain for the bias against non-Anglophone publications. It seems remarkable, nonetheless, that at least the non-Anglophone world is taken notice of: "Outside the Anglophone world, peer review is less dominant, although practices are tending to converge to this norm" (p. 4).

7 One should add that the whole Lvov-Warsaw School, which was founded by Twardowski, should be taken into account here. Cf. e.g. Woleński. 1989, Coniglione/Poli/Woleński eds. 1993, and the very useful edition of texts in German (Pearce/Woleński eds.1988).

What are we supposed to learn from this?[8] Well, we learn that logical positivism, or, as I prefer to call it, logical empiricism,[9] has completely British roots. It is characterized as "a natural outgrowth" of radical or British empiricism and logical atomism. As everybody will agree "radical or British empiricism" is somehow essentially British, and for "logical atomism" the *Dictionary* unsurprisingly refers to "Russell" who is described (p. 699) as "British philosopher, logician, social reformer, and man of letters, one of the founders of analytical philosophy". So I think it to be no exaggeration to conclude that logical empiricism is characterized in the *Dictionary* as a philosophical movement with exclusively British or Anglo-Saxon roots. I should add that the rest of the article does not use any proper names or indexicals that could give a hint to possible historical influences that are *not* of a British origin.[10] An innocent reader would conclude that "logical positivism" is part of the British, and of no other, philosophical tradition. Well, I think almost everybody outside the Anglophone world knows better, and knows better in the greatest detail. Nonetheless, I am afraid that Fumerton's article might be an indicator of what "globalization" has in store for philosophy in general and philosophy of science in particular. The wrapping of ideas has changed their content. In the case at hand it is, as Wesley Salmon once pointed out to me, probably the Anglophone wrapping of Logical Empiricism in Alfred Ayer's *Language, Truth and Logic* (1936). Ayer's book has been for some people in the Anglophone world the only source of information about logical empiricism.[11]

III. Globalized Parochialism

There are, particularly in the U.S., many more examples that show a coarse grained perception and sometimes even a remarkable ignorance of rest of the world connected with a sort of almost natural disinterest. Here a recent exchange on the HOPOS mailing list: there was some guessing who taught "the first seminar" in the philosophy of science. On March 26, 2011 at 3:15 p.m. Sydney Axinn, professor emeritus of Temple University, now what is beautifully called "Courtesy Professor" at the University of South Florida, wrote the following: "What is reported as 'the first … seminar in the philosophy of science …' was given at the University of Pennsylvania by Edgar A. Singer, Jr. […] Among Prof. Singer's students was C. West Churchman, who later became editor of the Philosophy of Science journal.

8 I more or less literally quote this passage from Wolters. 2003, 109f.

9 For the history of labeling logical empiricism see Uebel (forthcoming).

10 In the respective article the verifiability criterion of meaning is also traced back to British empiricism.

11 I should, however, add that a reader who follows the links given underneath the entry "logical positivism" is lead among other things to Thomas Uebel's splendid "Vienna Circle". – If one has a look at the "Board of Editorial Advisers" of the *Dictionary* one counts 26 NES and 2 NoNES …

Less noteworthy students included the author of this note." – A quarter of an hour later that day Alan Richardson, an excellent scholar of 19th and 20th centuries philosophy, commented: "It should be noted that the issue that I [...] was attempting to sort through was philosophy of science **in the USA** [emphases are mine] – so even if we were to agree that Singer's course was the first **US** course with the title 'philosophy of science' this is not to say that there weren't courses given elsewhere with that name or a direct translation before then. After all, Mach was Professor of the Philosophy of the Inductive Sciences in Vienna from 1895." – This little exchange seems to be significant for a widespread attitude of self-satisfaction in the Anglophone world, particularly in the US. I mean a habit to spontaneously locate in their part of the world everything worth to be thought and spoken about. Only on further reflection may come to mind that this might not be the case. Such habitually narrow scope is usually called parochialism. In contrast to good old parochialism, which hardly ever crossed the borders of the respective parish, the new Anglophone variety can claim to be *globalized parochialism*.

To be sure, I do not see any malicious intent here in the sense that people consciously and explicitly might want to exclude the rest of the world. The rest of the world they simply take little notice of.[12] I am sure that things were certainly not better if Finnish, French or German or any other language had become *lingua franca*. In my view globalized Anglophone parochialism has its roots in a general human habit to watch and judge things through the lens of one's own culture.[13] This universal habit has been enforced in the case at hand by two more components. *First*, there is in my view no doubt that the Anglophone and in particular the US preponderance in the philosophy of science and in some other fields is well deserved. *Second*, and connected with the first, given the general economic and military and in their own view also cultural supremacy of the US, hardly any American thinks that there might be out there positive developments worth to take notice of. This seems the more natural approach to take since the knowledge of foreign languages is rather limited not only among NES in general but also in particular among NES philosophers of science. For presidential candidates of the Republican Party knowledge of foreign languages raises even the suspicion of lack of patriotism among many voters.[14] And those who know foreign languages hardly ever read works written in those languages. A quick look in any bibliography of articles in *Philosophy of Science* or in books by NES authors gives statisti-

12 In Britain this attitude has been elevated to government policy: in 2004 pupils at age 14 were allowed to drop foreign languages. As was to be expected the youngsters drop them in fact on a huge scale, and replace e.g. French by "religious education". For more information see: http://www.independent.co.uk/news/education/education-news/the-language-crisis-in-british-schools-2061211.html.

13 Note that also NoNES are often – and often rightly – accused of "eurocentrism" by the RoW.

14 Cf. McWorther (2012). – Suspects are the two former Mormon missionaries Huntsmen (Chinese) and Romney (French).

cally very significant proof of this. Some NES, especially among our US friends, have even a sort of *apriori* knowledge that nothing worth to be taken notice of can come from outside the Anglophone world. It resembles very much the story told in the Gospel of St. John (I, 45f.), when Jesus collected his first disciples:

> The day following Jesus would go forth into Galilee, and findeth Philip, and saith unto him: Follow me. Now Philip findeth Nathanael, and saith unto him: We have found him, of whom Moses in the law, and the prophets, did write, Jesus of Nazareth, the son of Joseph. And Nathanael said unto him: Can there any good thing come out of Nazareth?

With this we have reached the present and turn to the question of how Anglophone globalization affects theory.

IV. "Globalization": its Impact on Philosophical *Theory*

As hinted to in the last section, the Anglophone preponderance in the philosophy of science is well deserved, notwithstanding the globalized parochialism that we find with some of its representatives. To their eternal praise I would even like to add that what is visible of European philosophy of science these days at least on a *European* level owes much to particularly one American institution: the Pittsburgh *Center for Philosophy of Science* that in the 52 years of its existence has probably done more for European philosophy of science than any European institution. Only the thus far two European networks in the philosophy of science designed and chaired by Maria Carla Galavotti are of similar importance. Many of us got to know each other via the Pittsburgh philosophical relay station. So, whatever critical things I have said about *some* Anglophone philosophers does not diminish the professional admiration and respect for *many others*, and particularly for the *Center*.[15]

When the Vienna and Berlin Circles fled Nazism in Austria and Germany there began what one might call the "professionalization" of philosophy of science, i.e. its exclusive concentration on more or less conceptual and technical problems. This meant considerably narrowing its focus. "Vienna", and to a lesser degree also "Berlin" had been different. Logical empiricism was embedded in both places in a major enlightenment movement that addressed society at large. The foundation of the *Verein Ernst Mach* (1928) and the so called Vienna Circle

15 The *Center* was founded in 1960 by Adolf Grünbaum. – One of its later directors, Gerald J. Massey, started in 1990 a special collaboration with the University of Konstanz, which included a still ongoing archival collaboration that brought to the *Philosophisches Archiv* at Konstanz a clone of the treasures of the Pittsburgh *Archives for Scientific Philosophy*. The holdings at Konstanz are particularly intended for the use of European scholars. – Furthermore the collaboration included the "Pittsburgh-Konstanz Colloquia in the Philosophy of Science", which thanks to Peter Machamer's untiring and generous efforts went to eight editions (the last in 2008).

Manifesto of 1929 are ample proof of the fact that the early logical empiricists saw their professional activity in a broader societal context, and tried to present their research in public lectures accessible to those interested in society at large.[16] This approach got completely lost, when Carnap & Co had reached the US. Until very recently it does not seem to have regained strength. It occurs to me that this also holds to a certain degree for other Anglophone countries.[17]

Here is what Michael Dummett, an excellent analytic philosopher from Oxford, and himself in his fight against racism a notable counterexample, testifies for the British case:

> very few [in Britain] think that there's any call on them to be involved in any practical sense, and partly it's a tradition in this country, I must say, and not only amongst philosophers. Well, I was very impressed recently, a few years back. I and various other Italian philosophers, and other British ones as I recall it, we all published articles in an Italian daily newspaper on philosophy. Now, that's unthinkable in this country, absolutely unthinkable! [...] In France, and to a lesser extent in Italy, intellectuals generally and philosophers in particular are expected to make remarks on political and social questions.

Dummett points here to a remarkable difference in the role that philosophy or philosophy of science, for that matter, play in various non-Anglophone European countries, as e.g. in France, Italy, Finland or Germany. Let me say a word to the Finnish case. The eminent Finnish philosopher Georg Henrik von Wright (1916–2003) has not only written much admired books in English. Besides this as it were "professional" work he has published most interesting contributions to general philosophical and cultural questions addressed to the general public that in my view are inspired by his clear and rigorous philosophical thinking. As Ilkka Niiniluoto pointed out to me von Wright thus became "in the 1980s and 1990s [...] the leading intellectual in Finland and in Scandinavian countries".[18] One example is his denouncing in 1967 of the Vietnam War.[19] Only a part of this his work, collected in books in Swedish and Finnish, has been translated into English, or any other language a substantial number of people could read. Thus, in a sense, sadly, even von Wright himself seems to have internalized the more technical "professionalization" of philosophy of science and its separation from culture at large that happened to Logical Empiricism in the US.

Another example of embedding philosophy of science in general culture is the Constructivism of the so called Erlangen School in Germany, founded by the mathematician Paul Lorenzen (1915–1994) and the philosopher Wilhelm Kamlah

16 The papers of Thomas Mormann, Donata Romizzi and Günther Sandner (forthcoming in the *Vienna Circle Yearbook*, see reference to Uebel) dealt with this aspect of logical empiricism.

17 See Fara/Salles. 2006, 10.

18 Personal communication February 11, 2012.

19 I take this from a moving memoir of von Wright's Finnish student, Lars Hertzberg. It can be found via a link in the Finnish Wikipedia entry "Georg Henrik von Wright".

(1905–1976). Erlangen constructivism is a bold program, which attempts a sort of operational foundation of mathematics, physics, philosophy and also politics. Some texts were translated into English.[20] To the best of my knowledge they were completely ignored. The same holds for the further development towards the so called "culturalistic" approach that has been put forward by Lorenzen's former student Peter Janich (2006). Here the focus is on the cultural implications and presuppositions that figure prominently in science and philosophy.

A third example is historical epistemology. Historical epistemology goes back to French thinkers like Gaston Bachelard (1884–1962) and George Canguilhem (1904–1995). Thanks to the efforts of Hans Jörg Rheinberger historical epistemology has gained some influence on a *European* scale. But it would be an exaggeration to say that it has entered the globalized scene of international, i.e. Anglophone philosophy of science. I cannot recall, for example, any article on this topic in recent editions of *Philosophy of Science*.

From these three examples – other people probably could easily add more – we can take the following lesson: the price of globalization and so called professionalization, for that matter, would be for Europeans to give up the longstanding and in my view very important embedding of philosophy and philosophy of science in the cultural life of their countries. One may add this as the *fifth* negative linguistic wrapping effect to those four mentioned in section II. Renouncing the cultural embedding is both a disadvantage for the respective national cultures *and* for philosophy. With the financial crisis progressing tax-paying citizens and their politicians are tempted to ask themselves why they should pay for the ivory tower business of philosophers. It is perhaps no coincidence that philosophy seems most endangered in Britain these days, where at least one department of philosophy was shut down recently.[21] Others might follow or have already followed in the meantime.

Besides this broad cultural approach to philosophy of science that fell victim to the Anglophone globalization, there are more professional European approaches in philosophy of science that were practically ignored by globalized parochialism. To name just three of certainly many more examples: first the work of the very interesting Italian philosopher of science Giulio Preti (1911–1972), to whom we owe among other things a fascinating pragmatist embedding of philosophy of science. Unfortunately, not a line seems to have been translated into English.[22] Furthermore I would like to mention the Polish philosopher Leszek Nowak (1943–2009), who seems to have launched the contemporary debate on idealization and has greatly contributed to it. He is, nonetheless, rarely quoted, although a substantial part of his work is published in English: He just seems to have had the wrong address:

20 A fine overview in English of the various approaches one finds in Butts/Brown eds. 1989. It does not cover, however, the political dimension.

21 I mean philosophy at Keele.

22 Those, who read Italian may consult among other things the presentation of Preti's work given in Parrini/Scarantino eds. 2004.

University of Poznań.[23] A third example is the theory of truth approximation that has been developed by Theo Kuipers (2000), Ilkka Niiniluoto (1987) and others. Although there are publications in first rate Anglophone journals and although it is vividly discussed among *European* philosophers of science it has not been taken much notice of on the global level.

These and many possible other examples of genuinely European approaches that did not enter the globalized international scene show – as *sixth* negative wrapping effect – one very important thing: The *agenda* of what counts in philosophy of science is not set in Europe but rather in the Anglophone world, particularly in the US. Everything that does not fit to cultural habits and traditions in those parts of the world has little chance to surface.[24] A rift even *within* Anglophone philosophy was pointed out to me by James R. Brown (Toronto):

> Many [philosophical] topics are perfectly international, but many ethical/political issues are local. Canadian philosophers who work on Canadian issues (e.g. bioethics within a system of socialized medicine) cannot publish in the [American, G.W.] 'top' journals, so suffer for it.

This development is further intensified by the so called "Philosophical Gourmet Report", which claims to deliver a ranking that "primarily measures faculty quality and reputation".[25] The top scorers in the "Gourmet Report" practically set the agenda for the rest of the world.

Top scorers of any sort not only set the agenda, they are also in a position to enforce it by their strong influence on grants and tenure. Here is what primatologist Carel van Schaik, who spent most of his career in the U.S., wrote me about this. Van Schaik certainly thinks of experiences in his own field, but things are probably not entirely different in philosophy:[26]

> I think there is one important strategic factor why people behave in this parochial way: grants and tenure. I write for the people whom I know, and sometimes respect, but who will judge me for major decisions, be it acceptance of journal articles, awarding of grants or evaluations of tenure! This great guy in China that nobody has ever heard of will not affect my chances of getting tenure, etc. This problem can also be solved when the evaluation part (grants, articles, tenure) also becomes more fully international. And once everyone is fluent in the lingua franca, this is an achievable goal! And if Europeans (or Asians!) do excellent work, one day the self-appointed dominants and keepers of the flame will have to notice.

23 An excellent overview for the Italian reader both of Polish philosophy of science and of Nowak's work is given in Coniglione 2010. Francesco Coniglione also first pointed out to me Nowak's importance.

24 Jim Brown, personal communication. – I am also grateful to Jim for explaining to me the negative effects of the "Gourmet Report".

25 See: http://www.philosophicalgourmet.com/.

26 Personal communication February 3, 2012.

NES not only set the agenda but sometimes also determine the *methods*. Here is an example coming from the political scientist Andreas Bieler (Nottingham).[27] Bieler in much of his work follows a method that might be characterized as analytical narrative. This method, as distinguished from the nomological hypothesis/confirmation method, seems to be very suited for large areas of history and the social sciences. When recently he submitted a paper to the journal *International Organization* with an almost exclusively American editorial board he received the following answer:

> At IO board meetings in recent years, we have discussed this matter in some depth. We look for articles that state hypotheses drawn from theories of international relations, discuss the literature from which these theories and hypotheses are drawn, propose empirical tests of the hypotheses, and then present findings that advance the relevant theoretical debates. Your piece does not fit this model of an article.

This methodological inferiority complex has recently very fittingly been designated 'The Social Sciences Physics Envy'.[28] I suppose that comparable tendencies of methodological imperialism hold true for philosophy of science, even if they are not pronounced as clearly as in the case at hand.

Unfortunately, the Anglophone perspective on philosophy has been more and more internalized even in non-Anglophone Europe: Invited speakers at European conferences include almost always NES, often the majority is NES, and sometimes one looks in vain for at least *one* speaker of the NoNES.[29] In Anglophone countries, however, *not* having NoNES as invited speakers is rather the rule than the exception. Here is an example from the [Notizie Filosofiche] mailing list (March 26, 2012). For a conference at the "Institute of Philosophy" in London "in collaboration with the University of East Anglia and supported by the Mind Association" on the topic "Philosophical Insights" I count 5 invited speakers from Britain and 4 from North America.[30] This selection indicates that the organizers do not expect any "philosophical insights" from NoNES.

It is interesting, not to say depressing, to see how Europeans deepen the fundamental asymmetry by almost slavishly following the newest fashions that are proclaimed in the "leading" journals in the field. – The key word "journal" brings me to the concluding section of my talk, which deals with journals, thus talking about the *seventh* negative linguistic wrapping effect.

---

27 Personal communication February 3, 2012.

28 New York Times April 1, 2012 (Kevin A. Clarke, David M. Primo).

29 A telling example of this conference policy is what is described in a circular in English as the "First International Conference of the German Society for Philosophy of Science >Gesellschaft für Wissenschaftsphilosophie (GWP) e.V.<" that is going to take place in March 2013. At this founding event of a *German* Society for Philosophy of Science that avoids the word "German" in its very name, one finds seven "keynote speakers": 4 NES, 1 Greek, and, after all, 2 Germans.

30 See: philosophy.sas.ac.uk/d/f/Philosophical_Insights_21230612.pdf

V. On the Perception Threshold in "Globalized" Philosophy of Science

Let me begin with a truism: In order to be perceived beyond the national scale one has to publish in English. I have got the impression that this in the meantime also holds for being taken notice of at home. The *European Science Foundation* has developed the so called *European Reference Index for the Humanities* (ERIH), which has appeared in two editions, the second in 2011.[31] It contains sections for journals both in "philosophy" and in "philosophy of science", and groups the journals in three classes: (1) National; (2) International 1; (3) International 2.

"INT1" is defined as endowed "with high visibility and influence among researchers in the various research domains in different countries, regularly cited all over the world". – "INT2", in turn, has only "significant visibility and influence" and it lacks being "cited all over the world".

I am not going to criticize here the enterprise "Reference Index" as such. I would rather like to have a look at the criteria, which led the commission: "Any journal accepted in the ERIH list has to meet stringent benchmark standards: peer review of submissions, an active international editorial board [...], openness to new authors [...]."

I would like to concentrate on peer review and the "active international board" of the leading journals in our field. As far as peer review is concerned, there is a vivid international discussion also in other domains, not only in philosophy, that has led to the result that papers written by NoNES that come from institutions outside the NES world have a significantly lesser chance to be accepted. The reason for this is simple: the editors and peer reviewers are mostly NES, and mostly American.

I have done some empirical research with three journals in the course of 20 years and have checked editorial boards and reviewers for *Philosophy of Science*, *The British Journal for Philosophy of Science*, and *Biology and Philosophy* in the respective editions of 1990, 2000 and 2010. Although this does not warrant a quantitatively exact result, the figures I have found seem representative and are sobering. Although all three journals are ranked INT1 in the ERIH-Index, the editorial boards of *Philosophy of Science* and *Biology and Philosophy* show no great changes in their overwhelmingly NES composition during the last twenty years. The editorial board of *Philosophy of Science* consisted in 1990 of 40 peo-

31 Cf. http://www.esf.org/research-areas/humanities/erih-european-reference-index-for-the-humanities/erih-foreword.html. Here one finds links to the various lists. – It is also explicitly stated (p. 2) that the "difference between the categories [...] is not of quality but of kind." – This is, of course, naïve nonsense because a publication in INT1 is judged better by everybody as one in INT2. And because publications in English count more, NAT papers can be simply forgotten. Several people from various countries told me that this has already become the procedure of hiring committees in their respective countries.

ple, among them two NoNES; in 2000 we find 55 people, among them 4 NoNES and 1 RoW, while in 2010 of 44 members 2 are NoNES. The number of NoNES referees is also always well below ten percent. In 2010, for example we find 182 referees, among them 13 NoNES. Similar things as for *Philosophy of Science* hold for *Biology and Philosophy*, while at the *British Journal for Philosophy of Science* *all* members of the editorial board have been NES so far, while perhaps 17 of the 252 reviewers of the year 2010 have been NoNES.[32]

I think it justified to say that NoNES and even more RoW are marginalized in these three journals, ranked "INT1" in the ERIH-Index. I cannot detect much of "international editorial boards". Consequently there is a *structural unfairness* in these journals to aspiring non-Anglophone contributors, and certainly in other journals that I did not check as well.[33] An exception is *Erkenntnis*, a journal that has had German editors so far. The board of *Erkenntnis* has consisted predominantly of NoNES but has always contained a fair number of NES.[34] But, for what reason ever, *Erkenntnis* is ranked INT1 in the "Philosophy of Science"-section of the ERIH-Index, while it is only INT2 in "Philosophy".

If one now puts some hope on the new *European Journal for Philosophy of Science,* the journal of the newly founded *European Philosophy of Science Association* (EPSA), the disappointment gives way to embarrassment. For, here we find a clear case of *self-discrimination*: if I have counted correctly the editorial board consists of 31 people. A majority of 19 works at non-Anglophone European Universities. This does not sound bad, although the NoNES/Anglophone ratio 19/12 seems somewhat out of balance, because NES are still represented at more than a third.[35] However, of the 29 articles published in the first four issues of the

32 There is a certain uncertainty in these figures. Not all referees are known to me. I have then categorized those with a clearly English name as NES. The other names unknown to me I have checked on the internet. This little uncertainty does, however, not influence the general result.

33 Van Parijs 2002 and 2007 discusses the unfairness question. – I thank Werner Callebaut of the *Konrad Lorenz Institute for Evolution and Cognition Research* for directing my attention to the articles of van Parijs and Clavero.

34 In the last 20 years the percentage of NES on the editorial board of *Erkenntnis* oscillates between ca. 25 and 45 percent. One might speculate, whether the downgrading of *Erkenntnis* in "philosophy"-section of the ERIH-Index has possibly to do with the fact that there is no majority of NES on the board. So what the authors seem to interpret as "active international editorial board" might be missing in the case of *Erkenntnis* …

35 The mostly Anglophone countries Britain and Ireland have a population of some 66,8 Million, while France has a population of about one million less, but together with the partly Francophone countries France, Belgium, Switzerland and Luxemburg it comes to clearly more French than English speakers in Europe. Germany with its population of ca. 81,7 million together with Germanophone Austria (ca. 8, 4 million) and the Germanophone part of Switzerland (ca. 5 million) adds up to around 95 million Germanophones, which is the by far largest linguistic community in Europe. I have not mentioned yet other large linguistic communities like Italians (60,1 million), Spaniards (46 million), and many, many others, including such small ones like Estonians

European Journal for Philosophy of Science, all but 4 come from Anglophone countries. There is no time to discuss this remarkable result.[36] I would only like to say that, for whatever reason, the *European Journal for Philosophy of Science* has not yet made any tangible contribution to further the cause of non-Anglophone European philosophers of science.

In concluding I would like to state that there is not much of genuine non-Anglophone European philosophy of science that has crossed the perception threshold of "globalized" philosophy of science. I am confident that the PSE programs and other European projects, above all the newly founded *European Philosophy of Science Association* and hopefully also the *European Journal for Philosophy of Science* will contribute to improving this unsatisfactory situation.

VI. What Might Be Done?

The admitted Anglophone superiority in the humanities does in my view not justify the degree of actual dominance, nor does it allow that the agenda id almost exclusively set by NES. How might the NoNES change this situation, which I regard as undeserved and unsatisfactory?[37] My own suggestions include the following:

(1) We ought to try to learn English as early and as well as possible. This is not only a question of personal will and determination but also of public structures, particularly schools. The most important example of a counterproductive public structure is dubbing. Although I know that people prefer dubbed movies and tv series to originals with subtitles I regard it as a great stupidity that in major European countries like France, Germany, Italy and Spain films are dubbed because the costs of dubbing are easily repaid. Whoever had the opportunity to live for a while in a non-dubbing country can testify what I experienced in recent years in Holland and Finland, namely that almost all younger people, from the supermarket cashier to the student of philosophy speak pretty good English. So my recommendation

(1,1 million speakers). – The European Union alone has a population of around 502,5 million people; the Anglophone part of it is slightly above 13%. It goes further down, when we add those European countries that are not part of the EU (like Switzerland, Norway, Ukraine, Belarus, and others), while leaving out the controversial question, whether parts of Russia or Turkey are parts of Europe. Particularly Turkey with its lively scene in philosophy of science could be counted as "European" (cf. Irzik/Güzeldere eds., 2005, perhaps also Russia. – The result of our linguistic census is that Anglophones, while representing (depending on the meaning of "Europe") at most about 10 percent of the European population are more than three times as much represented on the editorial board of the *European Journal for Philosophy of Science*. (Figures from the English language Wikipedia). – On the ambiguities and difficulties of "Europe" the German reader may see the splendid book Osterhammel, 2009, 143ff.

36 I have been informally told that above all there is a severe lack of acceptable submissions of papers by NoNES. This might hopefully change in the future.

37 Some people, e.g. Van Parijs 2002, and Clavero 2010 speak even of "injustice" here.

is: ban dubbing![38] Nobody will doubt that only people with excellent command of English are in a position to submit articles to Anglophone journals. In addition, in discussions most of us give a rather clumsy impression compared to our NES friends. This is no wonder because we have to struggle not only with questions and problems but also with the language, in which they have to be answered.

(2) We should be aware that it is not sufficient to only *produce* thoughts; one also has to *sell* them. Apart from linguistic asymmetry NoNES are on the whole rather bad vendors. In other words, compared particularly to their American colleagues, Europeans are in a statistically very significant way just lousy in presenting their work. Primatologist Carel van Schaik observed the following:

> I have recently visited some philosophical meetings in Europe, and I noted that many European philosophers present their work in a most unattractive way. If at a busy meeting with numerous presentations, one has to work very hard to figure out what the speaker is trying to say, one simply gives up and instead focuses on the talks that are easy to follow! So, another response [to my paper, G.W.] would be to make sure continentals produce better presentations, and so have more impact![39]

(3) My third proposal regards the university system. Independent teaching and research starts in most European countries after many years, sometimes decades, of serving – in some cases even of slaving – as assistants or whatever. This is most unproductive because the best work is usually done by people under 50. Professors should not aim at creating intellectual clones of themselves but rather independent, critical spirits, as is the rule in the Anglophone system, and a precondition for scientific creativity. The structural prerequisite for this is, however, the introduction of the tenure track system as it has been in use in the U.S. for a long time. Given the internal corruption, nepotism and clientelism in the universities of some European countries it should be excluded that somebody starts a tenure track carrier in the university, where he/she received his/her PhD. There should be created an open job market as in the U.S.

(4) Compared to American top universities European universities are as a rule grossly underfinanced. Take Harvard, a university with 21.225 students.[40] The "total operating expenses" as documented in the "Harvard University Financial Report Fiscal Year 2010" (p. 18) are $ 3,729,582,000.[41] This equals as of today (31 March 2012) € 2.794.583.251.The budget 2011 for the 9 universities[42] and sev-

---

38 Cf. Van Parijs 2007, 226ff.

39 Carel van Schaik (Zurich), personal communication (March 20, 2012).

40 Cf. the English Wikipedia entry "Harvard University".

41 Available at: http://www.google.de/url?sa=t&rct=j&q=operating%20budget%20harvard%20universiy&source=web&cd=1&ved=0CC4QFjAA&url=http%3A%2F%2Fcdn.wds.harvard.edu%2Ffad%2F2010_full_fin_report.pdf&ei=cbp1T53hHdGOswbA3ODDAw&usg=AFQjCNFo3hpftBp2vcsPqosigtR3-hPiYA&cad=rja.

42 Among them are large and old institutions with costly medical schools like Freiburg, Heidelberg and Tübingen.

eral other academic institutions in the German *Land* Baden-Württemberg is about 150 million Euros smaller: € 2.636.302.000. The number of students in Baden-Württemberg was, however, as of November 2010 at 294.362. This means that Baden-Württemberg spends less than 9.000 Euros/student compared to 131.664 at Harvard. This means that Harvard spends around 15 times as much per student than the comparatively well off *Land* of Baden-Württemberg does. I am afraid that the figures in other European countries ("Oxbridge" possibly excluded) are not much better. Certainly, not all American Universities are as rich as Harvard, but it seems clear that the operating expenses of the top institutions exceed by far everything what we might find in Europe. – Given this financial imbalance compared to U.S. top institutions it is astonishing what poor European universities, in fact, do still accomplish.

(5) Publishing in English is only a necessary condition for crossing the global perception threshold. Active networking has to be added. These two necessary conditions are, unfortunately, not jointly sufficient. Disappointments are unavoidable. For those feeling disappointed it might serve as a consolation that also in pre-globalization times philosophical quality did not always suffice to be perceived and acknowledged. There have always been unrecognized geniuses ….

It seems to me that there is no separate path to avoid publishing in English. Ongoing resistance in major European countries like France and Germany is in my view counterproductive and doomed to fail.[43]

(6) NoNES should stay away as much as we can from seeing the philosophical world through the lenses of globalized parochialism. There are interesting and sometimes fascinating developments in our countries that do not surface globally. To notice them presupposes attentiveness to what is going on in other NoNES countries instead of staring like a rabbit caught in the headlights at the newest revelations arriving from the world of globalized parochialism. As remarked earlier, it is Maria Carla Galavotti's European "Networks" that have achieved more than any other European Institution to open our eyes for what our NoNES friends have achieved and are achieving. Therefore projects on a European level are very much desirable. If only the bureaucratic procedures with the *European Science Foundation* (here I have ample experience myself) and probably other European institutions were not as tiresome, unproductive and nerve-wracking as they are!

43 According to a letter in the HOPOS mailing list (January 13, 2012) there is being launched a new bilingual (articles in French and English) journal *Lato Sensu – Revue de la Société de philosophie des sciences*. I very much doubt that articles written in French in this journal will be read by NES. I just checked the first 2012 issue of *Philosophy of Science*. Of the many references in 9 articles *not one* relates to a publication in a language other than English. I think this finding is fairly representative, and I am afraid that it will hold *lato sensu* also for publications in *Lato Sensu*. – Apart from this it is interesting to note that in the name of the French society the word "French" is missing, as *mutatis mutandis* in their German counterpart,

References

Butts, Robert E., James Robert Brown, eds. 1989. *Constructivism and Science: Essays in Recent German Philosophy*. Dordrecht: Kluwer

Clavero, Miguel. 2010. "Awkward wording. Rephrase": linguistic injustice in ecological journals. *Trends in Ecology and Evolution* 25: 552-553

Coniglione, Francesco, Roberto Poli, and Jan Woleński, eds. 1993. *Polish Scientific Philosophy: the Lvov-Warsaw School*, Amsterdam: Rodopoi

Coniglione, Francesco. 2010. *Realtà e astrazione. Scuola Polacca ed epistemologia post-positivista*. Rome: Bonanno

Fara, Rudolf, Maurice Salles. 2006. An interview with Michael Dummett: from analytical philosophy to voting analysis and beyond, London: LSE Research Online. Available at http://eprints.lse.ac.uk/archive/00000552

Fumerton, Richard A. 1999. Logical Positivism. In *The Cambridge Dictionary of Philosophy*, ed. Robert Audi, 514-516, 2nd edition. Cambridge: Cambridge University Press

Irzik, Gürol, Güven Güzeldere, eds. 2005. *Turkish Studies in the History and Philosophy of Science*, Dordrecht: Springer

Janich, Peter. 2006. *Kultur und Methode: Philosophie in einer wissenschaftlich geprägten Welt*. Frankfurt: Suhrkamp

Kuipers, Theo. 2000. *From Instrumentalism to Constructive Realism: on Some Relations between Confirmation, Empirical Progress, and Truth Approximation*. Dordrecht: Kluwer

McWorther, J. 2012. When Does Speaking a Foreign Language Get a Candidate in Trouble? In *The New Republic*, January 23

Mahoney, T. J. 2001. The Problems of English as a Foreign Language in Professional Astronomy. In *Organizations and Strategies in Astronomy*, ed. A. Heck, 185-192. Dordrecht: Kluwer)

Niiniluoto, Ilkka. 1987. *Truthlikeness*. Dordrecht: Reidel

Osterhammel, Jürgen. 2009. *Die Verwandlung der Welt. Eine Geschichte des 19. Jahrhunderts*. München: C. H. Beck

Parrini, Paolo, Luca Maria Scarantino, eds. 2004. *Il pensiero filosofico di Giulio Preti*. Milano: Guerini

Pearce, David, Jan Woleński, eds. 1988. *Logischer Rationalismus: philosophische Schriften der Lemberg-Warschauer Schule*. Frankfurt: Athenäum

Szaniawski, Klemens, ed. 1989. *The Vienna Circle and the Lvov-Warsaw School*. Dordrecht: Kluwer

Uebel, Thomas. 2011. Name ist Schall und Rauch? On Naming a Revolutionary Philosophy. In *Philosophy of Science in Europe – European Phi-*

losophy of Science and the Viennese Heritage, eds. Maria Carla Galavotti, Elisabeth Nemeth, and Friedrich Stadler (in preparation). Dordrecht: Springer (Vienna Circle Institute Yearbook)

Van Parijs, Philippe. 2002. Linguistic Justice. *Politics Philosophy Economics* 1: 59-74

Van Parijs, Philippe. 2007. Europe's Linguistic Challenge. In *The Language Question in Europe and Diverse Societies*, eds. Dario Castiglione, Chris Longman, 215—251. Oxford: Hart

Woleński, Jan. 1989. *Logic and philosophy in the Lvov-Warsaw school*, Dordrecht: Kluwer

Wolters, Gereon. 2003. Carl Gustav Hempel – Pragmatic Empiricist. In *Analytical and Continental Aspects of Logical Empiricism,* eds. Paolo Parrini, Merrilee H. Salmon, and Wesley C. Salmon, 109-122. Pittsburgh: University of Pittsburgh Press

Universität Konstanz
Fachbereich Philosophie
Universitätsstraße 10
78457 Konstanz
Gereon.Wolters@uni-konstanz.de

Veronika Hofer and Michael Stöltzner

Vienna Circle Historiographies

It is a great pleasure to congratulate Fritz Stadler on the occasion of the 20th anniversary of the Institute Vienna Circle. Back then the IVC had started out as a local endeavor to bring home a movement that was admired world-wide – not least by those philosophers who had set out to overcome it –, yet ignored locally – deliberately or simply as the consequence of an intellectual milieu that was, more often than not, inward-looking and highly selective in its historical attitude. To develop the IVC into an international center of scientific exchange, of publications, of historical research and archival documentation, required not only to launch a large number of local and international activities, exhibit an impressive tenacity in matters of administration and lobbying, but also the capacity to attract and integrate people for whom the IVC represented a – roundly Neurathian – boat headed at profoundly changing the local intellectual environment at the crossroads of philosophy of science and Austrian social and intellectual history of the early 20th century, and to make them devote significant time and effort into a project whose success was initially anything but certain.

The two of us first met at the IVC in 1996, when it was still located in the Museumstrasse 5, in a large flat – or, in today's parlance, a coworking-space – where in the late 1940s and 1950s the circle around the former Vienna Circle members Viktor Kraft and Béla Juhos had convened, continuing the tradition of scientific philosophy in a city whose academic philosophy had become openly hostile to it. Now the flat harbored scholars reflecting upon various elements of Vienna's intellectual history. One of us (M.S.) had just finished his master thesis in mathematical physics under Walter Thirring and worked his way into philosophy of science, learning from the Viennese physicists turned philosophers, among them Ernst Mach, Ludwig Boltzmann and Philipp Frank. The other (V.H.) was finishing her Ph.D. thesis on Ludwig von Bertalanffy under Fritz's supervision. One of us was a philosopher who became increasingly interested in the history of science, the other a philosophically-informed historian of science. The Vienna Circle, Logical Empiricism, Red Vienna, the history of science in Austria became the topic of countless debates among us – including the different strategies how to unravel this complex of thematic and personal interconnections into either historical narratives about science and its philosophy or philosophical outlooks that live up to the present standards of historiography. We are still debating, and concluded that the most adequate birthday present to the IVC – owing to the empirical orientation of the Vienna Circle – would be a personal field guide into the riches of historiographies by and of Vienna Circle that reflects upon the embedding of the

M.C. Galavotti et al. (eds.), *European Philosophy of Science – Philosophy of Science in Europe and the Viennese Heritage*, Vienna Circle Institute Yearbook 17,
DOI 10.1007/978-3-319-01899-7_21, © Springer International Publishing Switzerland 2014

movement – and its historiographies – into the history of modern science and discuss in which way this variety of historiographies has evolved over the past 20 years.

Gone are the days of the received view of the Vienna Circle; the view's obituaries have been shelved and its specter emerges less and less within contemporary analyses. Rather than continuing to overcome it, scholars have begun to inquire how it could have at all emerged and prevailed for so long. Moreover, during the past ten years or so, the field of Vienna Circle historiography has not only widened but brought a series of interesting studies of this historiography – including its employment by the movement itself and by its later historiographers – and reflections upon how such a historiography sits between the history of philosophy (of science) and the history of science. For scholars are increasingly taking into account that most Vienna Circle members were part of the scientific community of the day – some of them with major accomplishments – and that they coordinated key activities of the movement with leading scientists. Hence the Vienna Circle's understanding of scientific philosophy was not exhausted by importing scientific methods into philosophy – or its potential successor in academia: the logic of science. They also strove to influence the ongoing philosophical-methodological discourse within the sciences themselves and the broader debates concerning science's place in modern society. A substantial part of this discourse was led by scientists interpreting the history of their field of expertise and, more general, the development of modern science.

A case in point is the famous polemics between Max Planck and Ernst Mach during the years 1908–1910 that would serve as a reference point for an entire generation of scientist-philosophers. The issue was not merely the dichotomy between positivism and realism – which Planck's former student Moritz Schlick (1932) set out to diffuse by the principle of verification that, on its part, would become one of the pillars of the received view –, but also the difference between Mach's historico-critical analyses and Planck's merely philosophical employment of the history of physical theory devised to justify his view that the structural principles of our present physical world view have survived all recent scientific revolutions. More generally, we notice – especially among the Vienna Circle's Austrian members – a remarkably close connection to the critical tradition of history of science that Mach spearheaded at his time according to which science and its formative principles were in a permanent state of change, rather than nothing but a logical structure to be reduced to some basic elements.

We will outline our observation that there has been a significant development in Vienna Circle historiography over the past 20 years in an exemplary fashion rather than by a broad overview; neither are we aspiring at a general methodological discourse about the relationship between history and philosophy of science. To elucidate our thesis, we suggest to distinguish four modes of mutual reflections between historization and philosophical analysis. These four modes are not neatly separated and they are configured in different ways; broader historical narratives or philosophical analyses will typically involve more than one mode. First, many standard methods of the history of science and the history of philosophy have been applied to the historiography of the Vienna Circle. This has led to a large inventory of analyses including studies of its lesser-known members. Second, historiographies thus developed have been employed to justify philosophical or broader epistemological agendas. Among them was the contrast between a genuinely Austrian philosophy and Kantian readings of the Vienna Circle, a dichotomy that has largely disappeared from today's historiography. Third, the history of such employments has, more recently, become the object of investigation: starting from the role of self-historiography in the public presentation of the Vienna Circle itself, the later constructions of the received view and its demise, until the rediscovery of the Circle's history. Fourth, differences thus diagnosed have opened up new philosophical perspectives both as regards philosophical agendas and the history of science. The important point here is that the historiography of science and of philosophy of science intersect even though – in contrast to the days of the Vienna Circle – the disciplines of history of science and philosophy of science are fully established such that their mutual relationship becomes a topic of methodological reflection.

In Section 1, we focus on the creation of the hybrid identity of scientist-philosopher or scientist-historian and its relation to the emerging disciplines of history of science and philosophy of science. Section 2 analyzes the pretty complex self-historiography of the Vienna Circle. Section 3 shows that the status of the history of science within the Vienna Circle underwent significant changes which affected its historiography. In Section 4, we look at some approaches we have seen during the Institute's first decade, which mainly involve the first two modes. In Section 5, we provide more recent examples pertaining to the other two modes.

1. Hybrid Identities and the Emergence of Three Academic Disciplines

The plurality of historiographical approaches adopted by, and in the analysis of, the Vienna Circle and the complex dynamics of their implementation and employment should not come as a surprise. The Vienna Circle and its agenda of a scientific world conception emerged from within the sciences of the day, most prominently among them relativity theory and the new logical foundations of mathematics. The Circle's members and their allies set out to apply scientific scrutiny to the reflections on science and knowledge in general, a task which traditionally had been the prerogative of philosophy either by providing methodological warrant of the results of science – or in the weaker form popular around 1900 (cf. Ziche 2008) – of systematizing the knowledge reached by the sciences and the humanities. In this latter context, the expression "*wissenschaftliche Weltauffassung*" was used for the first time by authors who distanced themselves from the idealistic "*Weltanschauung*". Yet the Vienna Circle focused more consistently than Wilhelm Wundt or the neo-Kantians on the empirical sciences and their established methods because these promised more advanced and formally precise means of epistemological analysis and interpreted the scientific world-conception as a social utopia in which the sciences stood at the center of human life (cf. the concluding passage of the Manifesto). Though some Vienna Circle members held that no proper object of study, and no method of its own, would remain for philosophy, perhaps apart from the logic of science, others felt that precisely this new logic promised a revitalizing turn of philosophy, above and beyond its intimate connection with the sciences.

It is important to note that a substantial part of these debates was conducted by invoking the history of the sciences and the history of philosophy. Frank (1932), for one, tirelessly stressed that there was nothing in the revolutionary development of 20th century physics that would force scientists to abandon the empiricism of Galileo and Newton, while Schlick (1930) turned these developments into a justification of the ultimate renewal of philosophy that he advocated. The historical perspective was even more evident in the case of other sciences, among them Neurath's early strides between the German historical and the Austrian school in economics, the Viennese efforts to save Mach's empiricist heritage by purging it of psychologism and his peculiar Spencerian brand of evolutionism, and the early interest for the life sciences, where the specter of a return of metaphysics appeared imminent (cf. Hofer 2012).

This combination of historical and philosophical arguments, as well as the activity of the Circle's members within the scientific community, put Vienna Circle historiography at the crossroads of the history of science, a historically informed philosophy of science, and the history of philosophy. While today each of them represents an academic discipline in its own rights and with a distinct methodology – including the respective controversies –, in those days only the third was acknowledged as such.

Only after the Second World War would history of science and philosophy of science meet all sociological criteria of a discipline, among them specifically designated chairs, journals, conferences, and a methodological canon employed in academic education. It is true, long before the days of the Vienna Circle eminent books in the history of science had been published – most importantly by the physicists Ernst Mach and Pierre Duhem. There were also a few journals publishing such work. Moreover, scientists' popular addresses to the whole university, an academy, or a learned society, and their encyclopedic writings (e.g. for *Die Kultur der Gegenwart,* The Culture of the Present, 1903–1925) were based on a historical presentation of the subject. But the professionalization of the history of science – in terms of the mentioned sociological criteria – was only accomplished with the (Belgian-born) George Sarton in the U.S. after 1945.

The foundation of *Erkenntnis* in 1930 and the new professorships for Reichenbach and Carnap were important steps to establish the discipline of philosophy of science, but the process was completed only after 1945 and in the U.S. It is true, the idea of a "*wissenschaftliche Philosophie*" had been around since Hermann von Helmholtz's famous speech "On Human Vision" of 1855, and from 1877 until 1916 there existed the *Vierteljahrsschrift für wissenschaftliche Philosophie (und Soziologie* – after 1902), edited among others by Richard Avenarius and Ernst Mach. But the German concept "*Wissenschaften*" embraces the humanities, and the mutual relationships of the single "*Wissenschaften*" and their relationships with philosophy were amply discussed in the *Vierteljahrsschrift* and beyond. Helmholtz's take was highly influential, but Wundt instead embraced all *Geisteswissenschaften,* or at least their empirical parts, into the systematization accomplished by "*wissenschaftliche Philosophie*".

After the separation into *Geistes-* and *Naturwissenschaften* had won the day, Helmholtz and – for the Austrians – Mach became influential role models for the scientist-philosophers of the decades after 1900. These had typically made their entire academic career in the single sciences, but amply published on philosophical topics, in the *Vierteljahrsschrift*, in Wilhelm Ostwald's *Annalen der Naturphilosophie* (1901–1921) and in Arnold Berliner's *Die Naturwissenschaften* (after 1913). Often, they assembled their respective papers and addresses in separate volumes (cf. Planck 1922). Among those scientist-philosophers are several Vienna Circle members and allies, the physicist Frank, the mathematician Hans Hahn, the applied mathematician-statistician Richard von Mises, but also some of their philosophical opponents, among them the biologist-vitalist Hans Driesch. Although Schlick had early on pursued a philosophical career, his 1904 PhD in physics under Max Planck and his role in the struggles about relativity theory secured him a preferred access to the German physics community. Notice that the scientist-philosopher represented not only a sociological role model that was more widespread in Germany than in other countries, but also a hybrid intellectual identity that persisted even after the discipline of philosophy of science has been established. It could also be amended to a scientist-philosopher-historian, a perfect realization of the Machian role model. This hybrid identity and its historical dynamics provide the background for assessing the role of history of science and historiography in general among the Vienna Circle members.

2. The Variety of Self-Historiographies

In the movement's self-descriptions there existed a surprising diversity of historiographical approaches. It started right at the beginning of what Stadler (1997) has called the Vienna Circle's 'public phase' and persisted in different forms throughout the movement's entire history. The Manifesto – published by the *Verein Ernst*

Mach (1929) to honor Schlick and presented at Prague during the annual meeting of the German-speaking physicists and mathematicians – embedded the scientific world conception and its modernist agenda into an Austrian tradition starting with the local heroes Bernard Bolzano and Mach, thus drawing an essentially continuist picture of an undercurrent that better integrated modern science into the general world view than Kantianism to which – as Frank put it – most German scientists had "some sentimental ties" (1949, 40).

Schlick's (1930) own assessment, published in the first issue of *Erkenntnis,* revealed an interesting dialectic. For he simultaneously criticized the perpetual replacement of one outdated philosophical system by another and claimed that the revolutionary changes in logic permitted us to once and for all exit this never-ending spiral and turn philosophy into an activity of structural and logical analysis. But there was another side to Schlick's intellectual identity. To quite a few German scientist-philosophers, among them David Hilbert and Max Born, Schlick remained the principal authority in the philosophy of physics.[1] Here, Wittgenstein was one philosophical inspiration among others.

Neurath's post-Manifesto historiography, now in the context of the projected *Encyclopedia of Unified Science* and part of his endeavor to internationalize the movement, still mentioned that the Austrian tradition "avoided the intermezzo of Kant" (1935, 12) by directly reconciling rationalist logic (Bolzano) and empiricist science (Mach), but he broadened the base of what counted as scientific philosophy. The new 'big tent' included not only American pragmatists, but also French and Italian (experimental or scientific) rationalists, some of whom were more positive about the Kantian tradition. After long-winded multilateral negotiations a programmatic compromise was reached that was put on the back cover of the early *Encyclopedia* booklets. (cf. Stöltzner 1998) All agreed "in considering the unity of science as the ideal aim of their efforts, in eliminating any form of speculation other than that recognized in science, in stressing the importance of logical analysis in various fields, and in taking into account the historical development of scientific concepts and regulative principles."

Reichenbach, who had emigrated to Turkey in 1933, did not participate in the *Encyclopedia.* To an increasing extent he publicly emphasized the differences between his own method of "analysis of science" and the "Viennese school" (1936, 144), which he simply equated to Wittgenstein and Carnap's *Aufbau* project. Reichenbach's quest for priority was also a struggle about historiography because he located the origin of his method in his study on relativity theory (1920) even though, after correspondence with Schlick, he had abandoned his core conception of a relativized a priori. (Cf. Friedman 1999) Moreover, he claimed that his work on probability represented the greatest progress on the problem of induction since Hume (cf. Stadler 2011, Stöltzner 2013). Having emigrated to the U.S. in 1938, Reichenbach became a most influential pioneer for the discipline of philosophy of

1 Cf. the correspondence between Schlick and Born.

science. In the popular *Rise of Scientific Philosophy* (1951), he outlined a longue durée narrative of the philosophical renewal prompted by 19th and early 20th century science from which the Vienna Circle had almost disappeared.

Neurath's untimely death in 1945 bereft the Vienna Circle not only of its organizational motor and its programmatic helmsman, who was constantly balancing the various groups and thematic strands of the larger movement, but also of its chief chronicler. Neurath not only orchestrated the *Encyclopedia* project with historical briefs of the various national traditions involved, but presented the project as a renewal of the French *Encyclopédie* and placed it historically between rationalism and empiricism (cf. Zilsel and de Santillana 1941). Additionally, Neurath enlisted the Danish philosopher Joergen Joergensen for a booklet on *The Development of Logical Empiricism* that appeared in 1951, the same year Reichenbach's *Rise* was published. Joergensen introduced his contribution as a continuation of Zilsel's history of empiricism and outlined the history of the movement until 1940. "What unites its members ... is not so much definite views or dogmas as definite tendencies and endeavors." (1970, 847) When describing the Vienna Circle, Joergensen quoted long passages from Herbert Feigl's 1943 presentation and from the Manifesto, devoted ample space to its prehistory, providing a ten page digest of the *Tractatus* followed by a précis of Carnap's *Aufbau*. As did Neurath's presentations during the 1930s, Joergensen used the *Encyclopedia* congresses to portray the expansion and integrative activities of logical empiricism. The list is basically complete, except for the Italian Federigo Enriques and the journal *Scientia*. Interestingly, Joergensen accepted Reichenbach's 1936 claim to a genuine method of analysis of science, but emphasized that both his probability-logic and his theory of induction have not "won the general assent of the adherents of logical empiricism." (1970, 898) In his broad assessment of the divergences emerging in the late 1920s inside the Vienna Circle, Joergensen devotes ample space to Neurath's and Carnap's criticism of Wittgenstein's concept of philosophy, Carnap's principle of tolerance, the protocol sentence debate, the quest for unity of science, and physicalism. Coming to 1940, Joergensen considered the semantics of Carnap and the Lwow-Warsaw school as the most promising developments. Readers of Joergensen, accordingly, should not have been led to the received view.

In a postscript written in 1949, Carnap's student Norman M. Martin gave a short summary of the developments after 1940, mainly listing congresses and publications, and emphasizing progresses in semantic, the analysis of science, and the philosophy of probability. Interestingly, Martin not only counts Reichenbach's and Hempel's works among 'analysis of science', but also Frank (1941) and Felix Kaufmann (1944). Reichenbach's claim to a distinctive method was thus silently rebuked. The remainder of the postscript mainly lists books, among C. I. Lewis (1946) and Charles L. Stevenson (1944), thus making surprisingly little distinction between the members of the movement, its American partners in the *Encyclopedia* project, and its allies in a looser sense. Since another two years passed until the volume went to the press, the editors added a further note that simply mentioned new writings of Frank and Carnap, Kraft's (1950) history, and the activities of Frank's *Institute for the Unity of Science*. Both postscripts provided no deeper analysis of the debate or a characterization of the movement apart from the emphasis of semantics. In the same vein, the back covers of *Encyclopedia* booklets became more and more neutral, perhaps owing to the project's success in the U.S. educational system.

For the joint publication of all *Encyclopedia* booklets in 1970, Feigl and Charles Morris compiled a bibliography of the growing "literature of the logico-methodological foundations and the history of the sciences." (1970, 949) A majority of the entries concerned books published after 1945 or previous writings of the movement, and classics that had shaped their discussions. In contrast to the Manifesto, this bibliography does not serve to construe a group identity.

It seems tempting to follow Gerald Holton's diagnosis that by the time of Frank's death in 1966 the Vienna Circle in exile had run its course and "died of its members' success" (1993, 41), insofar as it had become fully integrated into the North-American debates. Frank had been involved in Harvard's general education program and was a key member of a Boston area community of scientists and philosophers, among them the young Holton. Carnap was a core figure of analytic philosophy and philosophy of science, whose works, not least the technical ones, set the standard for the now fully established discipline of philosophy of science. And so did Feigl, a member of the younger generation, who after his emigration to the U.S. in 1930, was active in publicizing the movement in the New World (cf. Blumberg and Feigl (1931) on "Logical Positivism", and the above-mentioned Feigl (1943), now under the rubric of "Logical Empiricism"[2]). The goal was to build bridges, not to set the distinctive characteristics of the European tradition apart from the American scene.

From the perspective of current Vienna Circle historiography, it remains nevertheless surprising that – after they had established themselves in the U.S. in the 1950s – neither Carnap nor Frank ever came out against the emerging misrepresentations of the factual history of their discussions during the 1920s and 1930s, either to reassert Neurath's historiography or to amend it in a suitable fashion that reflected the philosophical progress and the changed intellectual context – a context in which the opponent was no longer Kantian metaphysics or the anti-scientific tendencies of the interwar period. It is true, Carnap and Frank arranged translations of some of their previous works, to the extent that they considered them still adequate. It is also true, that the criticism against metaphysics continued in a revised form (cf. Uebel 2011). Yet the introductions to Frank's philosophical essays (1941, 1949) and Carnap's (1963) autobiography for the Schilpp volume rather read like the recollections of natural scientists who tend to present the development of their own ideas purged from positions they had long abandoned

2 For the terminology, see Uebel (2013). As a matter of fact, the first overview of the new scientific philosophy was published by Kurt Grelling (1928), a member of Reichenbach's group.

unless such a change of mind appears important for contextualizing their present views or contributes to ongoing debates.[3] Both adopted a rather considered or even defensive style of historiography and were more interested in the current state of the art. This tone was substantially different from the revolutionary thrust of the Manifesto, Schlick's "Turn in Philosophy" (1930), or the call for a new *Encyclopedia*; but also from Reichenbach's (1951) historiographical posture that came close to a disciplinary Founders Myth narrated by a core figure of American philosophy of science.[4] Kraft's (1950) book was soon translated into English, but it did not have much of an impact in the U.S. George Reisch's (2005) investigations about the Vienna Circle during the Cold War indicate that during the oppressive McCarthy years Frank and Carnap had little incentive to set the record straight and carve out the modernist and leftist European roots of philosophy of science. Today we know that Carnap had actually written up a much broader depiction of his early thinking for his intellectual autobiography, but had to shorten his text.[5]

The movement's fading self-historiography and the increasingly stereotypical narratives by others, viz. the received view, were not without effects on the subsequent philosophical debates. But this was not mere negligence but also a consequence of the fact that history of science had undergone substantial changes as regards the role of philosophical concepts in science. The Neurathian self-historiography of the movement of the 1930s would not have met the standards of the new discipline. Moreover the role models of scientist-philosopher – or scientist-philosopher-historian in the Machian tradition – had largely come out of fashion or were transformed into public science – a fact which made Frank a well-respected, but unique character.

3. Changes in the Role of the History of Science

From the very beginning, Neurath sought to include the history of science into the *Encyclopedia,* but the designated author Enriques died shortly after Neurath. The remaining editors Carnap and Morris contacted George Sarton and I. Bernard Cohen who declined, then contemplated a booklet on "Sociology of Science" (by Louis Wirth) instead, until the assignment finally went to Thomas S. Kuhn (1962), thus eventually from the avowed continuist Enriques to the author of *The Structure of Scientific Revolutions*. In contrast to other volumes, Kuhn's booklet was substantially expanded and revised for the final edition of the *Encyclopedia.* It is not without irony that, in the 1970s, *Structure* was commonly read as the foundational document of a new epistemology that overcame logical empiricism and its heirs,

3 Cf. the footnote about nominalism and realism in Carnap (1950).

4 For the role of Founder Myths in the history of scientific disciplines, cf. Bensaude-Vincent (1983).

5 Carus (2007) uses this original version to portray the early Carnap against the backdrop of the late German *Bildungsbürgertum*.

among them – surprisingly from the standpoint of Vienna Circle historiography – Kuhn's actual combatants Karl Popper and Imre Lakatos. Moreover, the facts that (a) we find the continuity-discontinuity dichotomy already behind the Mach-Planck debate – Planck held that abstract principles typically survive the revolution –, that (b) Carnap had endorsed Kuhn's book (Reisch 1991), and that (c) Kuhn actually developed the paradigm concept out of his analysis of scientific dogma (Reisch 2012), make one wonder whether the Vienna Circle had a shared and coherent understanding of the history of science and what role, if any, the history of science played within the post-1945 scientific world conception. This is not irrelevant for a movement for which philosophy had to be part and parcel of science itself.

Already the early Neurath worked on the history of optics – or rather on ways how to better write its historiography combining Mach and modern logic – and that Zilsel provided a detailed analysis of the social origins of modern science (cf. Neurath 1915, 1916; Zilsel 1942). Moreover, Neurath used his historiographical skills to establish and justify the identity of the movement, both in the form of an Austrian philosophical tradition and a narrative reaching back to the Enlightenment. But after Neurath's and Zilsel's untimely death little was left from this aspect of the Machian heritage.[6]

In the bibliography by Feigl and Morris, the section on "History of Science" seems to be pretty wavering, given that the discipline had meanwhile been fully established. For they not only assembled books on the history of science, but also on the history of philosophy of science and on the movement of logical empiricism. As for the former, one finds, among others, books by George Sarton, Herbert Butterfield, Max Jammer, Abraham Wolf, Thomas S. Kuhn, and – to list authors outside the Anglo-American hemisphere – Claude Bernard, Eduard Jan Dijksterhuis, Alexandre Koyré, and Ettore Carruccio, a former assistant of Enriques. Enriques himself appears only with his *Problems of Science* (1914), but in the section "General Philosophy of Science", and not with his writings on the history of mathematics and logic. Mach and Duhem's classical historical writings appear under "Foundations of Physics", so does Holton's (1958). On the other hand, we find the *Encyclopedia* entries of Joergensen, Santillana and Zilsel, and Kuhn, the Manifesto, Neurath (1935) and Kraft's Vienna Circle book (1950). Neurath's and Zilsel's writings on the history of science however are not listed, perhaps because they were not monographs. Similarly, Holton's historical papers are missing, probably because they had not yet been assembled into his *Thematic Origins* (1973). From the perspective of today's history of science, the absence of Canguilhem and Bachelard is striking.

The bibliography also failed to list Frank's works among the contributions to the history of science. Despite the involvement of many scientists in the activities of the Harvard Institute for the Unity of Science, Frank kept a comparably low pro-

6 For its prospects within contemporary history of science, cf. Section 5.

file outside the Boston area. But in his post-war writings, he effectively developed the role model of scientist-philosopher into a broader hybrid identity in which both the history of science and the emerging sociology of science found a place. Frank's open-minded discussion of "The Variety of Reasons for the Acceptance of Scientific Theories" (1956) appears outside the forthcoming frontlines of the philosophical discourse about the rationality of the history of science. This did not favor its later reception and rediscovery by Vienna Circle historiography. A consequence of this neglect was that the relationship between the history and the philosophy of science has mainly been discussed in terms of the debates between Kuhn and Feyerabend, on the one side, and Popper and Lakatos, on the other – a setting that became increasingly unattractive for historians of science.

Frank's emphasis of the open character of the historical development of science was forgotten in the face of the battles between Kuhnian incommensurability and Lakatosian rational reconstruction – battles in which the received view associated the Vienna Circle with a non-historical, logic-centered brand of rational reconstruction. An important methodological point is that Frank – and to a certain extent Holton as well – developed his explanations and criticisms of the various aspects of the history of science without reference to a general explanatory framework. Following Max Weber, Frank did not only reject the distinction between understanding and causal explanation, but he also followed Mach in admitting a plurality of research methods and justifications within a broader empiricist attitude. Today's historians, more often than not, are instead indebted to overarching investigative frameworks providing specific anthropological or social dispositions for the scientific actors, and epistemic things, among them Foucault's archeology of knowledge, Bourdieu's habitus or Rheinberger's epistemic cultures.

4. Rediscovering the Vienna Circle

When the IVC was founded, its primary research program was to go "back to the sources" and overcome the received view that took Logical Empiricism as a long and justly abandoned set of philosophical dogmas. Using this misconception as their default jumping board, most scholars laid the blame at the doors of Ayer (1936) and Quine (1951). Aspiring at an explanation of how this view had come to dominate, was not yet on the agenda. Thus most of his early historiography can be classified as the first mode. The objective was to demonstrate that if only one read the original texts and duly contextualized them by archival material, the Vienna Circle was different and astonishingly pluralistic. Not only was the Circle rooted in a broad variety of contemporary traditions, but its actual members pursued diverging agendas within a broader strategic alliance held together by common goals and Neurath's stewardship. Some of these analyses, especially those by philosophers, explicitly set out to overcome the received view from within – as Thomas Uebel (1992) has aptly put it.

With a certain delay, established and novel approaches in the history of science were brought to bear on the Vienna Circle, among them the history of networks, the history of institutions, the history of ideas, and the history of scientific cultures. They showed that the plurality was not only one of abstract philosophical theses, but the consequence of a culture of vivid debate and driven by a social agenda. Only in the 1990s did Vienna Circle historiography meet Sarton's vision. Thus the received view had not only brought a defective history of philosophy; it also neglected the complex identity of the Vienna Circle members as scientist-philosophers and intellectuals deeply invested in the various projects of Red Vienna. No doubt, being streamlined into traditional philosophers with a logical-analytical tack, they were an easier target for philosophers.

After having demonstrated the pluralism of the original Circle, scholars typically moved to the second mode of reflection and incorporated this fact into a variety of different philosophical agendas. The interaction between philosophers' and historian's historiographies was not always frictionless. As regards the research of the 1990s, we basically identify four major lines, three philosophical ones and a broader cultural-historical agenda.[7] First and already since the 1970s, Rudolf Haller (1986) and others had begun to embed the Vienna Circle into a broader Austrian Philosophy that had started with Bolzano and Brentano, was strongly influenced by Wittgenstein, and extended into contemporary analytic ontology.[8] In the 1990s, Austrian Philosophy simultaneously developed into a distinct philosophical agenda (cf. Smith 1997) and was historicized in a more consistent fashion, not least by Stadler (1997). The difference of agendas effectively led to a dissolution of the broader concept of Austrian philosophy after 2000. While philosophers continued to identify it with a set of positive and negative philosophical creeds, thus at bottom reconstructing a philosophical school to justify a contemporary conception of analytic metaphysics (cf. Monist 2000), the analysis of the factual historical tradition in Austria and the Habsburg lands revealed a more nuanced picture that accommodated diverse philosophical agendas.

Second and against this Habsburgian strand of interpretation, Michael Friedman (1999), Alan Richardson (1999), and others articulated the Kantian and neo-Kantian roots of the movement, at least of its German members, and turned them into a historical motivation for an epistemological agenda in the philosophy of science that saw itself as a continuation of late 19th century scientific philosophy (cf. Friedman 2001, Parrini 1995; see Section 5).

Third, the work of Neurath underwent a remarkable renaissance (cf. Uebel 1992, Nemeth/Stadler 1996, Cartwright et al. 1996) that emphasized not only the

7 Needless to say, we cannot do justice to the steadily growing literature on the Vienna Circle, among them the volumes published by the Institute Vienna Circle. As said in the introduction, we are proceeding in an exemplary fashion and not by showcases.

8 Cf. the book series "Studien zur Österreichischen Philosophie" at Rodopi.

break with the traditionally universalist and theory-oriented philosophy of science, but brought questions of the social role of science, both historically and thematically, to the fore-front of the debate. There has been substantial work on the *Encyclopedia* and the concept of unity of science, and more recently about Neurath's philosophy of economics. The Neurath reception also prompted a renewed discussion of Zilsel's thinking, which had already been taken up by sociologists of science. (cf. Wolfgang Krohn's introduction to Zilsel 1976)

Fourth, Neurath has become not only a reference point for present social epistemology and for the history of Red Vienna, but has to a growing extent also attracted historians of science who understand the Vienna Circle as a part of a broader history of modernism. Part of this literature (cf. Galison 1990) has to be seen against the backdrop of recent developments in the history of science that are guided by broader notions of epistemology and culture.

In retrospect, the list of institutions involved in this rediscovery at the time when the IVC was launched, appears impressive. It includes classical archives, among them the Vienna Circle Stichting in Haarlem; eminent philosophy departments that had founded archives, among them the Graz *Dokumentationsstelle*, the Minnesota Center, the Pittsburgh Archives for Scientific Philosophy, and the Philosophical Archive at the University of Konstanz; publication projects, among them the Vienna Circle Collection at Reidel-Kluwer (since 1973) and the Suhrkamp series (since 1976); and finally a constantly growing number of congresses that were often prompted by anniversaries. Nevertheless, in the early 1990s the community appeared still somewhat fragmented into various networks and determined by strong bilateral ties, such as Graz-Minnesota or Konstanz-Pittsburgh collaborations. That this was to change was, to our mind, not least a consequence of the double centenaries of Carnap and Reichenbach that led to a unprecedented number of publications with a considerable intersection of invitees: collections (Stadler 1993, Haller/Stadler 1993, Danneberg et al. 1994, Salmon/Wolters 1994), special numbers of *Erkenntnis* (Spohn 1991) – re-founded in 1975 to continue the tradition, not primarily to reflect upon it – and *Synthese* (Sarkar 1992). Notably, the Viennese volumes amended the program with the centenary of Edgar Zilsel and the city of Prague. But also the foundation of the HOPOS working group – now a learned society – played an important role. By the year 2000, Vienna Circle historiography had finally become an international research field that has brought since steady progress in the two first modes mentioned.

5. Reflections on and Historizations of Historiography

Let us finally emphasize some recent developments that can be classified under the third and fourth mode of reflection. Our first example concerns the historiographical aspects of the Carnap-Quine debate – a debate that for North American analytic philosophy became as important a point of reference as had been the Mach-Planck debate during the European debates of the 1920s and 1930s. As Richard Creath has argued, Quine had not only an enormous influence on the emergence and dissemination of the received view "that Carnap is some sort of British empiricist whose driving concern in the *Aufbau* was the ontological reduction of science to an absolutely certain domain of sense data." (Creath 2007, 334) More generally, his "historical writing plays a crucial role in his argument for his non-historical views." (ibid., 332) This is especially true for the classic papers "Two Dogmas" (1951) and "Epistemology Naturalized" (1969), where Quine developed his own argument – that would undergo substantial revisions in later versions – against a carefully crafted picture taking Carnap "as the culmination of the development of empiricism." (Creath 334) On the basis of recent historical scholarship about the Vienna Circle, Creath shows that in actual fact the seven basic doctrines Quine is typically associated with, can be squarely placed in the Vienna Circle tradition.

The novel point of historiography in the third mode is not that philosophers, among them Quine and Creath, reinterpret the history of philosophy so that it suits their argument. They have always done so. Rather does Creath reveal the historicity of philosophers' historiography by putting Quine's then victorious narrative of the Vienna Circle against a historiography that was conspicuously absent after 1945, the once flourishing self-historiography of Vienna Circle. Such a perspective goes beyond overcoming the received view on the factual level by means of a historiography that convinces philosophers to take another look.

Alan Richardson's (2007) study of the barely known history of the 'received view' in the 1960s provides another example of this kind of historiography. It focuses on how Kuhn's *Structure* "helped to make widespread and to stabilize the very image of logical empiricism it sought to argue against." (2007, 365) Despite unfamiliarity with the history of the movement,

> Kuhn not only succeeded in dislodging the particular topics associated in his mind and the minds of his readers in the 1960s with logical empiricism (a timeless, universal logic of induction or of explanation, say); he also succeeded in changing the methods and tools of philosophy of science. (ibid., 367)

Recent historiography (starting from Reisch 1991) has shown that Carnap was not wrong in relating his linguistic frameworks to Kuhnian paradigms. While "contemporary researchers such as Michael Friedman … can invoke logical empiricism and find resources within it exactly when concerning themselves with the historical dynamics of science", the logical empiricists of the 1960s believed that Kuhn's call for a role for history "did not importantly disrupt their own" (both Richardson 2007, 368) technical projects and they did not combat the widely shared view that Kuhn's work had shown the poverty of their own approach. Rather were it Popper and Lakatos, themselves staunch critics of logical empiricism, engaged Kuhn head-on and charged him of denying the rationality of the scientific enterprise. However both Popper and Lakatos entertained conceptions of history of science that were fundamentally at odds with the Vienna Circle, even more so if one considers the Machian heritage in Frank and Neurath's thinking. This shows again the importance of history of science for today's Vienna Circle historiography.

Let us turn first to Friedman's (2001) program to secure some rationality for Kuhnian revolutions by means of historically relativized constitutive a priori principles. These principles concern the mathematical framework of a theory, e.g. non-Euclidian geometry, and – at a more specific level – scientific principles, e.g. general covariance, that make the formulation of particular scientific laws at all possible. Since not all those levels break at the same time, these principles provide bridges across fractures occurring on other levels. Moreover they are embedded into a philosophical meta-discourse that makes new paradigms thinkable before a scientific revolution occurs. Friedman's program is motivated by the discussions about relativized a priori principles of the early 1920s and Carnap's (1950) linguistic frameworks that he views as a way how to modernize the former, prematurely abandoned conception.

The decisive point is that Friedman not only employs his own contributions to Vienna Circle historiography as justification of his systematic philosophical agenda, but combines them with a reconfiguration of the history of philosophy of the 1920s. This renders it an example of historiography in the fourth mode. In *A Parting of the Ways* (2000), Friedman argues that the 1929 "International University Course" at Davos, Switzerland, figuring lectures by Cassirer and Heidegger, represented the branching point between analytic and continental philosophy. In the decade before, to Friedman's mind,

> logical empiricism, Husserlian phenomenology, neo-Kantianism, and Heidegger's new 'existential-hermeneutical' variant of phenomenology were … engaged in a fascinating series of philosophical exchanges and struggles, all addressed to the revolutionary changes that were then sweeping both the *Naturwissenschaften* and the *Geisteswissenschaften*. … [T]hey still spoke the same philosophical language, and they actively engaged one another on a common set of philosophical problems. (2000, xi)

Davos convinced the young Carnap, who sat in the audience, that this unity had broken apart and a radical departure from the philosophical heritage was warranted, a conclusion reinforced by the political developments. Carnap (1931) simply denied that a philosophy such as Heidegger's possessed any meaningful object.

It is not uncommon for historians of science to consider conferences as watersheds for long-time developments. Take the *Naturforscherversammlungen* of 1895, when the conflict between Boltzmann and Ostwald's energeticism figured center-stage, or 1920, when Einstein and Philipp Lenard fought about the nature of modern physics. But it is rare in the history of philosophy because such a historization embeds philosophical programs into the broader scientific and historical context. While in the eyes of some philosophers this introduces an element of historical contingency, Friedman's strategy to accept a historical change of rationality, of the meaning of the most basic concepts of science, in itself does not contradict the rationality of science as long as there is an understanding of the dynamics of reason that combines its inner-scientific and reflective (philosophical) elements. The alternatives would be either an ahistorical conception of scientific rationality or a meta-theory of the history of science, be it a Hegelian or Darwinian one. Both of these alternatives falsely suggest a homely feeling of epistemic security that misses the basic openness of the scientific process and the historical nature of the scientific world conception.

Taking both sides of Friedman's recent work together, one can no longer neatly separate the historiography of a science in which philosophical reflections play a key role, e.g., as regards early relativity theory, from the historiography of a philosophy that is largely inspired by the scientific developments of the day. Such a comprehensive approach continues in an important sense the hybrid identity of scientist-philosopher-historian of the days of Mach and the early Vienna Circle into the post-1945 epoch when both history of science and philosophy of science became fully established disciplines. It involves a rather complex historiography and a critical reflection upon the disciplines' diverging methods, which is precisely the point we want to address with our distinction of historiographical modes.

Let us provide a second example of historiography in the fourth mode by showing that Machian historiography and its transformation by the Vienna Circle have influenced present-day history of science in a more complex way than the alleged demise of logical positivism and the post-Kuhnian developments might suggest. We will do so by referring to Holton's (1993) historical perspective on the anti-science phenomenon and Daston and Galison's (2010) history of objectivity. It is well-known that the Vienna Circle purged the Machian heritage from phenomenalism and the empiricist foundation of mathematics and portrayed him as the standard bearer of a renewed Enlightenment. Yet it was precisely Mach's historico-critical reconstruction of key concepts, such as force, that provided the methodological basis to understand the development of a scientific world conception also as a struggle for the interpretation of the history of science. Although this brought science closer to other social and political practices, it did not abandon the focus on the basic concepts as the bearers of the historical dynamics of reason – to borrow Friedman's term. Mach allowed the advocates of a new Enlightenment to overcome the heroic and teleological narratives prevailing in 19th century historiography of science, including the convergent realism that had been triggered by the mechanical world view but survived its demise in a more abstract fashion, for instance in Planck's conception of the physical world view.

The members of the so-called First Vienna Circle set out to renew Mach's critical historiography by replacing its foundation in Spencerian evolutionism with modern logic (cf. Neurath 1915, 1916), which they considered as a much stronger tool for historical critique and conceptual renovation. Frank viewed "The Significance of Mach's Physical Epistemology for the Intellectual Life of the

Present"[9] in having adapted the Enlightenment project, locating the latter's main tenet in "the protest against the misuse of merely auxiliary concepts" (1917, 70) as an absolute foundation of physics and philosophy. For such misuse bore the danger of conceiving any change in the foundations of physical theory as a bankruptcy of the scientific world conception as a whole.

In the following year, Oswald Spengler's (1918) *Decline of the West* took exactly this route, predicting that modern science and occidental Faustian rationality as a whole would repeat the course of the Apollonian science of antiquity and, once again, give way to an epoch of belief. Modern physics, to Spengler's lights, exhibited clear symptoms of exhaustion: excessive emphasis on abstract theory, the abandonment of absolute space and time, and the abandonment of strict causality. Most scientists criticized that Spengler had missed that the progress of modern science consisted precisely in overcoming such a mechanistic world view. However, Spengler's cultural morphology challenged the Vienna Circle's agenda of a renewed Enlightenment and its Machian historiography at a more basic level by offering a grand and seductive counter-narrative to modernism at-large.

Neurath's *Anti-Spengler* (1921) approached this challenge by charging Spengler of an ill-conceived pseudo-rationalism that settled for simplistic explanations and denied the openness of both modern science and modern society. Neurath's aim was to save Enlightenment by regaining the individual and the collective as historical actors in science against the fatalism of morphology. Neurath and Frank (1932) continued the criticism against several heirs of Spengler, including Rudolf Steiner's anthroposophy and Jakob von Üexküll's renewal of Goethe's qualitative and phenomena-oriented science, and against the conservative holism and organismic conceptions of society advocated by Othmar Spann and others. This defense of the Enlightenment project as a whole, for which the criticism against metaphysics was pivotal, also provided the historical context for the Vienna Circle's fight against a static metaphysical holism – in contrast to Mach's open empirical holism – and against Hans Driesch's vitalist version of biology.

Frank and Neurath's goal was not to defend an already well-entrenched, rule-like methodology of science and a timeless conception of scientific rationality against pseudo-scientific trends but to secure the openness of scientific rationality in action. There was no rationality to be reconstructed for its own sake but history to be made. Historiography stood not outside the sciences, it was in an important sense part of their internal dynamics. Its employment was critical, not teleological, defensive, but always complemented by the Neurathian style of utopia, the fight for modernism and the scientific world conception as a social practice that still had to be won.

Holton (2006) has often acknowledged the importance of Frank and their collaboration within Harvard's general education program for his own conception of

9 The translation in Frank (1949) was titled "The Importance for Our Times of Ernst Mach's Philosophy of Science".

historiography. His collection *Science and Anti-Science* (1993) combines a historical perspective on the Machian tradition in the Vienna Circle with an analysis of the contemporary anti-science phenomenon. Interestingly, he still traces the latter back to Spengler, thus effectively embedding the tradition of Neurath and Frank into present-day debates. Significantly, Holton does not commence from the Kuhn debate or continue the dichotomy rationality versus historicity, but counts on formation from early age on, broad adult education, and public exposure of anti-science. He does so not in a demarcationist fashion but as an expression of the ethos of science that he exemplifies at the principal author of the Declaration of Independence, Thomas Jefferson, who "was no less a patriot for being comfortable with scientific thinking. Again and again, he hoped to conquer superficial variety and differences by embracing themata, fundamental to science, namely generalization and unification." (1993, 113) Jefferson's research program, as Holton sees it, took something from each of Bacon and Newton, combining them into a new style that "locates the center of research in an area of basic scientific ignorance that lies at the heart of a social problem." (ibid., 115)

Let us finally turn to Lorraine Daston and Peter Galison's (2010) book *Objectivity* that is dedicated to their teacher and friend Holton. On their account, the core elements of scientific rationality, among them experiment, accuracy, and objectivity "emerged in a specific time and place, mutated over time, and connected thickly with concrete practices as well as disciplinary values". (Daston 2001, 216) In their book, they distinguish various types of objectivity that have developed at different times, but are still present in contemporary thinking, making the concept multilayered and not free of internal antagonisms.

> What process of historical fusion soldered the metaphysical, the methodological, and the moral into the amalgamated concept of scientific objectivity? How was each distinct component of the amalgam formed, and what affinities among components made their bonding first thinkable and then apparently inevitable? It is not enough to say simply that history has united what logic would have put asunder. History's unions may be less constrained than logic's, but even history cannot arbitrarily recombine elements – +otherwise we would have chimeras instead of concepts. A history of objectivity must explain why some ideas and practices melded with one another and others slide away. (2010, 379)

We read this passage also as a criticism not only of a merely fact-gathering historiography, but also of narratives that remain on the surface of scientific practices and entirely replace the analysis of the history of scientific concepts acting in them by social or cultural context, deriving their justification from the poverty of a logic-focused reconstruction of an allegedly true historical rationality. Daston emphasizes the significance of Machian historiography as a watershed and as a role model. Its objective was not to "record the steady growth of knowledge, but rather to clarify conceptual puzzles in contemporary science, … revealing the past roots of present incoherences." (2001, 210)

Inasmuch as their general approach develops a promising historiography in the fourth mode, Daston and Galison (2010) do not apply it to the Vienna Circle in a fully consistent fashion. To begin with, they largely subsume the new philosophy of science under the structural objectivity emerging from within 20th century physics. They rightly emphasize that "Carnap's neutralist stance toward structure involved more than logical quantifiers. For him and the Vienna Circle colleagues, it was also a moral stance, a way of life, in conscious defiance of traditional philosophy." (2010, 291) And Galison (1990) has provided a highly influential account of Carnap's modernist agenda, which has fathered much historiography in the second mode (see Sect. 4). But their understanding of Carnap's neutral attitude, "which he soon elevated to an ontological (and political) 'principle of tolerance'" (2010, 292) comes out too static and does not include its increasingly pragmatic tack, let alone other Vienna Circle actors, among them Neurath and Frank. This disturbs the subtle equilibrium between the icy slopes of logic and the political advancement of the scientific world conception mentioned in the concluding passage of the Manifesto, an equilibrium of epistemic and political agendas that recent Vienna Circle historiography has brought out in many different case studies.

Carnap was, no doubt, a modernist, yet what were – in Daston and Galison's terms – the epistemic virtues at work in his reconstruction of science? If Friedman's (2000) reading of the Davos debate is correct, Carnap's epistemic virtues had significant consequences for the history of 20th century philosophy. If Friedman's *Dynamics of Reason* (2001) is right about the promise of linguistic frameworks, one might wonder whether it was precisely Carnap's oft-mentioned principle of tolerance that intended to specify the place for history. Granted, it might sound uninspired to subsume all history under pragmatics. But in a historical perspective that harks back to the days of the Vienna Circle *in actu* – not in exile –, Carnap could view himself on one side in a complex movement. Bringing Machian historiography of science, and its lasting influence in the work of Neurath and Frank, into sharper relief additionally suggests that precisely the interaction between – in Daston and Galison's terms – structural and judgmental objectivity was an issue of concern for the Vienna Circle and its understanding of the history of science.

That the subtle equilibrium set by the Manifesto was lost after 1945, had significant effects on Vienna Circle historiography. This has been convincingly shown by the recent scholarship on the history of the received view, to which one might add yet another twist of historical reflection and ponder about the historical conditions of such an influential take on the historiography of science as Daston and Galison's. It seems that Frank's influence (through Holton) on their picture of the Vienna Circle was significantly weakened by the Quinean reading of Carnap – their positive reception of Mach's critical historiography notwithstanding. Unfortunately, Frank is still one of the less scrutinized figures of the Vienna Circle. But we feel confident that he would not have found many reasons to disagree with the following assessment of Daston.

We live in a science-saturated society: science and science-based technology not only shape the contours of cities, the terms of war and peace, the balance of plenty and famine; they also set the standards as to what counts as knowledge and who qualifies as a knower. (2001, 202)

References

Alfred Ayer (1936): *Language, Truth and Logic*, London: Gollancz.

Albert Blumberg, Herbert Feigl (1931): "Logical Positivism. A New Movement in European Philosophy", *Journal of Philosophy 28*, 281-296.

Bernadette Bensaude-Vincent (1983): "A Founder Myth in the History of Sciences? The Lavoisier Case", in: Loren Graham, Wolf Lepenies, Peter Weingart (eds.), *Function and Uses of Disciplinary Histories*, Dordrecht: Reidel, pp. 53-78.

Rudolf Carnap (1931): "Überwindung der Metaphysik durch logische Analyse der Sprache", *Erkenntnis* 2, 219-241.

Rudolf Carnap (1950): "Empiricism, Semantics, and Ontology", *Revue international de philosophie 4,* 20-40.

Rudolf Carnap (1963): "Intellectual Autobiography", in Paul A. Schilpp (ed.), *The Philosophy of Rudolf Carnap*, La Salle IL: Open Court, pp. 1-84.

Andre W. Carus (2007): *Carnap and twentieth-century thought*, Cambridge: Cambridge University Press.

Nancy Cartwright, Jordi Cat, Lola Fleck, and Thomas E. Uebel (1996): *Otto Neurath: Philosophy between Science and Politics*. Cambridge: Cambridge University Press.

Richard Creath (2007): "Vienna, the City of Quine's Dreams", in: Richardson/ Uebel (2007), pp. 332-345.

Lutz Danneberg, Andreas Kamlah, Lothar Schäfer (eds.) (1994): *Reichenbach und die Berliner Gruppe*, Braunschweig/Wiesbaden: Vieweg.

Lorraine Daston (2001): "The Historicity of Science", in: Glenn W. Moss (ed.), *Historicization-Historisierung*, Göttingen: Vandenhoeck & Ruprecht, pp. 201-221.

Lorraine Daston, Peter Galison (2010): *Objectivity,* New York: Zone Books.

Federigo Enriques (1914): *Problems of Science,* Chicago: Open Court (Italian original *Problemi della Scienza*, Bologna: Zanichelli, 1906).

Herbert Feigl (1943): "Logical Empiricism", in Dagobert D. Runes, *Twentieth Century Philosophy,* New York: Greenwood Press, pp. 371-416.

Herbert Feigl and Charles W. Morris (1970): "Bibliography and Index", in Otto Neurath et al. (eds.), *Foundations of the Unity of Science,* Chicago: The University of Chicago Press, vol. II, pp. 947-1023.

Philipp Frank (1917): "Die Bedeutung der physikalischen Erkenntnistheorie Machs für das Geistesleben der Gegenwart", *Die Naturwissenschaften* 5, 65-72.

Philipp Frank (1932): *Das Kausalgesetz und seine Grenzen,* Vienna: Springer.

Philipp Frank (1941): *Between Physics and Philosophy*, Cambridge, MA: Harvard UP.

Philipp Frank (1949): *Modern Science and Its Philosophy*, Cambridge, MA: Harvard UP.

Philipp Frank (1956): "The Variety of Reasons for the Acceptance of Scientific Theories", in: Frank (ed.), *The Validation of Scientific Theories*, Boston: Beacon Press, pp. 3-28.

Michael Friedman (1999): *Reconsidering Logical Positivism*, Cambridge: Cambridge University Press.

Michael Friedman (2000): *A Parting of the Ways: Carnap-Cassirer-Heidegger*, Chicago: Open Court.

Michael Friedman (2001): *Dynamics of Reason,* Stanford, CA: CSLI.

Peter Galison (1990): "Aufbau/Bauhaus: Logical Positivism and Architectural Modernism." *Critical Inquiry* 16, 709–52.

Kurt Grelling (1928): "Philosophy of the Exact Sciences: Its Present Status in Germany", *The Monist 38,* 97-119.

Rudolf Haller (1986): "Gibt es eine Österreichische Philosophie?", in: *Fragen zu Wittgenstein und Aufsätze zur Österreichischen Philosophie*, Amsterdam: Rodopi, pp. 31-43.

Rudolf Haller, Friedrich Stadler (eds.) (1993): *Wien–Berlin–Prag. Der Aufstieg der wissenschaftlichen Philosophie*, Vienna: Hölder-Pichler-Tempsky.

Veronika Hofer (2013): "Philosophy of Biology in Early Logical Empiricism", forthcoming in Hanne Andersen et al. (eds.), *New Challenges to Philosophy of Science*, Dordrecht: Springer.

Gerald Holton (1973): *Thematic Origins of Scientific Thought. Kepler to Einstein,* Cambridge: Harvard University Press.

Gerald Holton (1993): *Science and Anti-Science,* Harvard University Press, Cambridge, MA.

Gerald Holton (2006): "Philipp Frank at Harvard University: His Work and His Influence", *Synthese* 153, 297–311.

Joergen Joergensen (1970): *The Development of Logical Empiricism*, Chicago: The University of Chicago Press; in Otto Neurath et al. (eds.), *Foundations of the Unity of Science,* Chicago: The University of Chicago Press, vol. II, pp. 845-946.

Felix Kaufmann (1944): *Methodology of the Social Sciences,* New York: Oxford University Press.

Viktor Kraft (1950), *Der Wiener Kreis, der Ursprung des Neopositivismus; ein Kapitel der jüngsten Philosophiegeschichte,* Wien: Springer; English translation by Arthur Pap: *The Vienna Circle, the origin of neo-positivism; a chapter in the history of recent philosophy*. New York, Philosophical Library, 1953.

C. I. Lewis (1946): *Analysis of Knowledge and Valuation,* La Salle, IL: Open Court.

Monist (2000): "Austrian Realism: From Aristotelian Roots to the Vienna Circle" (ed. by Herbert Hochberg), *The Monist* 83, 1.

Elisabeth Nemeth and Friedrich Stadler (eds.) (1996): *Encyclopedia and Utopia. The Life and Work of Otto Neurath (1882–1945)*. Dordrecht: Kluwer.

Otto Neurath (1915): "Prinzipielles zur Geschichte der Optik", *Archiv für die Geschichte der Naturwissenschaften und der Technik* 5, 371-389.

Otto Neurath (1916): "Zur Klassifikation von Hypothesenssystemen", *Jahrbuch der Philosophischen Gesellschaft an der Universität zu Wien 1914 und 1915*, Leipzig: Hölder, pp. 39-63.

Otto Neurath (1921): *Anti-Spengler*, München: Callwey.

Otto Neurath (1935), *Le Développement du Cercle de Vienne et l'avenir de l'empirisme logique,* Paris: Hermann.

Paolo Parrini: *Conoscenza e Realtà. Saggio di Filosofia Positiva*, Roma e Bari, Laterza, 1995; English version: *Knowledge and Reality: An Essay in Positive Philosophy,* Dordrecht: Kluwer 1998.

Max Planck (1922), *Physikalische Rundblicke,* Leipzig, Barth.

Hans Reichenbach (1920): *Relativitätstheorie und Erkenntnis apriori.* Berlin: Springer.

Hans Reichenbach (1936): "Logistic Empiricism in Germany and the Present State of its Problems", *Journal of Philosophy* 33, 141-160.

Hans Reichenbach (1951): *The Rise of Scientific Philosophy,* Berkeley: University of California Press.

George Reisch (1991): "Did Kuhn Kill Logical Empiricism?", *Philosophy of Science* 58, 264-277.

George Reisch (2005): *How the Cold War Transformed Philosophy of Science: To the Icy Slopes of Logic,* Cambridge: Cambridge University Press.

George Reisch (2012): "The Paranoid Style in American History of Science", *Theoria* 27: 323-342.

Willard Van O. Quine (1951): "Two Dogmas of Empiricism", *Philosophical Review*, 60, 20-43.

Willard Van O. Quine (1969): "Epistemology Naturalized", in *Ontological Relativity and other Essays,* New York: Columbia University Press, pp. 69-90.

Alan Richardson (1998): *Carnap's Construction of the World: The Aufbau and the Emergence of Logical Empiricism*. Cambridge: Cambridge University Press.

Alan Richardson, Thomas E. Uebel (eds.) (1997): *The Cambridge Companion to Logical Empiricism*, Cambridge: Cambridge University Press.

Wesley Salmon, Gereon Wolters (eds.) (1994): Logic, Language, and the Structure of Scientific Theories, Pittsburgh & Konstanz: University of Pittsburgh Press, Universitätsverlag Konstanz.

George de Santillana, Edgar Zilsel (1941): *The Development of Rationalism and Empiricism,* Chicago: Chicago University Press.

Sahotra Sarkar (ed.) (1992): *Carnap: A centenary appraisal*, *Synthese* 93,1-295.

Moritz Schlick (1930): "Die Wende der Philosophie", *Erkenntnis* 1, 4-11.

Moritz Schlick (1932): "Positivismus und Realismus", *Erkenntnis* 3, 1-31.

Barry Smith (1997): "The Neurath-Haller Thesis: Austria and the Rise of Scientific Philosophy", in: Keith Lehrer and Johann C. Marek (eds.), *Austrian Philosophy. Past and Present. Essays in Honor of Rudolf Haller*, Dordrecht: Kluwer, pp. 1-20.

Oswald Spengler (1918): *Der Untergang des Abendlandes. Erster Band: Gestalt und Wirklichkeit,* Wien–Leipzig: Wilhelm Braumüller.

Wolfgang Spohn (ed.) (1991): *Erkenntnis Orientated: A Centennial Volume for Rudolf Carnap and Hans Reichenbach*, Dordrecht: Kluwer (= Vol. 35 of Erkenntnis 1991).

Friedrich Stadler (ed.) (1993): *Scientific Philosophy. Origins and Developments*, Dordrecht: Kluwer.

Friedrich Stadler (1997): *Studien zum Wiener Kreis. Ursprung, Entwicklung und Wirkung des Logischen Empirismus im Kontext*, Frankfurt am Main: Suhrkamp.

Friedrich Stadler (2011): "The road to *Experience and Prediction* from within: Hans Reichenbach's scientific correspondence from Berlin to Istanbul", *Synthese* 181, 137-155.

Charles L. Stevenson (1944), *Ethics and Language,* New Haven: Yale University Press.

Michael Stöltzner (1998): "Federigo Enriques e l'Enciclopedia Neurathiana", *Rivista di Storia della Filosofia* LIII, 463-494.

Michael Stöltzner (2013): "Did Reichenbach Prefigure the Possibility of Quantum Mechanics?", forthcoming in N. Milkov, V. Peckhaus (eds.), *The Berlin Group and the Philosophy of Logical Empiricism*, Dordrecht: Springer.

Thomas E. Uebel (1992): *Overcoming Logical Positivism from Within: The Emergence of Neurath's Naturalism in the Vienna Circle's Protocol Sentence Debate*, Amsterdam–Atlanta: Rodopi.

Thomas E. Uebel (2011): "Beyond the formalist criterion of cognitive significance: Philipp Frank's later antimetaphysics", *HOPOS: The Journal of the International Society for the History of Philosophy of Science* 1: 47-72

Thomas E. Uebel (2013): "'Logical Positivism' – 'Logical Empiricism': What's in a Name?", forthcoming in *Perspectives on Science.*

Verein Ernst Mach (ed.) (1929): *Wissenschaftliche Weltauffassung. Der Wiener Kreis,* Vienna: Artur Wolf Verlag.

Paul Ziche (2008): *Wissenschaftslandschaften um 1900. Philosophie, die Wissenschaften und der nichtreduktive Szientismus*, Zurich: Chronos.

Edgar Zilsel (1942): "The Genesis of the Concept of Physical Law", *Philosophical Review* 51, 245-279.

Edgar Zilsel (1976): *Die sozialen Ursprünge der neuzeitlichen Wissenschaft*, edited by Wolfgang Krohn, Frankfurt am Main: Suhrkamp.

Veronika Hofer
Institute for the History of Medicine
Währingerstraße 25
1090 Wien
Austria
veronika.m.r.hofer@gmail.com

Michael Stöltzner
Department of Philosophy
University of South Carolina
Columbia, SC 29208
USA
stoeltzn@mailbox.sc.edu

18th Vienna Circle Lecture
November 16, 2010

DAGFINN FØLLESDAL

HUSSERL AND GÖDEL ON MATHEMATICAL OBJECTS AND OUR ACCESS TO THEM[1]

I am very grateful for the invitation to give this lecture. The Vienna Circle was of great importance for philosophy in the Nordic countries, mainly through my late colleague Arne Næss and through Eino Kaila in Finland. The Circle also provided the intellectual setting for Kurt Gödel's early development. Gödel developed an interest in philosophy already when he was in the gymnasium, and although he was interested in other philosophers than those who were popular in the Vienna Circle, his contact with the Circle acquainted him with a combination of philosophy on the one hand and mathematics and science on the other, which stimulated his philosophical reflections.

Gödel had a strong interest in philosophy throughout his life. Although it was his contributions to logic that gave him fame, he devoted considerably more time to philosophy than to logic. He also emphasized the importance of philosophy for work in logic. He pointed out, for example, that my compatriot and advisor Thoralf Skolem had all the technical results that were needed for the completeness theorem long before Gödel proved this theorem in 1930. However, Skolem did not have the philosophical perspective on logic and the concepts that made sense of the problem and its solution.

Jean van Heijenoort discovered this when he edited *From Frege to Gödel* (1967). He wrote to Gödel and inquired about it, and Gödel answered, on 14 August 1964:

> As for Skolem, what he could justly claim, but apparently does not claim, is that, in his 1922 paper he implicitly proved: 'either *A* is provable or ~*A* is satisfiable' ('provable' taken in an informal sense). However, since he did not clearly formulate this result (nor, apparently had made it clear to himself), it seems to have remained completely unknown, as

1 I am grateful to Charles Parsons, Solomon Feferman, Richard Tieszen, Michael Friedman, Jens Erik Fenstad, David Smith, Ronald McIntyre and Christian Beyer for valuable discussions of Husserl and Gödel over many years. My co-researcher Kai Hauser and I would also express our thanks to the Templeton Foundation for their generous support of this and other work.

M.C. Galavotti et al. (eds.), *European Philosophy of Science – Philosophy of Science in Europe and the Viennese Heritage*, Vienna Circle Institute Yearbook 17,
DOI 10.1007/978-3-319-01899-7_22,

follows from the fact that Hilbert and Ackermann in 1928 do not mention it in connection with their completeness problem.[2]

Also Gödel's later work was philosophically motivated. He focused on problems that he regarded as philosophically fundamental, and he had the ability to solve them.

Gödel's relation to the Vienna Circle has been explored and analyzed in a number of fine studies, by Friedrich Stadler, Eckehart Köhler and several others here in Vienna and many, many others, some of whom I will mention as I go on.

In this lecture I will focus on Gödel's connection with Edmund Husserl. Gödel heard about Husserl in Vienna, Carnap had studied with Husserl, Schlick wrote on him and Felix Kaufmann applied phenomenology to the social sciences and also to mathematics.[3] Husserl even came to Vienna to lecture in 1935, but Gödel did apparently not go to hear him.[4] It was only much later that Gödel started a serious study of Husserl.

The separation of philosophy into "continental" and "analytic" that started approximately in 1930, had as a consequence that few phenomenologists were interested in Husserl's work on the philosophy of mathematics and logic, and few people with an interest in mathematics and logic read Husserl. Long before the separation started, the prominent mathematician Hermann Weyl (1885–1955) developed a strong interest in phenomenology. He spent the years 1904–13 in Göttingen and attended some of Husserl's lectures. Husserl presided on his doctoral examination in 1908. Weyl's wife Helene, whom Weyl met when she came to Göttingen to study with Husserl, contributed to strengthening his interest in phenomenology.[5] After the first World War Oskar Becker (1889–1964), who had a doctorate in mathematics from Leipzig, came to Freiburg to study with Husserl and finished his Habilitation dissertation, on *The Phenomenological Foundations of Geometry*, in 1923.[6]

Gödel did not become interested in phenomenology until 1959. He was originally strongly attracted to Leibniz's ideas. He then became more and more engaged in Kant, and then in 1959 he started an intensive study of Husserl. In this lecture we shall look into the relationship between Gödel and Husserl and see how Gödel used Husserl to throw light on some of the fundamental philosophical issues in mathematics, particularly the two classical questions: Are there mathematical entities? And if so, how can we know about them?

---

2 Heijenoort, p. 510. See also Takeuti, p. 22.

3 See Stadler 1997 and 2003 and Huemer for more on Kaufmann.

4 Van Atten/Kennedy, p. 427.

5 For more material on Weyl I recommend Ryckman 2004 and Feferman 1998, 2000 and also John Bell and Herbert Korté's long article on Weyl in the *Stanford Encyclopedia of Philosophy*.

6 There is an interesting correspondence between Weyl and Becker, edited by Paolo Mancosu and T. A. Ryckman, see Bibliography.

GÖDEL AND HUSSERL

The separation of philosophy into "continental" and "analytic" was largely based on misconceptions of what was on the other side. This in turn made it seem preposterous to "analytic" philosophers to contend that Gödel, whom they regarded with great respect, should have learned anything from Husserl, whom "continental" philosophers regarded as their forefather. Georg Kreisel seems to have been one of the first to learn about Husserl's influence on Gödel. He was invited by Gödel to spend 1955–57 at the Institute for Advanced Study. At that time Gödel had not yet become interested in Husserl. However, Kreisel kept in contact with Gödel, and when Kreisel came to Stanford in 1962 he knew that Gödel had begun a very thorough study of Husserl. Kreisel was not surprised by this and set out to learn more about Husserl. When I came to Stanford in 1966 we talked frequently about Husserl, and later gave a joint seminar on the philosophy of mathematics with emphasis on Husserl. Also Robert Tragesser discussed these matters with Kreisel, and he participated in a summer-long seminar on these issues in 1975. This led to his book *Phenomenology and Logic,* 1977, the first book devoted to the relation between Gödel and Husserl.

Beginning in October 1971 Hao Wang had regular conversations with Gödel, and he was puzzled by Gödel's high appreciation of Husserl. Wang reports these conversations and discusses them in several of his books and articles. He touches on them in *From Mathematics to Philosophy* (1974) and discusses them more fully in *Reflections on Kurt Gödel* (1987) und *A Logical Journey: From Gödel to Philosophy* (1996). Already on November 10, 1971, Husserl came up in these conversations, and from then on he was a central figure. Although Gödel in these conversations repeatedly expressed his agreement with Husserl, Wang claimed that Gödel's views were importantly different from those of Husserl, particularly when it came to ultimate justification and certainty. Wang writes in *Beyond Analytic Philosophy* (1985):

> ... he [Husserl] spent most of his energy in the struggle for securing a 'true beginning.' the indiscriminately applied pejorative term 'foundationalism' is a quite appropriate label for his work. (p. 32)
>
> ...he looks for the absolutely certain or synthetic a priori foundation once and for all, while I feel such a quest for an unattainable goal is doomed to failure, however useful and interesting its by-products might be. (pp. 37-38)

Against this, Wang quotes Gödel as saying "For me there is no absolute knowledge, everything goes only by probability."[7] Yet, Gödel insisted that he was in almost total agreement with Husserl. Wang found it difficult to accept this. However, he thought very highly of Gödel as a philosopher, and he did his best to understand Gödel's enthusiasm for Husserl. He invited me to lectures and discus-

7 Wang 1996, p. 170.

sions on Husserl at Rockefeller University, where he was teaching, and he benefitted from having Charles Parsons and later also Richard Tieszen at Columbia.

In my view, the apparent differences between Gödel and Husserl are mainly a matter of how to interpret Husserl. I think that Wang is wrong in his interpretation and Gödel right. First, lest one be misled by Husserl's foundationalist way of talking, note the following question from the *Crisis:*

> Can I begin with a truth – a definitive truth? … If I already had such 'immediately self-evident' truths, then I could perhaps mediately derive new ones from them. But where do I have them?[8]

The only case where we might possibly have a definitive truth may be the case of our own existence. However, Husserl points out that even our self-experience is in any case inadequate, partial and obscure.[9] Scholars who interpret Husserl as a foundationalist often point to the emphasis he puts on "adequate evidence" or "adequate givenness." However, Husserl does not claim that we can ever achieve this kind of perfection:

> For it might turn out that such adequate givenness is a mere "idea," in a sense analogous to the way in which we call pure red an "idea": seen red is, we think, always only red in more or less imperfect purity, belonging to incremental scales through which we may come closer to pure red, although we finally nevertheless remain (more or less) far from it.[10]

There are numerous further statements in Husserl's writings that go against interpreting Husserl as a foundationalist. I will not quote these here, but refer to my articles 1988 and 2005.

Husserl's frequent use of 'a priori' is often taken as evidence that he was a foundationalist. Thus, for example, in all his writings Husserl characterizes phenomenology as a study of the *a priori.* This makes it natural to assimilate him to Kant and Kant's foundationalism. However, Husserl means something different with 'a *priori*' than does Kant. For Husserl, the *a priori* is that which we anticipate, that which we expect to find when we experience the world. Phenomenology studies and attempts to chart these anticipations, but as we know, our anticipations often go wrong, our experiences turn out differently from what we expected, and again and again we have to revise our views and our expectations.[11] We shall get back to these anticipations later.

Although there are important differences, there are many similarities between Husserl's phenomenology and Kant's philosophy. In his conversations with Wang, Gödel repeatedly came back to this. Wang writes: "The whole phenomenologi-

---

8 *Krisis,* § 73, Husserliana VI, 269.24-29 = Carr, 335.

9 *Cart. Med.*, § 9, Husserliana I, 62.9-20 = Ströker, 24 = Cairns, 22-23.

10 *Husserl 1959. Erste Philosophie,* 31. Vorl., Husserliana VIII, 33.8-34.1.

11 Follesdal 1988, p. 115.

cal method is, according to Gödel, a precise formulation of the core of Kantian thought"[12] And in a manuscript from late 1961 or shortly thereafter Gödel writes:

... just because of the lack of clarity and the literal incorrectness of many of Kant's formulations, quite divergent directions have developed out of Kant's thought – none of which, however, really did justice to the core of Kant's thought. This requirement seems to me to be met for the first time by phenomenology, which, entirely as intended by Kant, avoids both the death-defying leaps of idealism into a new metaphysics as well as the positivistic rejection of all metaphysics. But now, if the misunderstood Kant has already led to so much that is interesting in philosophy, and also indirectly in science, how much more can we expect it from Kant understood correctly?[13]

This manuscript remained unpublished, so did everything else that Gödel wrote on Husserl and phenomenology. Why? We will consider this question briefly before we go on to discuss Gödel and Husserl's views on the philosophy of mathematics. Here the Vienna Circle comes in as part of the explanation. From the Vienna Circle Gödel was accustomed to discussions where one did not just present one's views, but focused on the arguments given for the views. In addition to providing support for the conclusions, the arguments also help clarifying one's position; seemingly small differences in the formulation of the position may make decisive differences to the arguments. Gödel was well aware of the importance of good arguments, and he was never satisfied with what he did in philosophy. An example is the contribution he promised to write for the Carnap volume in *The Library of Living Philosophers*. He worked on it for six years, and produced six different drafts, but none of these satisfied him. He finally wrote to Paul Edward Schilpp, the editor of the series:

The fact is that I have completed several different versions, but none of them satisfies me. It is easy to allege very weighty and striking arguments in favor of my views, but a complete elucidation of the situation turned out to be more difficult than I had anticipated, doubtless in consequence of the fact that the subject matter is closely related to, and in part identical with, one of the basic problems of philosophy, namely the question of the objectivity of concepts and their relations. On the other hand, in view of widely held philosophical prejudices, it may do more harm than good to publish half done work.[14]

The last sentence here shows that Gödel thought that his views on the objectivity of concepts and their relations would encounter massive resistance. The fact that he had consented to present these views in a volume for Carnap indicates that he may have hoped to bring around Carnap, who has justly been widely praised for his openness and willingness to listen to counterarguments.[15] However, the

12 Wang 1996, p. 357.
13 Gödel 1961, p. 387.
14 Goldfarb 1995, p. 324.
15 See, for example, Kaplan 1971.

last three words of the quotation make it clear that Gödel was not satisfied with his work. As Warren Goldfarb mentions in his Introduction to Gödel's drafts on Carnap in Gödel's *Collected Works,*[16] Gödel gives no account of how we gain knowledge of the mathematical realm. Gödel says that we do it by our faculty of mathematical intuition, but he gives no further detail on this. Gödel continued working on this issue. A main aim of this lecture is to give an account of how we gain this sort of knowledge.

The same dissatisfaction made Gödel abandon a lecture that he was supposed to give as a newly elected member of the American Philosophical Society. His *Nachlass* contains a full manuscript for the lecture, but he never gave it. In this case, Gödel did not have to fear a hostile audience. There would have been only one philosopher in the audience, W. V. Quine, who is likely to have been the one of those who proposed Gödel for membership, probably together with Saunders Mac Lane, who was the only other logician in the Society.[17] Perhaps also Einstein was a co-nominator, he thought very highly of Gödel. However, in spite of the positive audience Gödel decided not to give the lecture. He gave no reason for this. There is no indication that Gödel even replied to the invitation. Cautious as he always was, he may have decided to find out whether he had something satisfactory to say before he accepted the invitation.

A very good discussion of Gödel's caution has been given by Solomon Feferman.[18] He points out that when Gödel talks about "the philosophical prejudices of our time," he probably has in mind not only the logical empiricists, but also the Hilbert school. These were the colleagues he had to convince.

As a final example of Gödel's caution let me mention a recent discovery by Mark van Atten and Juliette Kennedy. In a draft for the Supplement to the Cantor article they found a final paragraph that starts:

> Perhaps a further development of phenomenology will, some day, make it possible to decide questions regarding the soundness of primitive terms and their axioms in a completely convincing manner.[19]

Why did Gödel decide to leave this paragraph out? It seems to me that here again he was dissatisfied with his lack of arguments and probably also felt that it was out of place to state his opinion without arguing for it.

---

16 Goldfarb 1995, pp. 333-334.

17 Saunders Mac Lane received his doctorate in logic with Bernays and Weyl in Göttingen in 1934. He later nominated Gödel for the National Medal of Science. Mac Lane even accepted the award for him from President Ford when Gödel was too ill to accept it in person.

18 Feferman 1984, p 180 of the reprint in Feferman 1998.

19 Van Atten, Mark, and Kennedy, Juliette, p. 466.

Gödel's study of Husserl

Gödel started to study Husserl in 1959.[20] He found Husserl quite congenial, and he soon became quite absorbed by Husserl's writings. He owned all of Husserl's main works,[21] and his underlining and comments in the margin indicate that he studied them carefully. The comments are mostly in Gabelsberger shorthand, but thanks to Cheryl Dawson, who has very kindly transcribed them, we can see that Gödel's comments are mostly approving, amplifying some of Husserl's points. However, sometimes he is critical, notably in his notes to the *Logical Investigations* (1900–01) and to some sections of Husserl's last work, the *Crisis* (one part published in 1936, the rest after Husserl's death). Generally, Gödel is most appreciative of the *Ideas* (1913), the *Cartesian Meditations* and the other works written after Husserl's "idealist" conversion around 1907. This is remarkable, since most "analytic" philosophers have normally found Husserl's early work more palatable than his later idealist work. However, I am in full agreement with Gödel on this. Ever since I taught my first course on Husserl, at Harvard in 1961–62, I have required the students to read the *Ideas* and the *Cartesian Meditations*, since I regard these two together as the best introduction to Husserl's mature phenomenology.

Gödel had expressed views on the philosophy of mathematics similar to those of Husserl long before he started to study him. What he found in Husserl was not radically different from his own view; what impressed him seems to have been Husserl's general philosophy, which provided a systematic framework for a number of his own earlier ideas on the foundations of mathematics.

From Gödel's manuscripts and particularly through his notes in his copies of Husserl's works we can see what Gödel was attracted to in Husserl, and how he

20 Wang 1981, p. 651 and Wang 1987, pp. 12 ff.

21 Gödel owned the 1968 edition of Husserl's *Logische Untersuchungen* (Max Niemeyer Verlag, Tübingen, first edition 1900/01), which is unchanged from the second edition (1913: Vol. I and Vol. II, Part 1, and 1921: Vol. II, Part 2). He owned the *Husserliana* edition of *Ideen*, Book I (1950), *Cartesianische Meditationen* und *Pariser Vorträge* (2. edition, 1963), *Die Krisis der europäischen Wissenschaften und die transzendentale Phänomenologie* (2. edition, 1962). He also owned Quentin Lauer, *Phenomenology and the Crisis of Philosophy* (Harper Torchbooks, Harper and Row, New York, 1965), which is an English translation of Husserl's two essays "Philosophy as Rigorous Science" (1911) and "Philosophy and the Crisis of European Man" (1935). Gödel also owned both volumes of the second edition (1965) of Herbert Spiegelberg, *The Phenomenological Movement* (Phenomenologica 5) Nijhoff, The Hague. All of these books are heavily annotated by Gödel, with the exception of *Logical Investigations*. In the latter were found, however, several pages of Gödel's shorthand notes, referring to page numbers in the text, indicating that Gödel probably had based his study of that text on a borrowed copy. The work was out of print for many years before the 1968 edition was published. The notes indicate that the text Gödel had used was the second edition, which first appeared in 1913-1921 and was reprinted in 1922, 1928 and 1968. References to Husserl in the following will be, as usual, to page and line in the Husserliana editions.

understood Husserl. It seems to me that Gödel's understanding of Husserl is extraordinary, placing him among the foremost interpreters of Husserl.

Husserl: Perception and Intuition

The first publication where Gödel brings in what he has learned from Husserl is the revised version of "What is Cantor's continuum problem?" (1964) Paul Benacerraf and Hilary Putnam wanted to include this article in *Philosophy of Mathematics: Selected Readings.* However, Gödel insisted that he wanted to revise the article and add a supplement, since his philosophical views had changed since he wrote the article. The changes reflect the impact that the study of Husserl had on his views. Among the many passages that reveal this influence, I find the following particularly useful, since it seems cryptic, but becomes understandable when read on the background of Husserl's theory of perception. It thereby provides a motivation for our using Husserl as a key to understanding Gödel. Gödel writes:

> That something besides the sensations actually is immediately given follows (independently of mathematics) from the fact that even our ideas referring to physical objects contain constituents qualitatively different from sensations or mere combinations of sensations, e.g., the idea of object itself … . Evidently, the 'given' underlying mathematics is closely related to the abstract elements contained in our empirical ideas. It by no means follows, however, that the data of this second kind, because they cannot be associated with actions of certain things upon our sense organs, are something purely subjective, as Kant asserted. Rather they, too, may represent an aspect of objective reality, but, as opposed to the sensations, their presence may be due to another kind of relationship between us and reality.[22]

This passage expresses in a condensed and cryptic way a key to Husserl's theory of perception, which is the basis for his and Gödel's philosophy of mathematics.

Traditional theories of perception regard perception as largely a matter of reception. The object perceived affects causally our sense organs and through processes in our nervous system and brain we come to have some sort of representation of the object in our mind. Aristotle's view is an example:

> By a 'sense' is meant what has the power of receiving into itself the sensible forms of things without the matter. This must be conceived of as taking place in the way in which a piece of wax takes on the impress of a signet-wing without the iron or gold.[23]

Kant's Copernican revolution turned this around. Kant looked upon perception as much more of an active process, where the subject gives structure to what

22 Gödel 1947/1964, p. 268. Supplement added for the reprint in Benacerraf and Putnam, p. 271-272.

23 *De Anima,* II 12, 424a18f.

we perceive. Husserl followed up this Kantian idea and studied this active structuring in great detail.

The duck/rabbit figure may be of help for understanding Husserl. It is nowadays well known to everybody, mainly through Wittgenstein, who gives credit to the Gestalt psychologist Jastrow (1900) for the figure. (See also Jastrow 1899). Actually the duck/rabbit figure first appeared in the German humor Magazine *Fliegende Blätter,* published in Munich in 1892,[24] the year after Husserl published his *Philosophy of Arithmetic,* where in Chapter XI he sets forth the basic ideas of the later Gestalt Psychology. Frege, in his critical review of this book exempts Chapter XI from his criticism and writes "I should like to direct the attention of psychologists especially to Chapter XI … ." However, the main developments in Gestalt psychology came later, from around 1910 on.

I should emphasize that Husserl never used the duck/rabbit figure. Ever since I started teaching Husserl fifty years ago I have found the figure pedagogically helpful as an aid to understanding Husserl. This may have misled some writers who use the phrase "Husserl's duck/rabbit figure." This is wrong. Nor is there any indication that the duck/rabbit figure was inspired by Husserl. This kind of thinking was very much "in the air" when Husserl studied. He was deeply influenced by his teacher Brentano and his ideas of intentionality, and expresses his gratitude at several occasions. Brentano led him to Hume, whom Husserl studied very carefully and regarded very highly. Another student of Brentano's and of Brentano's student Meinong, Christian von Ehrenfels, presented somewhat different ideas on the same issue, in the article "Über Gestaltqualitäten" (1890), one year before Husserl finished his *Philosophy of Arithmetic.*

Both Ehrenfels and Husserl give credit to Ernst Mach. It may be particularly appropriate to mention this here in the Vienna Circle, which was originally named "*Verein Ernst Mach"* in honor of Ernst Mach. Husserl owned and studied carefully Mach's Beiträge zur Analyse der Empfindungen (1886), and this is also the work Ehrenfels refers to.[25] However, in a letter to Ehrenfels after Ehrenfels had sent him his article, Mach mentions that he had already put forth the main ideas in an

---

24 October 23, 1892, p. 147. See Kihlstrom 2004 and McManus et al. 2010.

25 Husserl writes in a footnote in *Philosophie der Arithmetik*: The foregoing investigations had been worked out for almost a year before the penetrating work by Chr. Ehrenfels, "Ueber Gestaltqualitäten", *Vierteljahrsschrift f. wiss. Philos.*, 14, 1890, S. 249-292) appeared, in which the „figural Moments" investigated above only incidentally, in the interest of an explanation of the indirect apprehension of groups , are sujected to a comprehensive investigation . Unfortunately, the essay mentioned is not accessible to me as I prepare these pages for the press, so that I must forego a more full acknowledgment of it. As he states at the very opening of his presentation, Ehrenfels' research was instigated by E. Mach's *Beiträge zur Analyse der Empfindungen,* Jena 1886. Since I had read this work by the gifted physicist right after its appearance, it is quite possible that I was partly influenced in the process of my thought by reminiscences from that reading. (Husserl 1891, Husserliana XII, pp. 210-211, here quoted from Douglas Willard's translation, Husserliana: Collected Works, p. 223 n.)

earlier paper. Kevin Mulligan and Barry Smith has found out that this earlier paper is almost certainly his "Bemerkungen zur Lehre vom räumlichen Sehen," which came out already in 1865.[26]

Now to the figure. Here is the original figure from 1892:

When we look at this picture, we may see a duck or a rabbit. What reaches our eye is the same in both cases, so the idea that perception is merely reception must be wrong. The difference must be something coming from us. We structure what we see, and we can do so in different ways. The impulses that reach us from the outside are insufficient to uniquely determine which object we experience; something more gets added. Our consciousness becomes so structured that it is *as of* a duck or *as of* a rabbit. This structuring activity of our consciousness Husserl called the *noesis.* Each act has a *noesis*. The noeses are experiences, that is temporal events. When, looking at the figure, I see a duck, then flip over to seeing a rabbit and then flip back to seeing a duck, all these three acts of seeing will have distinct, that is numerically different, noeses. However, the first and the third noeses will be very similar. If they are perfectly similar, they both instantiate the same *noema.* The noemata are abstract structures, instantiated by the noeses. Peirce's distinction between type and token may be helpful here: the noeses are *tokens*, the noemata *types*.

It happens very rarely that two acts have perfectly similar noeses. In the duck/rabbit case, when I first see the figure as a duck and then discover that it could also be seen as a rabbit, and then flip back to seeing it as a duck again, the second see-

---

26 Mulligan, Kevin, and Smith, Barry, 1988, p. 125.

ing it as a duck will be different from the first. I learned from seeing it as a rabbit that I should not be too certain that the figure is of a duck. As soon as there is a difference between the noeses, the corresponding noemata will be different. Husserl devotes much attention to the noetic-noematic correlation. Usually, one can describe one's experience using either term.

Our consciousness structures what we experience. How it structures it depends on our previous experiences, the whole setting of our present experience and a number of other factors. If we had grown up surrounded by ducks, but had never even heard of rabbits, we would have been more likely to see a duck than a rabbit in the above figure; the idea of a rabbit would probably not even have occurred to us.

According to Husserl, all our experience could in principle be structured in different ways; what reaches our senses is never sufficient to uniquely determine what we experience. Only in a few rare cases, such as in the duck/rabbit example, can we go back and forth at will between different ways of structuring our experience.

Usually we are not even aware of any structuring going on; objects are simply experienced by us as having a structure. Like another of Brentano's students, Sigmund Freud, Husserl was very interested in the multifarious activities of our consciousness that we are not normally aware of. We shall now take a look at these various activities.

First we have the anticipations involved in perception. We have different anticipations when we see a duck and when we see a rabbit. In the first case we anticipate, for example, feathers that we will feel when we touch the object. In the latter case we expect to find fur. When we get the experiences we anticipate, the corresponding component of the noema is said to be *filled*. In all perception, there will be some filling: the components of the noema that correspond to what presently "meets the eye" are filled, and similarly for the other senses. We also have "anticipations" with respect to the future and the past. If I see a big tree in front of me and turn away from it for a moment, I expect it to be there when I turn back. Likewise, in such a case I expect the tree to have been there before I turned to it, and, for example, that it would appear in a photo taken in that same location the day before. We also have anticipations concerning value etc. These are important in ethics, but I will not discuss them here, since they are less relevant to the philosophy of mathematics.

Such anticipation and filling is what distinguishes perception from other modes of consciousness, for example imagining or remembering. If we merely imagine things, then our noema can be of anything whatsoever, an elephant or a locomotive standing here beside me. In perception, however, my sensory experiences are involved; the noema has to fit in with the experiences I have when sensory organs are impinged upon.

This eliminates a number of noemata which I could have had if I were just imagining. In my present situation I cannot have a noema corresponding to the perception of an elephant. This does not reduce the number of perceptual noemata I can have just now to one, for example, of you sitting there in front of me. It is a central point in Husserl's phenomenology that in every situation I can have a variety of different perceptual noemata that are compatible with the present impingements upon my sensory surfaces.

In the duck/rabbit case this was obvious; we could go back and forth at will between having the noema of a duck and having the noema of a rabbit. In most cases, however, we are not aware of this possibility. Only when something untoward happens, when I encounter "recalcitrant" experience that does not fit in with the anticipations in my noema, do I start seeing a different object from the one I thought I saw earlier. My noema "explodes," to use Husserl's phrase, and 1 come to have a noema quite different from the previous one, with new anticipations. This is always possible, says Husserl. Perception always involves anticipations that go beyond what presently "meets the eye," and there is always a risk that we may go wrong, regardless of how confident and certain we might feel. Misperception is always possible.

No intermediaries

It is now time to leave the duck/rabbit figure, since it has one feature that may easily lead us astray when we want to understand perception. Wittgenstein used the figure to illustrate his point that all seeing is seeing as. Seeing the duck/rabbit figure we may see it as a duck or as a rabbit. However, perception is not normally of pictures. In order to come closer to Husserl we should modify the example and consider not a picture, but a silhouette of the real animal against the sky. When we face such a silhouette against the sky, we may see a duck or a rabbit.

What is it in normal perception that can be seen in one way or in another way? Certainly not pictures. Philosophers have often appealed to sense data to answer this question. Sense data are supposed to be something unstructured that is grasped by our sensory organs and then structured into objects.

Husserl does not introduce any such intermediaries. There is nothing that is *given* that is structured in our acts. What constrains us, is not data that we observe and give structure to. What constrains us, are experiences which we typically have when our sensory organs are affected. Husserl calls these experience *hyle,* using the Greek word for matter. We may have hyle also when our nervous system is disturbed, for example by fever or drugs. But typically we have them when our sensory organs are impinged upon. For perception to take place, our hyletic experiences have to harmonize with our noeses.

In acts of perception the hyle play a constraining role. While in acts of imagination our acts can have a wide variety of noemata and corresponding objects, independently of what goes on at our sense organs, acts of perception are constrained to having noemata that harmonize, or are compatible with the hyle. This notion of compatibility is rather intricate, the hyle are experiences, they do not have properties that correspond to the properties of the object nor to properties of the noema of the object. The hyle have no colors and no shapes. They do have a temporal duration, but this duration does not correspond to the duration of the object, which usually outlasts our experience of it. Also, hyle cannot be reidentified from one act to another, different act, as has sometimes been claimed.[27]

The thetic component, existence, reality

This notion of harmonizing is important for understanding Husserl's theory of perception and its connection with his motion of reality. I mentioned earlier that in my present situation I cannot have a noema corresponding to the perception of an elephant. I can, however, have a noema corresponding to imagining an elephant. The two noemata differ in one important ingredient, the *thetic* component (and also in various other ways, that I shall not discuss here). The noema in an act of perception attributes *existence* to its object, it is experienced as *real*. The noema in an act of imagination does not attribute reality to its object, and there is no need for hyletic experiences or other constraints that have to fit in with our noetic experience.

The thetic character of acts is highly important in Husserl's idealism. The study of the thetic character gives us insight into what it means for the world and its objects to be:

> Phenomenological idealism does not deny the factual [*wirklich*] existence of the real [*real*] world (and in the first instance nature) as if it deemed it an illusion … Its only task and accomplishment is to clarify the sense [Sinn] of this world, just that sense in which we all regard it as really existing and as really valid. That the world exists … is quite indubitable. Another matter is to understand this indubitability which is the basis for life and science and clarify the basis for its claim.[28]

The constraints are crucial for our experience of the world as real. What are these constraints? That the object impinges on our sensory organs are, as we noted, one constraint. We have a conception of the world where things that I see have to be in a certain location relative to my eyes. It does not suffice that they are in front

27 For example by Gurwitsch, 1966, p. 256. That hyle cannot be reidentified from act to act is stressed by Husserl several places, for example in the manuscripts *Analysen* zur *passiven Synthesis*, Husserl 1918-1926, Husserliana XI, 363.

28 Husserl 1930, "Nachwort zu meinen 'Ideen zu einer reinen Phänomenologie und phänomenologischen Philosophie," pp. 549-570. Husserliana V, pp. 138-162. English version used as Preface to the Gibson's translation of *Ideas*, Allen & Unwin, London, 1931. Here my translation from the German version in Husserliana, V, 152.32-153.5. For more on Husserl's idealism, see Follesdal 1998.

of me. If I am turned toward a big tree and there is a rabbit behind it, I do not have hyle associated with the rabbit, even if I should happen to know that there is a rabbit there. There is therefore a comprehensive interplay between my conception of the world and of myself in this world and the hyletic experiences I undergo.

Intersubjectivity

Also I take others to be similar to myself in important respects, for example in not being able to see through trees. Husserl devotes great attention to the way in which we take ourselves to be living in a shared world, which we experience from different perspectives, both because of our different location in that world and because of our different past experiences that give rise to different anticipations as to what the world is like and also, to some extent, because of differences in the acuity of our senses. Although our conceptions of the world may differ to some extent, we experience ourselves as living in a common world and we can learn from one another what this world is like. Perception, which might seem so much an individual activity, is dependent on intersubjective adaptation. Husserl studied this adaptation in great detail and concluded that "even what is straightforwardly perceptual is communal."[29]

More on anticipation

Our anticipations usually relate to general features of the world. For example, when we see a figure like this:

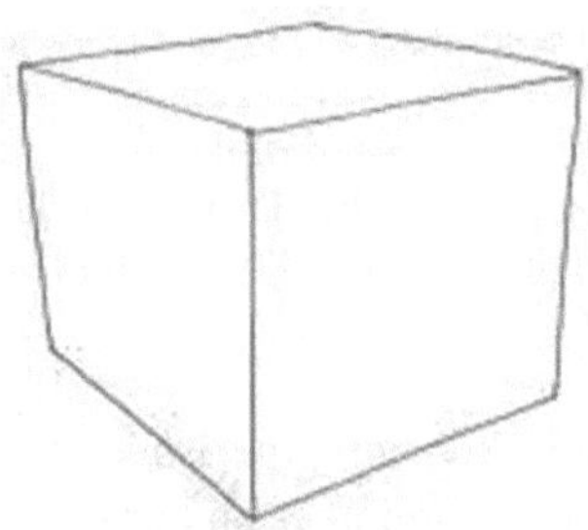

We may take it to be a cube, or maybe three flat areas put up to form a corner, or maybe just nine lines meeting in various points, or any of a number of other configurations There is an interdependence between these first anticipations and our further anticipations. If we take it to be a cube, we anticipate that it will change its shape in specific ways if we move it or if we move around it. We may not be familiar enough with geometry to be able to specify our anticipations very accurately, but on the basis of the sedimentations left in us from previous experience

29 Husserl 1936, Husserliana VI, § 47, 166.19-22. Carr's translation p. 163.

of things moving in space we pretty quickly spot deviations from a cube's normal appearances, if there should be any.

Our anticipations need not be based on experience with this particular cube, any cube will do. The cube-shape is then the object of our act, and we learn how that shape appears differently from different perspectives. Husserl calls such objects, in this case the cubic shape, *essences* (*Wesen* and also *eidos*). This particular cube is a physical object, its shape is a general object, an essence. Words that end on '–ness- and '–ity' are usual names for essences in English: triangularity, roundness, greenness, etc.

Very many of our activities in daily life pertain to essences and not to individual physical objects. Consider what Quine and Davidson call "triangulation" in language learning, for example a mother who teaches a child the word 'dog' by pointing to a dog and saying 'dog'. The child is then rewarded for uttering 'dog' next time a dog appears. Apparently, the example involves three physical objects: mother, child and dog. However, then the father comes along, puts the child on his lap and says when a dog appears: 'Fido'. The child may be a little confused, but thinks that perhaps there are two words for dogs. Next time a dog appears, the child says 'Fido', but is not rewarded. This new dog was not Fido, but another dog. The word 'dog' is a general term, while 'Fido' is a singular term. In order to learn the difference the child must be able to individuate the contents of the world, it has to master the difference between the two contrasting pairs 'like-unlike' (or similar-different) and 'identical-distinct':

	identical	distinct
like (similar)		
unlike (different)		

Two drops of water may be similar, but being two, they are not identical. I myself am different now from when I was a child, but I am identical to the individual I was then. Individuation and counting come as a package, we cannot have one without the other. Distinguishing mass terms, like water, from terms with divided reference, like glass, is part of that same package. Quine discusses this in *Word and Object* and comes back to it in several later works, pointing out, for example, that individuation is also closely connected with our conception of space and time:

> Distinguishing a body from an utterly similar one at a later time is what calls for the next big development. It requires acquisition of our whole schematism of space and time and the unobserved trajectories of bodies within it.[30]

As we shall see, Husserl and Gödel make very similar observations, and Quine in interview with Giovanna Borradori in 1994, said "I recognize that Husserl and I, in very different ways, addressed some of the same things."[31]

The determinable X

We are now ready to proceed to a further important component of the noema: the determinable X. The determinable X, which Husserl also calls the object pole of the noema, is a component of the noema that corresponds to the act being directed towards an object. We are not just experiencing a heap of features, but features of objects. We structure the world into objects, that have properties, bear relations to one another, etc. The determinable X has two characteristic features, both of which have to do with individuation:

First, it constitutes the object pole, around which the other components of the noema are grouped, such that our experience is of an object, with all its properties, an object that has more to it than what meets the eye, and that may remain the same object although its properties change. The object is experienced as the same through all changes in its properties and in our perspective:

> It *points ahead* to possible perceptual multiplicities which, merging continuously into one another, join together to make up the unity of one perception in which the continuously enduring thing is always showing some new "sides" (or else an old "side") in a new series of adumbrations.[32]

Secondly, the determinable X also gives sense to our notion of several objects being numerically distinct although they may be very similar. When we encounter an object that is very similar to one that we encountered before, we may still raise the question "Is this the same object, or a distinct one?" And conversely, an object may change its properties, look very different from what it did before, and still be the same object. The notion of the determinable X makes us distinguish the two pairs of opposites that we just mentioned: identity versus distinctness and similarity versus difference. Things may be distinct in spite of similarities. And things may be identical in spite of changes.

The determinable X is hence what makes sense of what goes wrong in cases of mistaken identity. "What cannot all step into the unity of an identifying conscious-

---

30 Quine 1995, p. 36.

31 "Twentieth-Century Logic," Quine 2008, p. 64.

32 Husserl 1913, § 44, Husserliana III,1, p. 91.27-32 . Here quoted from Kersten's English translation, p. 91, slightly modified.

ness, what cannot a person mix up " Husserl writes in a manuscript from 1911.[33] In this manuscript he discusses what he calls the "twin world" problem and other examples which show the inadequacy of his earlier theory of demonstratives. Husserl discusses in this manuscript the following problem:

> But how is it, if on two heavenly bodies two people in completely similar appearances of the surroundings are considering "the same" objects and make "the same" judgments about them? Does not the "this" then in these two cases have a different meaning?[34]

Husserl made several interesting observations concerning this and similar examples, without arriving at a satisfactory solution to the problems they raise. His idea of the determinable X enabled him to make sense of the problem, which Hilary Putnam sixty years later thought of independently, under the heading "twin earth" and which has later been widely discussed.[35]

Essences

Now on the essences: When we stand looking at an object, for example a traffic sign:

We may have this individual physical object as object of our act. That is, our act may be directed towards that individual physical object, that sign that we have seen in that particular location for years, and which has gradually started to fade.

However, the object of our act may also be an essence, for example the triangular shape of the object. It may be directed towards triangularity or towards equiangularity or perhaps equilaterality or some other essence that is instantiated by this figure. If the essence towards which the act is directed is triangularity we may replace the traffic sign with a nicely trimmed Christmas tree and our act will still have the same object:

33 Husserl 1908, Husserliana XXVI,213.5-6. my translation.

34 Husserl 1908, Husserliana XXVI,211.44-212.2, my translation.

35 Putnam 1973. For discussions of Husserl's view on indexicals, see Beyer 1996, pp. 172 ff.; Smith, David 1989; Künne 1990, and Schumann 1993. See also McIntyre 1982.

For a different kind of essence let us look at the following three figures:

Which of them are you regarding as similar to one another? Many will decide for the cup and the glass, they are both suitable for drinking. Some few, mostly students of topology, will point to the cup and the donut, because they can be turned into one another through continuous movements, not closing any holes, nor opening any new ones. They are "one-hole" objects, unlike the glass. Both answers are right. Among the many essences that are instantiated by both the cup and the glass are the practically useful feature of being drinking vessels, while the cup and the donut share the theoretical feature of one-hole-ness.

Essential insight = eidetic intuition

In our daily lives we are much more frequently concerned with essences than with individual physical objects. The individuality of the object is important for us mainly when we relate to objects to which we have strong emotional ties, such as persons whom we feel close to or objects that relate to such persons, for example an engagement ring or a clock that was made by my grandfather, or in the earlier example the dog Fido. We normally relate to human beings as particular individuals, but even humans are often treated as just instances of a kind. Even relations that are traditionally regarded as profoundly human and expressing deep emotions may be turned into relations to a member of a kind, for example to anyone who is available for sexual satisfaction.

Husserl regards perception, where the object of our act is an individual physical object, as only a special kind of a more general kind of acts, which give us insight into the object of the act. Husserl calls these acts intuitive acts, or intuitions [*Anschauung*]. There are hence two kinds of intuition: perception, where the object of an act is a physical object, and essential insight [*Wesensschau*], where the object is an essence.

Intuitions differ from the many kinds of other acts by their thetic component: what we intuit is regarded by us as real, not as merely imagined. Intuition gives us insight into the object of the act. When we intuit an object we anticipate various features of the object. Sometimes our anticipations get confirmed by our further experiences, other times they are not fulfilled and we have to revise our view of what the object is like. As we noted earlier, reality and constraints go together. Just as perception differs from fantasy, so does intuition in general.

We are now in a position to make sense of the cryptic passage from Gödel which we quoted earlier, and we see that it gives us in a nutshell the answers to the two questions concerning the foundations of mathematics that we started with. First, are there non-material objects?

> That something besides the sensations actually is immediately given follows (independently of mathematics) from the fact that even our ideas referring to physical objects contain constituents qualitatively different from sensations or mere combinations of sensations, e.g., the idea of object itself … .

That is: perception is not just a matter of a physical object affecting our sensory organs. There is much more structure to perception. We perceive physical objects, but the notion of a physical objects involves a comprehensive structure that includes arithmetic, geometry and other branches of mathematics. The notion of a physical object contains, for example, the idea of individuation: features are grouped together as belonging to the same individual. Individuals are distinct from one another although they may be similar, and they may keep their identities through changes. This structuring into objects, their properties and relations between them is crucial to perception. These abstract elements are incorporated into what we perceive; there is no perception without such elements. They are features of the world we experience, they are not objects of perception, but objects of eidetic intuition. We anticipate these features, but this does not make them any more subjective than the physical objects, which are themselves part of this structure. Now the second question: How do we know about them?

Evidently, the 'given' underlying mathematics is closely related to the abstract elements contained in our empirical ideas. It by no means follows, however, that the data of this second kind, because they cannot be associated with actions of certain things upon our sense organs, are something purely subjective, as Kant asserted. Rather they, too, may represent an aspect of objective reality, but, as opposed to the sensations, their presence may be due to another kind of relationship between us and reality.

So, we experience these "abstract elements contained in our empirical ideas" not because they are physical objects that affect us causally, but because they are elements of that same objective reality. They are anticipated in our experience of the world, and the further course of our exploration of the world may lead us to give up or revise some of these anticipations. So even "the abstract elements contained in our empirical ideas" are subject to constraints and revision.

As we noted earlier, the experience of constraints is crucial for our regarding something as real. We may believe and anticipate, but our further experience may correct us. To experience something as real involves that we experience that its existence and properties are not fully up to us. Mathematics is not like chess. In chess we are constrained by rules we ourselves make. But we are free to change the rules, and this has been done up through the history of chess, just as rules has been changed in soccer and other sports. One may object that the situation is similar in mathematics, for example in geometry, where one has developed new geometries. It turned out that non-Euclidean geometry became useful in physics. However, from a mathematical point of view, each geometry is an exploration of a structure.

The same holds in all branches of mathematics. Some structures turn out to be very interesting from a purely formal view. For example, in algebra, the simple structure called 'group' opened a new world, and there are several other ideas in mathematics that similarly have led to new insights and the discovery of highly interesting structures.

Gödel often came back to the connection between the constraints in mathematics and the existence of mathematical objects:

> … the activity of the mathematician shows very little of the freedom a creator should enjoy. Even if, for example, the axioms about integers were a free invention, still it must be admitted that the mathematician, after he has imagined the first few properties of his objects, is at an end with his creative ability, and he is not in a position also to create the validity of his theorems at will. If anything like creation exists at all in mathematics, then what any theorem does is exactly to restrict the freedom of creation. That, however, which restricts it must evidently exist independently of the creation.[36]

36 Gödel 1951, p. 314.

More examples of essences

Husserl regarded topology as a mathematical discipline that is particularly rich in examples of essences that are understandable by everybody. Here is an example from a branch of topology, the topology of knots, that was developed after Husserl suggested this, but that I find particularly apt to illustrate what Husserl had in mind. Look at the following two knots:

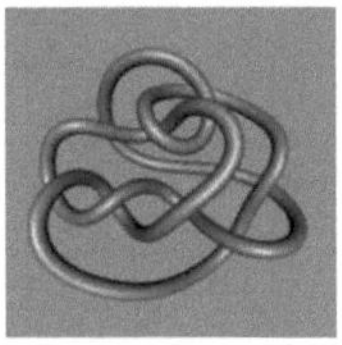

Is there any similarity between them? The study of such structures was started off by the mathematician Kurt Reidemeister (1893–1971), who as a student in Freiburg im Breisgau followed lectures by Husserl after Husserl came there in 1916. Reidemeister came in 1922 to Vienna as an Associate Professor. Through Hans Hahn, Gödel's later teacher, he came in contact with the ideas of the Vienna Circle. He became a junior colleague of Wilhelm Wirtinger, who had begun to study the topology of knots and encouraged Reidemeister to devote himself to this field.

In 1925 Reidemeister became a full professor in Kant's city Königsberg, where he made some of his most important contributions to the topology of knots. In September 1930 Reidemeister co-organized four coinciding conferences in Königsberg, where both Hilbert and Gödel lectured and several members of the Vienna Circle were invited. Gödel presented his incompleteness theorem, which he had just proved.

In 1933 Reidemeister was forced to leave his professorship. He was not a Jew, but he was in strong opposition to the Nazi movement. He received a position in Marburg, where he remained until after the war. In 1955 he became professor in Göttingen, where I took his courses, and could experience his ability to find illuminating ways of looking at the mathematical structures.

Knots come in all shapes, as we know from fishing trips or knitting or dealing with our shoe laces. Often they seem quite chaotic. However, Reidemeister created order in chaos. What he did, is what we strive for in all intellectual pursuits, typically in philosophy and in the sciences, but also in so many other fields: he tried to find a way of looking at the field that made us see the structure, see how things "fall into place." He showed in 1927 that two knots are the same when and only when the one can be changed into the other through a finite series of what we now call Reidemeister Movements. These are three very simple types of movements:

Type 1:

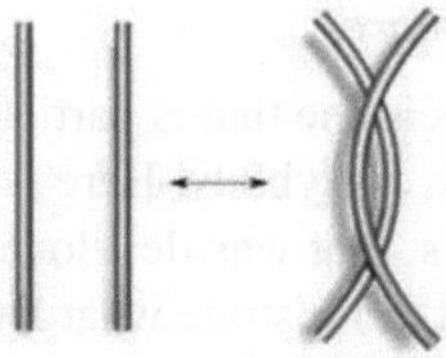

Type 2:

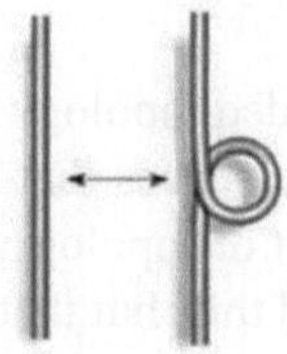

Type 3:

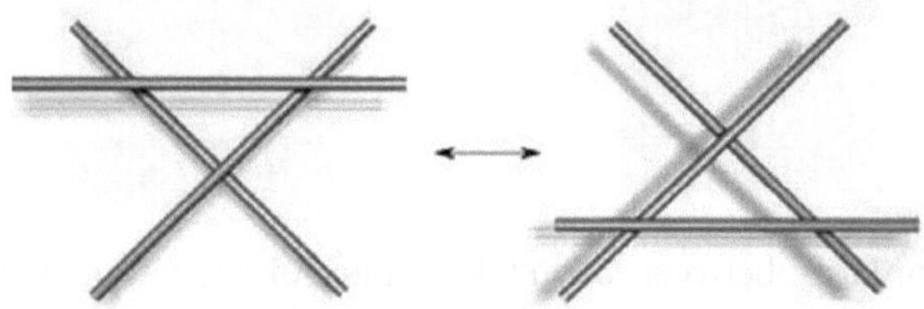

Reidemeister's contribution to he topology of knots is a good example of the exploration of what Gödel, following Husserl, called the "abstract elements contained in our empirical ideas," or also "Wesensschau." We are studying structures, and although the knots may be man-made, the structures are not. The knot example hence throws some light on the two basic questions in the philosophy of mathematics that I mentioned in the beginning:

– Are there non-material objects?

– If so, how can we know about them?

Although we started from some material objects, knots, we very soon began to disregard the individuality of these objects and focused on their structure. The objects studied in mathematics are not the material objects, but the structures they exemplify. A material object will normally exemplify many abstract features that can be studied by mathematicians and sometimes give rise to new branches of mathematics. Husserl mentions several such examples and in particular he envisioned several new branches of topology.

We need not confine ourselves to structures that are exemplified in the physical world. Often modifications, generalizations etc. of the structures we start with lead to new structures that are not found in the physical world, but that turn out to have interesting features that engage mathematicians. While some branches of mathematics are not instantiated in the physical world, others, like arithmetic, are instantiated everywhere. Arithmetic is based on the feature mentioned by Gödel and explored by Husserl, Quine and many other philosophers: individuation.

The idea is hence that what we study in mathematics is as much a feature of what we experience as are material objects.

Here are some more passages from Gödel that support this Husserl-inspired interpretation. First, the objective existence of mathematical objects:

The question of the objective existence of the objects of mathematical intuition (which, incidentally, is an *exact replica* of the question of the objective existence of the outer world) … .[37]

Perception and mathematical intuition:

But despite their remoteness from sense experience, we do have something like a *perception* of the objects of set theory, as is seen from the fact that the axioms force themselves upon us as being true. I don't see any reason why we should have less confidence in this kind of perception, i.e., in mathematical *intuition*, than in sense perception, which induces us to build up physical theories and to expect that future sense perceptions will agree with them.[38]

Constraints. Here we have the passage that was quoted earlier:

… the activity of the mathematician shows very little of the freedom a creator should enjoy. Even if, for example, the axioms about integers were a free invention, still it must be admitted that the mathematician, after he has imagined the first few properties of his objects, is at an end with his creative ability, and he is not in a position also to create the validity of his theorems at will. If anything like creation exists at all in mathematics, then what any theorem does is exactly to restrict the freedom of creation. That, however, which restricts it must evidently exist independently of the creation.[39]

Inductive methods in mathematics

As we have seen, Husserl and Gödel draw several parallels between our experience of the physical world and our experience of the abstract world studied in mathematics. As we should expect, this leads to similarities also in their method. Here is a concise and important passage about this from Gödel's Gibbs lecture (1951):

If mathematics describes an objective world just like physics, there is no reason why inductive methods should not be applied in mathematics just the same as in physics. The fact is that in mathematics we still have the same attitude today that in former times one had toward all science, namely we try to derive everything by cogent proofs from the definitions (that is, in ontological terminology, from the essences of things). Perhaps this method, if it claims monopoly, is as wrong in mathematics as it is in physics.[40]

---

37 Gödel 1947/1964, p. 268. Supplement added for the reprint in Benacerraf and Putnam, p. 272. My emphasis.

38 Gödel 1947/1964, p. .268. Supplement added for the reprint in Benacerraf and Putnam, p. 271. My emphasis.

39 Gödel 1951, p. 314.

40 Gödel 1951, p. 313.

Already seven years earlier, in his contribution to the Russell volume in the Library of Living Philosophers, Gödel wrote the following about classes and concepts:

> It seems to me that the assumption of such objects is quite as legitimate as the assumption of physical bodies and there is quite as much reason to believe in their existence. They are in the same sense necessary to obtain a satisfactory system of mathematics as physical bodies are necessary for a satisfactory theory of our sense perceptions and in both cases it is impossible to interpret the propositions one wants to assert about these entities as propositions about the "data," i.e. in the latter case the actually occurring sense perceptions.[41]

Note that the article on Russell and the Gibbs lecture antedate Gödel's study of Husserl. This confirms that Gödel had ideas similar to those of Husserl long before he started to study him. Presumably, it was the systematic framework that Husserl provided for these ideas that impressed him.

So, Gödel maintained, like Husserl, that mathematics, like physics, should proceed by hypotheses: we set forth a set of hypotheses that seem to fit in with the boundary conditions, we study how they fit in and derive consequences from them to see whether these consequences also fit in. We shall come back later to what kind of evidence we can have for such mathematical theories. But let us now first focus on a related issue, the role of concepts in mathematics.

Concepts in mathematics

In scientific theories our concepts are modified when the theory develops, and they can be radically changed if the theory undergoes changes. Examples abound in all sciences: atom, mass, gene, etc. A good choice of concepts helps to improve the theory, make it simpler, more comprehensive, unifying it with other theories, etc. One sometimes talks about the concepts changing, other times about words denoting different concepts from time to time. Husserl and Gödel talk about concepts being made clearer through the progress of theories. Given the parallel between mathematics and the sciences, this can be illustrated by examples both from the sciences and from mathematics. Let us consider a couple of examples from mathematics.

First, an example from Bernard Bolzano (1781–1848). In the early part of the nineteenth century mathematicians took it for granted that continuous functions are differentiable except in particular points. Ampère even believed that he had proved this, in a paper from 1806. However, around 1830 Bolzano gave an example of a function which is everywhere continuous, but nowhere differentiable.[42]

Husserl admired Bolzano and praised his work, although he did not know this particular result of Bolzano's. Because his political views challenged Emperor Franz and the other authorities in Vienna, Bolzano was from 1820 on not permit-

41 Gödel 1944, p. 137.

42 For a history of the subject, see Thim.

ted to publish in Austria, not even in logic and mathematics. This result and several others of his important insights therefore lay unknown until the 1920s. when they were discovered by people going through his manuscripts.

However, Husserl knew this result and many others from his teacher Weierstraß, who proved them independently thirty years later. Bolzano arrived at his example by using his "method of variation," which Husserl knew from the Bolzano's *Wissenschaftslehre*, which was published in 1837. Husserl made use of this method to study various essences, among them some topological notions that he used as examples. The method consists in finding examples that satisfy the concept one wants to clarify, for example continuity, and then vary the examples to see what variations are permitted. Most concepts, including those that we study in mathematics, can be instantiated also by imagined objects. We can imagine red objects, round objects, and functions of various kinds. It is much easier to vary an object in imagination than in the material world around us. In mathematics, as well as in other fields, we therefore often vary objects in our imagination. Bolzano imagined various functions and curves that depict them, and varied their properties in imagination. Through this "method of variation" he came up with a function that is continuous, but nowhere differentiable. Actually, the function he came up with is simpler than that of Weierstraß, testifying to Bolzano's ability to see exactly what is needed. Here, as in many other fields, he saw the issues very clearly and was able to communicate to his readers both this particular insight and the general structure of which it is a part.

It is worth reflecting on what happens when in mathematics or in a science one achieves such a "clarification" of concepts, which usually is part of a simplification of the whole field, and sometimes opens up a new way of looking at this field. Bolzano, Husserl and Gödel talk as if there are concepts waiting to be discovered. Others talk about "conceptual change" and say that the word 'continuous' got a new meaning through Bolzano's work. There is no doubt that the word 'continuous' got a new meaning through Bolzano's work. However, in this case, as in most similar cases, it would be more appropriate to say that the word got a more precise meaning. Ampère and others who took it for granted that every continuous function was differentially in most points, would not say that Bolzano cheated, that he used the old word in a new sense. Such changes of meaning some times take place both in the sciences and in daily life. However, what Bolzano did, was much more interesting. Mathematicians who had believed that continuous functions were differentiable in most points, would acknowledge that Bolzano had proved them wrong.

Note that we should not take it for granted that when people use the same word, they are expressing the same concept. Solomon Feferman has pointed out that the word 'continuum' which is used in the formulation of the continuum hypothesis (not to be mixed up with the notion of 'continuity' discussed by Bolzano and Weierstraß), may be used for many different concepts:

On the face of it, there are several distinct forms of the continuum as a mathematical concept: in geometry, as a straight line, in analysis as the real number system (characterized in one of several ways), and in set theory as the power set of the natural numbers and, alternatively, as the set of all infinite sequences of zeros and ones. Since it is common to refer to *the* continuum, in what sense are these all instances of the same concept? When one speaks of the continuum in current set-theoretical terms it is implicitly understood that one is paying attention only to the cardinal number that these sets have in common. Besides ignoring the differences in structure involved, that requires, to begin with, recasting geometry in analytic terms. More centrally, it presumes acceptance of the overall set-theoretical framework.[43]

On the other hand, we should not take it for granted that the intuitive notions that develop when we study the physical world around us, correspond to the more exact definitions that emerge when we work out the mathematics for these notions. Hermann Weyl, who as we noted, was inspired by Husserl, wrote about this:

To the criticism that the intuition of the continuum in no way contains those logical principles on which we must rely for the exact definition of the concept "real number," we respond that the conceptual world of mathematics is so foreign to what the intuitive continuum presents to us that the demand for coincidence between the two must be dismissed as absurd. Nevertheless, those abstract schemata which supply us with mathematics must also underlie the exact science of domains of objects in which continua play a role.[44]

Clarification of concepts in this way goes hand in hand with improvements of scientific hypotheses. As science progresses, we adapt its theories and concepts to one another so as to best satisfy the scientific norms of simplicity etc. This reflects, I think, that the situation is as Plato decribed it, that conceptual clarification seeks to carve nature at its joints.

Another example of how insight in structures and insight in clarification of concepts go hand in hand, is the clarification of the concept of general recursiveness (or Turing computability) in the early thirties. Gödel praises its great importance:

It seems to me that this importance is largely due to the fact that with this concept one has for the first time succeeded in giving an absolute dfinition to an interesting epistemological notion, i.e., one not depending on the formalism chosen.[45]

Later, Gödel in a similar way praised Abraham Robinson for his introduction of non-standard analysis:

… there are good reasons to believe that non-standard analysis, in some version or other, will be the analysis of the future. One reason is the just mentioned simplification of proofs,

---

43 Feferman 2009, p. 169.

44 Weyl, 1918, p. 108

45 Gödel 1946, p. 150..

> since simplification facilitates discovery. Another, even more convincing reason, is the following: Arithmetic starts with the integers and proceeds by successively enlarging the number system by rational and negative numbers, irrational numbers, etc. But the next natural step after the reals, namely the introduction of infinitesimals, has simply been omitted. I think it will be considered a great oddity in the history of mathematics that the first exact theory of infinitesimals was developed 300 years after the invention of the differential calculus.[46]

Gödel here gives two reasons for his praise of non-standard analysis: it simplifies proofs and it is a next natural step after the reals. This leads us to our last section, a brief survey of the kinds of evidence Gödel recognizes in mathematics.

Gödel on evidence in mathematics

Gödel's view that we have something like a perception of mathematical objects does not mean that mathematics can achieve certainty. Neither perception nor eidetic intuition are infallible sources of evidence. They always involve anticipations concerning aspects of their objects that have not yet been explored, and which may turn out to be wrong. As we have noted, Husserl acknowledges that errors are always possible, even in mathematics and logic.

Gödel sketches *four* different methods one can use to get insight into the mathematical realm:[47]

1. *Elementary consequences*

Just as we test highly general physical theories by trying to derive observable predictions from them, Gödel noted in his Russell article, as you remember, that mathematical objects

> are in the same sense necessary to obtain a satisfactory system of mathematics as physical bodies are necessary for a satisfactory theory of our sense perceptions[48]

Gödel also points out that very recondite axioms can have elementary consequences. Thus, discussing various axioms of infinity in "What is Cantor's Continuum Problem," Gödel observes that

> … it can be proved that these axioms also have consequences far outside the domain of very great transfinite numbers, which is their immediate subject matter: each of them, under the assumption of its consistency, can be shown to increase the number of decidable propositions even in the field of Diophantine equations.[49]

---

46 Gödel 1974, p. 311.

47 The following two pages are excerpted from my 1995 article "Gödel and Husserl."

48 Gödel 1944, p. 128.

49 Gödel 1947/1964, p. 261. Supplement added for the reprint in Benacerraf and Putnam, p. 264.

2. *"Success"*

In "What is Cantor's Continuum Problem" Gödel points to another reason we may have for thinking that a proposed axiom is true:

> A probable decision about its truth is possible also in another way, namely, inductively by studying its "success." Success here means fruitfulness in consequences, in particular in *verifiable* consequences, i.e. consequences demonstrable without the new axiom, whose proofs with the help of the new axiom, however, are considerably simpler and easier to discover, and make it possible to contract into one proof many different proofs. The axioms for the system of real numbers, rejected by the intuitionists, have in this sense been verified to some extent … .[50]

So, in this case what matters is not the derivation of new theorems, but that the axiom permits us to derive old theorems in a more elegant way.

Such simplification, which makes proofs "simpler and easier to discover, and makes it possible to contract into one proof many different proofs," is often achieved also by finding new ways of looking at old material. Here is an example from my own experience: While I was studying mathematics in the 1950s, we used an old German textbook in functional analysis, which I found very unsatisfactory. It did not provide a good overview of the field, proofs were divided into a large number of cases which had to be dealt with separately, etc. Then, in the middle of the term, our professor told us to replace the old book with a new one that had just come out: Frédéric Riesz and Béla Sz.-Nagy, *Leçons d'analyse fonctionelle*.[51] Shifting to the new book opened the field for us students, we saw the structure of the field, how everything was interconnected and fell into place; the proofs were simple and perspicuous. They gave us evidence, to use Husserl's and Gödel's word. Certainly, the proofs in the old German book provided evidence, too, we could see how each step was according to rules. However, they did not give us the kind of intuitive insight that for Husserl and Gödel is the ultimate evidence for the rules. In this example no new axioms were involved, but the evidence was of the same kind as what Gödel discusses in connection with ne axioms.

3. *Clarification*

Sometimes, we find that the axioms we have been setting up so far, are satisfiable by two or more quite differently defined classes of objects. Then, by reflecting upon the concepts we are trying to capture by these axioms, we may find that one of these classes does not fit the concept. In "What is Cantor's Continuum Problem" Gödel remarks that

50 Gödel 1947/1964, p. 261. Supplement added for the reprint in Benacerraf and Putnam, p. 265.

51 Riesz/Sz.-Nagy 1952

… it may be conjectured that the continuum problem cannot be solved on the basis of the axioms set up so far, but, on the other hand, may be solvable with the help of some new axiom which would state or imply something about the definability of sets.[52]

4. *Systematicity*

In the manuscript from 1961 where Gödel discusses Husserl's phenomenology, he mentions a fourth way in which we can arrive at new axioms, viz. by arranging the axioms in a systematic manner, which enables us to discover new ones:

… it turns out that in the systematic establishment [*Aufstellen*] of the axioms of mathematics, new axioms, which do not follow by formal logic from those previously established, again and again become evident. It is not at all excluded by the negative results mentioned earlier [incompleteness] that nevertheless every clearly posed mathematical yes-or-no question is solvable in this way. For it is just this becoming evident of more and more new axioms on the basis of the meaning of the primitive notions that a machine cannot imitate.[53]

Husserl, who had a "reflective equilibrium" approach to justification,[54] attaches great significance to systematization as a way to clarify concepts. It is an important factor in his conception of justification.

Husserl and Gödel's discussion of evidence in mathematics is relevant to not only questions of new axioms, but also to the analysis of proofs and how they provide evidence. One problem that often confronts editors, is to decide whether a proof is new or not. As an editor, I once received a manuscript where an important open problem was solved. I sent it to a referee whom I regarded as the best expert in this field, and he immediately wrote back and told me that he has solved this problem two weeks earlier and sent his manuscript to the other editor of the journal. I checked with my co-editor, and he confirmed that he had received the manuscript, postmarked two weeks earlier than the one I had received. The proofs were, as one would expect, not line by line identical, but they were based on the same crucial ideas, and only the first one could be published, but with a note telling about the second proof, which clearly was arrived at independently.

An example of the opposite kind, where a second proof is clearly different from the first, is Dana Scott's proof of the independence of the continuum hypothesis. This hypothesis, which was set forth by Georg Cantor in 1878, states that in the hierarchy of infinities the integers are immediately followed by the continuum, that is by the real numbers – there is nothing in between. The problem of proving or disproving the continuum hypothesis was the first of the 23 famous problems that Hilbert presented at the International Congress of Mathematicians at Paris in 1900. In 1940 Gödel proved that the continuum hypothesis is consistent with the standard axioms of set theory. He worked for years trying to show that it is in fact

52 Gödel 1947/1964, p. 262. Supplement added for the reprint in Benacerraf and Putnam, p. 266.

53 Gödel 1961, p. 385.

54 See Follesdal 1988.

independent of these axioms, that is, that it can also not be disproved from these axioms. This was proved by Paul Cohen in 1963, using a new, very fruitful idea of "forcing."

Many of the ideas in this proof arose through discussions between Cohen and Feferman, who were colleagues at Stanford; Feferman reported that his contribution to these discussions was mainly to act as a sounding board and to help verify Cohen's arguments (cf. Moore 1988, p. 156). In 1967 Scott published a paper, "A Proof of the Independence of the Continuum Hypothesis," in which he used so called Boolean-valued models to provide an analysis of the independence of the continuum hypothesis that was very unlike that of Cohen, based on different evidence and insight. (As stated by Scott, the method of Boolean valued models was due independently to Robert M. Solovay as an alternative means to obtain the independence results previously found by the method of forcing.) As is typical of Scott's work, this new analysis and proof is very simple and crystal clear and thereby helps us to better see the structure of the field.

Once the independence of the continuum hypothesis has been established the next question is "Is it true or false?" Very interesting work is being done on this issue by Hugh Woodin and others, who consider new axioms, which unify the field and from which the continuum hypothesis or its negation follows. These considerations exemplify all the four kinds of evidence that were mentioned above.

If one does not accept the platonist view that Husserl and Gödel argued for, one will not speak of the truth or falsehood of axioms or of the continuum hypotheses, but regard them as conventions or as having some other status. It then becomes a challenge to explain what status they have, and also to explain why some elements in perception are not real, while other elements in perception presumably are. Husserl and Gödel's analysis of perception as a special case of intuition is a key to their arguments concerning the existence and knowability of mathematical objects.

Let me finally mention that considerations that are used to determine whether proofs in mathematics are original or just old ones in new clothing are akin to those one uses to determine what is importantly similar and what is different in two pieces of music. This is not astonishing – both mathematics and music are disciplines in which the notion of structure is important. I once taught a seminar on phenomenology and music together with Lawrence Ferrara, Professor of Music at New York University, who in addition to teaching music and philosophy also has taught music copyright at Columbia University Law School and Harvard Law School and is a copyright consultant for courts and artists, particularly famous pop artists, who are often plagiarized and where it is difficult – and often important – to determine what is creative and original and what is derivative.[55]

55 A major part of Ferrara 1991 is devoted to phenomenology and music.

References

Benacerraf, Paul / Putnam, Hilary, eds.

1964 *Philosophy of Mathematics: Selected Readings.* Englewood Cliffs, New Jersey: Prentice-Hall.

Beyer, Christian

1996 *Von Bolzano zu Husserl – Eine Untersuchung über den Ursprung der phänomenologischen Bedeutungslehre* (Phaenomenologica 139). Dordrecht: Kluwer.

Davis, Martin, ed.

1965 *The undecidable. Basic papers on undecidable propositions, unsolvable problems and computable functions.* Hewlett, N.Y.: Raven Press.

Dreyfus, Hubert L. ed.

1982 *Husserl, Intentionality and Cognitive Science.* Cambridge, Mass.: MIT Press.

Ehrenfels, Christian von

1890 "Über Gestaltqualitäten," in *Vierteljahrsschrift für wissenschaftliche Philosophie* 14, pp. 249-292.

Feferman, Solomon

1988 "Weyl vindicated: *Das Kontinuum* 70 years later," in *Temie prospettive della logica e della filosofia della scienza contemporanee*, I, Bologna (1988), 59-93. Reprinted in Feferman 1998, pp. 249–283.

1998 *In the Light of Logic.* Oxford, Oxford University Press.

2000 "The significance of Hermann Weyl's *Das Kontinuum*," in Hendricks et al., (eds.), *Proof Theory.* Dordrecht: Kluwer.

2009 "Conceptions of the Continuum," *Intellectica* 51 (2009/1), 169-189

2012 Three Problems for Mathematics: Lecture 2: Is the Continuum Hypothesis a definite mathematical problem?, slides for inaugural Paul Bernays Lectures, ETH, Zurich, Sept. 12, 2012.

Feferman, Solomon, ed.

1986 (with J.W. Dawson, Jr., S. C. Kleene, G. H. Moore, R. M. Solovay, and J. van Heijenoort, eds.), *Kurt Gödel. Collected Works, Vol. I. Publications 1929-1936.* Oxford: Oxford Univ. Press.

1990 (with J.W. Dawson, Jr., S. C. Kleene, G. H. Moore, R. M. Solovay, and J. van Heijenoort, eds.), *Kurt Gödel. Collected Works, Vol. II, Publications 1938-1974.* Oxford: Oxford Univ. Press.

1995 (with J.W. Dawson, Jr., W. Goldfarb, C. Parsons, and R. M. Solovay, eds.), *Kurt Gödel. Collected Works, Vol. III, Unpublished essays and lectures.* Oxford: Oxford Univ. Press.

Ferrara, Lawrence
1991 *Philosophy and the Analysis of Music: Bridges to Musical Sound, Form, and Reference* (Contributions to the Study of Music and Dance) New York: Greenwood Press, Praeger.

Føllesdal, Dagfinn
1988 "Husserl on evidence and justification," in Robert Sokolowski, ed., *Edmund Husserl and the Phenomenological Tradition: Essays in Phenomenology* (Proceedings of a lecture series in the Fall of 1985.) (Studies in Philosophy and the History of Philosophy, Vol. 18). Washington: The Catholic University of America Press, 1988, pp. 107-129.
1995 "Gödel and Husserl," in Jaakko Hintikka, Hrsg., *Essays on the Development of the Foundations of Mathematics.* Dordrecht: Kluwer, pp. 427-446.
1998 "Husserl's idealism," in Marcelo Stamm, ed., *Philosophie in synthetischer Hinsicht* (Festschrift für Dieter Henrich). Stuttgart: Klett-Cotta, 1998, pp. 189-202.
2005 "Ultimate Justification in Husserl and Wittgenstein," in Johann Christian Marek and Maria Elisabeth Reicher, eds., *Experience and Analysis. Erfahrung und Analyse* (Proceedings of the 27th International Wittgenstein Symposium, August 8 - 14, 2004).. Wien: hpt et öbv, 2005, pp. 127-142.
2010 "Husserls Begriff der Lebenswelt," in: Carl Friedrich Gethmann, Hrsg.: *Lebenswelt und Wissenschaft* (Proceedings of XXI. Deutscher Kongress für Philosophie, Essen 2008). *Deutsches Jahrbuch Philosophie* 2. Hamburg: Felix Meiner Verlag, pp. 284-306.

Gödel, Kurt
1944 "Russell's mathematical logic", in Paul Arthur Schilpp, ed., *The Philosophy of Bertrand Russell* (The Library of Living Philosophers), Open Court, La Salle, Ill., 1944, p. 137, reprinted in Benacerraf and Putnam 1964 and in Feferman 1990, pp. 119-141, with introductory note by Charles Parsons, pp. 102-118.
1946 "Remarks before the Princeton bicentennial conference on problems in mathematics," first published in Davis 1965:84. Reprinted in Feferman et al., eds., 1990, pp. 150-153, with introductory note by Charles Parsons, pp. 144-150.
1947/64 "What is Cantor's continuum problem?" *American Mathematical Monthly* 54 1947) , pp. 515-525, errata, vol. 55, 151. Expanded version in: Paul Benacerraf / Hilary Putnam, eds. *Philosophy of Mathematics: Selected Readings*, Englewood Cliffs, New Jersey: Prentice-Hall, pp. 258-273. Both versions reprinted in Feferman et al. eds., 1990, pp. 176-187 and 254-270, with introductory note by Gregory H. Moore, pp. 154-175.
1951 "Some Basic Theorems on the Foundations of Mathematics and their Implications" (Dec 1951), Gödel's Gibbs lecture to American Mathematical

Society, in Feferman et al, eds, 1995, pp. 304-323, with introductory note by George Boolos, pp. 290-304.

1953-59, III and V
"Is mathematics syntax of language?" [unfinished contribution to Schilpp, 1963] in Feferman, et al., eds., 1995, pp. 334-362, with introductory note by Warren Goldfarb, pp. 324-334.

1961 "The modern development of the foundations of mathematics in the light of philosophy," in Feferman, et al., eds., 1995, pp. 374-387.

1974 "Remark on non-standard analysis," in Feferman, et al., eds., 1990, p. 311, with introductory note by Jens Erik Fenstad, pp. 307-310.

Gurwitsch, A.

1966 *Studies in Phenomenology and Psychology*, Evanston, Ill.: Northwestern University Press.

Hauser, Kai

2001 "Objectivity over Objects: A Case Study in Theory Formation," *Synthese* 128, pp. 245-285.

2002 "What New Axioms Could Not Be," *Dialectica* 56, pp. 101-117.

2002 "Is Cantor's Continuum Problem Inherently Vague?" *Philosophia Mathematica*, pp. 257-285.

2005 "Is Choice Self-Evident?" *American Philosophical Quarterly* 42, pp. 237-261. Abstract in *Review of Metaphysics* 59, p. 466

2004 "Was sind und was sollen neue Axiome," in G. Link, editor, *One Hundred Year of Russell's Paradox*, Berlin: De Gruyter, pp. 93-117.

2006 "Gödel's Program Revisited. Part I: The Turn to phenomenology," *Bulletin of Symbolic Logic*, 12, pp. 529-590.

Huemer, Wolfgang

2003 "Logical empiricism and phenomenology: Felix Kaufmann," in Stadler, 2003, pp. 151-61.

Husserl, Edmund

1891 *Philosophie der Arithmetik. Psychologische und logische Untersuchungen.* Erster Band. Halle-Saale: C.E. M. Pfeffer (Robert Stricker), 1891. Husserliana XII. Den Haag 1970.

1908 *Vorlesungen über Bedeutungslehre.* Sommersemester 1908. Husserliana XXVI. Den Haag, 1987.

1913 *Ideen zu einer reinen Phänomenologie und phänomenologischen Philosophie. Erstes Buch: Allgemeine Einführung in die reine Phänomenologie.* Halle, Max Niemeyer. Husserliana III, I. Den Haag 1976.

1918–1926 *Analysen* zur *passiven Synthesis*. Aus Vorlesungs- und Forschungsmanuskripten, 1918–1926, ed. Margot Fleischer, Husserliana XI. Den Haag 1966.

1923/24 *Erste Philosophie.* Zweiter Teil: Theorie der phänomenologischen

Reduktion. Husserliana VIII. Den Haag 1959.

1930 "Nachwort zu meinen 'Ideen zu einer reinen Phänomenologie und phänomenologischen Philosophie.'" *Jahrbuch für Philosophie und phänomenologische Forschung*, 11. Halle a.d.S. (1930), 549-570. Reprinted in Husserliana V, Den Haag 1971, pp. 138-162. Used as Preface to the Gibson's translation of *Ideas*, London: Allen & Unwin, 1931.

1936 *Die Krisis der europäischen Wissenschaften und die transzendentale Phänomenologie. Eine Einleitung in die phänomenologische Philosophie.* Husserliana VI. Den Hague 1954. English translation by David Carr: The Crisis of the European Sciences and Transcendental Phenomenology, Evanston, Ill.: Northwestern University Press, 1970,

1950– *Husserliana. Edmund Husserl: Gesammelte Werke,* herausgegeben vom Husserl Archiv (Leuven). The Hague: Martinus Nijhoff; later Dordrecht: Kluwer; now Berlin: Springer.

Jastrow, J.

1899 "The mind's eye." *Popular Science Monthly,* 54, 299-312.

1900 *Fact and fable in psychology*. Boston: Houghton Mifflin.

Jensen, Ronald

1972 "The fine structure of the constructible hierarchy," *Annals of Math. Logic* 4, pp. 229-308.

Kaplan, David

1971 "Homage to Rudolf Carnap," in *PSA 1970*, vol. 8 of *Boston Studies in the Philosophy of Science*, ed. Roger C. Buck and Robert S. Cohen. Dordrecht: Reidel, pp. xlv-xlvii.

Kihlstrom, John F.

2004 Joseph Jastrow and His Duck – Or Is It a Rabbit? http://socrates.berkeley.edu/~kihlstrm/JastrowDuck.htm Unpublished

Köhler, Eckehart

2002 "Gödels Platonismus," in Köhler et al., eds., 2002b, pp. 341-386.

2011 "Collaborating with Hao Wang on Gödel's Philosophy." In Parsons/Links, 55-72.

Köhler, Eckehart et al., eds.

2002a *Kurt Gödel. Wahrheit und Beweisbarkeit*, Band 1: Dokumente und historische Analysen, Wien: obv & hpt.

2002b *Kurt Gödel. Wahrheit und Beweisbarkeit.* Band 2: Kompendium zum Werk. Wien: obv & hpt.

Koellner, Peter

2009 "On reflection principles." *Annals of Pure and Applied Logic*, 157 (2-3), pp. 206-19.

Kreisel, Georg
1965 *Mathematical logic, Lectures on modern mathematics* (T. L. Saaty, editor). New York: Wiley.
1980 "Kurt Gödel, 28 April 1906 – 14 January 1978." *Biographical Memoirs of Fellows of the Royal Society* 26, pp. 148-224. Corrections, *ibid.* 27, p. 697, and 28, p. 718.

Künne, Wolfgang
1990 "The Nature of Acts – Moore on Husserl," in: D. Bell and N. Cooper, eds., *The Analytic Tradition – Meaning, Thought and Knowledge*, Oxford: Blackwell, pp. 104-116.

Mach, Ernst
1865 "Bemerkungen zur Lehre vom räumlichen Sehen," in: *Zeitschrift für Philosophie und philosophische Kritik*, neue Folge, 46, pp. 1-5.
1886 *Beiträge zur Analyse der Empfindungen*, 1st ed., Jena: Fischer.

Maddy, Penelope,
1998 "Believing the axioms" I, II. *The Journal of Symbolic Logic,* 53, 481-511, 736-764.
2000 "Does Mathematics Need New Axioms?" *Bulletin of Symbolic Logic* 6, 413-422.
2009 *Second Philosophy.* Oxford: Oxford University Press.

Mancosu, Paolo, and Ryckman, T.A.
2002 "Mathematics and Phenomenology: The Correspondence between O. Becker and H. Weyl," *Philosophia Mathematica*, 3d Series, vol. 10, 130-173, bibliography pp. 195–202.

McIntyre, Ronald
1982 "Intending and Referring," in Dreyfus 1982.

McManus, I. C., Freegard, Matthew, Moore, James, Rawles, Richard
2010 "Science in the Making: Right Hand, Left Hand. II. The duck-rabbit figure." *Laterality*, 15 (1/2), pp. 166-185.

Mulligan, Kevin, and Smith, Barry
1988 "Mach and Ehrenfels," in Smith, Barry, ed., 1988.

Parsons, Charles
2008 *Mathematical Thought and Its Objects*. New York: Cambridge University Press.
2012 *From Kant to Husserl: Selected Essays*. Cambridge, Mass.; Harvard University Press.

Parsons, Charles, and Links, Montgomery, eds.
2011 *Hao Wang: Logician and Philosopher*. Milton Keynes: College Publications.

Putnam, Hilary
1973 "Meaning and Reference." *Journal of Philosophy* 70, 699-711.
1975 "The meaning of meaning." In K. Gunderson, ed., *Language, Mind and Knowledge*, Minnesota Studies in the Philosophy of Science, VII, Minneapolis, Minnesota: University of Minnesota Press. Reprinted in Putnam, *Philosophical Papers, Vol. 2: Mind, Language and Reality*. Cambridge: Cambridge University Press, pp. 215-271.

Quine, W. V
1995 *From Stimulus to Science*, Cambridge, Mass. Harvard University Press.
2008 *Quine in Dialogue.* Douglas B. Quine and Dagfinn Føllesdal, eds., Cambridge, Mass.: Harvard University Press.

Riesz, Frédéric and Béla Sz.-Nagy
1952 *Leçons d'analyse functionelle.* Budapest: Académie des Sciences de Hungrie. Deuxième edition, 1953.

Ryckman, Thomas
2004 *The Reign of Relativity: Philosophy in Physics 1915–1925*. Oxford Studies in Philosophy of Science. Oxford: Oxford University Press.

Schilpp, Paul Arthur
1963 *The Philosophy of Rudolf Carnap* (The Library of Living Philosophers). La Salle, Illinois: Open Court Publishing.

Schumann, Karl
1993 "Husserl's Theory of Indexicals," in F. M. Kirkland and D. P. Chattopadhyaya, eds., *Phenomenology – East and West*, Dordrecht: Reidel.

Scott, Dana
1967 "A proof of the independence of the continuum hypothesis." *Mathematical Systems Theory* 1, pp. 89–111.

Smith, Barry, ed.
1988 *Foundations of Gestalt Theory*. Munich and Vienna: Philosophia Verlag

Smith, David
1989 *The Circle of Acquaintance*. Dordrecht: Reidel.
2007 *Husserl*. London and New York: Routledge.

Stadler, Friedrich, ed.
1997 *Phänomenologie und Logischer Empirismus. Zentenarium Felix Kaufmann (1895–1959)*, ed. Friedrich Stadler. Wien–New York: Springer.
2003 *The Vienna Circle and Logical Empiricism*. (Vienna Circle Institute Yearbook, Vol 10). Dordrecht: Kluwer.
2010 *The Vienna Circle in the Nordic Countries. Networks and Transformations of Logical Empiricism.* (with Juha Manninen) (Vienna Circle Institute Yearbook, Vol. 14). Wien–New York: Springer.

Tait, W.W.

2005 *The Provenance of Pure Reason: Essays in the Philosophy of Mathematics and Its History*. Oxfor: Oxford University Press.

Takeuti, Gaisi

2003 *Memoirs of a Proof Theorist: Gödel and Other Logicians*. Singapore: World Scientific Publishing Co., Revised Edition

Thim, Johan

2003 *Continuous, Nowhere Differentiable Functions*. Master's Thesis. Luleå University of Technology, Department of Mathematics: http://epubl.ltu.se/1402-1617/2003/320/LTU-EX-03320-SE.pdf. Unpublished.

Tieszen, Richard

1989 *Mathematical Intuition: Phenomenology and Mathematical Knowledge.* (Synthese Library) Dordrecht: Kluwer, now Berlin: Springer.

2005a *Phenomenology, Logic, and the Philosophy of Mathematics.* Cambridge: Cambridge University Press.

2005b "Consciousness of Abstract Objects," in *Phenomenology and the Philosophy of Mind*, D. Smith and A. Thomasson, eds. Oxford: Oxford University Press.

2011 *After Gödel: Platonism and Rationalism in Mathematics and Logic.* Oxford: Oxford University Press.

Tragesser, Robert S.

1977 *Phenomenology and Logic*. Ithaca: Cornell University Press.

1984 *Husserl and Realism in Logic and Mathematics* (Modern European Philosophy). Cambridge: Cambridge University Press.

van Atten, Mark, and Kennedy, Juliette

2003 "On the Philosophical Development of Kurt Gödel," *The Bulletin of Symbolic Logic*, Vol. 9, No. 4 (Dec., 2003), pp. 425-476.

Wang, Hao

1974 *From mathematics to philosophy*. New York: Humanities Press.

1985 *Beyond Analytic Philosophy. Doing Justice to What We Know.* Cambridge, Mass., MIT Press. Paperback edition 1987.

1987 *Reflections on Kurt Gödel.* Cambridge, Mass., MIT Press. Paperback edition 1990.

1996 *A Logical Journey: From Gôdel to philosophy.* Cambridge, Mass.: MIT Press.

Weyl, Hermann

1918 *Das Kontinuum. Kritische Untersuchungen über die Grundlagen der Analysis*. Leipzig: Veit.

Woodin, Hugh

1999 *The axiom of determinacy, forcing axioms, and the nonstationary ideal.*

Berlin: de Gruyter.

2011 *The realm of the infinite, Infinity, new research frontiers*. Cambridge: Cambridge University Press.

Department of Philosophy
Stanford University
450 Serra mall
Stanford, CA 94305-2155
dagfinn@csli.stanford.edu

MASSIMO FERRARI

LOGICAL EMPIRICISM IN HISTORICAL PERSPECTIVE. RECENT WORKS ON MORITZ SCHLICK

MATHIAS NEUBER, *Die Grenzen des Revisionismus. Schlick, Cassirer, und das ‚Raumproblem'*, Schlick-Studien, Bd. 2, Wien–Berlin: Springer, 2011

BRIAN MCGUINNESS, *Wittgenstein und Schlick. Mit einer „Erwiderung" von Mathias Iven*, Schlickiana, Bd. 3, Berlin: Parerga Verlag 2010

FYNN OLE ENGLER, MATHIAS IVEN (eds.), *Moritz Schlick: Ursprünge und Entwicklungen seines Denkens*, Schlickiana, Bd. 5, Berlin: Parerga Verlag 2010

Moritz Schlick, the leading figure of the Vienna Circle and one of the most eminent spokesmen of scientific philosophy, was for a long time forgotten by scholars working on logical empiricism. The critical edition of his writings[1], new attention to his contributions both to the epistemological interpretation of Einstein's relativity theory and to the discussion within the Circle, and last but not least an increasing interest in the intellectual developments of Rudolf Carnap, Otto Neurath, Hans Reichenbach, Philipp Frank and other protagonists of the golden age of logical empiricism between the two World Wars have radically changed this situation.[2]

---

1 Within the first section of Moritz Schlick's critical edition (*Kritische Gesamtausgabe*) up to now have been published the following volumes: vol. 2, *Über die Reflexion des Lichtes in einer inhomogenen Schicht. Raum und Zeit in der gegenwärtigen Physik*, edited by Fynn Ole Engler and Mathias Neuber, Wien–New York: Springer 2006; vol. 3, *Lebensweisheit. Fragen der Ethik* edited by Mathias Iven, Wien–New York: Springer 2006; vol. 6, *Die Wiener Zeit. Aufsätze, Beiträge, Rezensionen*, edited by Johannes Friedl and Heiner Rutte, Wien–New York 2008; vol. 1, *Allgemeine Erkenntnislehre*, edited by Hans Jürgen Wendel und Fynn Ole Engler, Wien–New York: Springer 2009. Forthcoming is the vol. 5, *Rostock, Kiel, Wien – Aufsätze, Beiträge, Rezensionen 1919–1925*, edited by Heidi König and Edwin Glassner, Wien-New York: Springer 2012.

2 It is here impossible to give a precise account of recent works about Schlick and Logical Empiricism. We would like to refer, e.g., to Paolo Parrini, *L'empirismo logico. Aspetti storici e prospettive teoriche*, Roma: Carocci 2002; Thomas Ryckman, *The Reign of Relativity. Philosophy in Physics 1915-1925*, Oxford: Oxford University Press 2005; Michael Stölzner / Thomas Uebel (Eds.), *Wiener Kreis. Texte zur wissenschaftlichen Weltauffassung von Rudolf Carnap, Otto Neurath, Moritz Schlick, Philipp Frank, Hans Hahn, Karl Menger, Edgar Zilsel und Gustav Bergmann*, Hamburg: Meiner 2006; Jacques Bouveresse / Delphine Chapuis-Schmitz / Jean Jacques Rosat (Eds.),

M.C. Galavotti et al. (eds.), *European Philosophy of Science – Philosophy of Science in Europe and the Viennese Heritage*, Vienna Circle Institute Yearbook 17,
DOI 10.1007/978-3-319-01899-7_23,

The four volumes of the *Schlickiana* edited by Parerga in Berlin offer a very instructive document of recent scholarship on Schlick. Fynn Ole Engler and Mathias Iven have edited the papers presented to a Schlick conference held in Rostock in September 2007.[3] Mathias Iven has provided an extremely well documented biographical sketch of Schlick's early intellectual adventures up until 1907.[4] As to the two volumes to be discussed in the present Review Essay: the contributions to the second Schlick Symposion in Rostock (September 2009) can be found in the fifth volume of the series, edited again by Engler and Iven; and finally Brian F. McGuinness has collected articles on Schlick and Wittgenstein, Gilbert Ryle, Friedrich Waismann, Piero Sraffa and Rozalia Rand giving an account of (to some extent) marginal aspects of history of analytical philosophy between Austrian and Great Britain.

By the way, the publication of some other volumes or papers and the translations that have recently appeared are evidence that Schlick is considered today in a new light, both in biographical-historical and in systematic perspective.[5] Neglected

---

L'empirisme logique à la limite. Schlick, le langage et l'expérience, Paris: CNRS Éditions 2006; Alan Richardson/Thomas Uebel (Eds.), *The Cambridge Companion to Logical Empiricism*, Cambridge: Cambridge University Press 2007; Michael Friedman/Richard Creath (Eds.), *The Cambridge Companion to Carnap*, Cambridge: Cambridge University Press 2007; Massimo Ferrari, "Moritz Schlick, la teoria della relatività e l'immagine del mondo della fisica moderna", in: *Giornale critico della filosofia italiana* 86, 2007, pp. 423-449; Steffen Kluck, *Gestaltpsychologie und Wiener Kreis. Stationen einer bedeutsamen Beziehung*, Freiburg–München: Alber 2008; Carsten Seck, *Theorien und Tatsachen. Eine Untersuchung zur wissenschaftstheoriegeschichtlichen Charakteristik der theoretischen Philosophie des frühen Schlicks*, Paderborn: Mentis Verlag 2008.

3 Fynn Ole Engler/Mathias Iven (Eds.), *Moritz Schlick. Leben, Werk und Wirkung*, Berlin: Parerga 2008 (*Schlickiana*, vol. 1).

4 Mathias Iven, *Moritz Schlick. Die frühen Jahre (1882–1907)*, Berlin: Parerga 2008 (*Schlickiana*, vol. 2).

5 We refer first of all to François Schmitz, *Le Cercle de Vienne*, Paris: Vrin 2009 (the introductory chapter, pp. 21-72, is entirely devoted to Schlick) and to the French translation of the *Allgemeine Erkenntnislehre* (see *Théorie générale de la connaissance*, traduit de l'allemand et présenté par Christian Bonnet, Paris: Gallimard 2009). Among recent French translations of Schlick's works are moreover to remember *Questions d'éthique*, traduit par Christian Bonnet, Paris: Presses Universitaires de France 2000 and *Forme et contenu. Une introduction à la pensée philosophique*, traduit par Delphine Chapuis-Schmitz, Marseille: Agone 2003. See also the excellent anthology published by Christian Bonnet/Pierre Wagner (Eds.), *L'âge d'or de l'empirisme logique. Vienne-Berlin-Prague 1929–1936*, Paris: Gallimard 2006 and the little book by Mélika Oulbani, *Le Cercle de Vienne*, Paris: Presses Universitaires de France 2006. Remarkable contributions to Schlick are also available in: Jacques Bouveresse/Pierre Wagner (Eds.), *Mathématique et expérience. L'empirisme logique à l'épreuve (1918–1940)*, Paris: Odile Jacob 2008 and Anne Siegetsleitner (Ed.), *Logischer Empirismus, Werte und Moral. Eine Neue Bewertung*, Wien-New York: Springer 2010. For a balanced presentation of Schlick's philosophical personality see Thomas Mormann, "Zwischen

aspects of his work (e.g., Schlick's early interests in ethics and philosophy of life) have been examined in the context of German philosophical culture at the turn of 20th century. At the same time, his later commitment to Vienna and Austrian cultural tradition are no longer been considered as simple biographical details.[6] By employing a German expression, we can say that the *Schlick-Forschung* has gradually acquired a prominent place in the agendas of historical inquiry in Europe as well as in the United States.

As Friedrich Stadler and Hans Jürgen Wendel point out in the foreword to the first volume of the *Schlick-Studien*, which was published in 2009, we can speak more broadly of a "new forum" that will enable us to appreciate Schlick's very articulated philosophical developments and achievements.[7] Whereas this first volume contains a rich variety of papers that we will still have an opportunity to mention, the second volume of the *Schlick-Studien* is here particularly worthy of discussion. Matthias Neuber offers indeed a rich analysis of Schlick's realistic interpretation of space-theory and brings it in relation to Ernst Cassirer's neo-Kantian position, reconstructing in this way a stimulating debate he defines as "the limits of revisionism". What is meant by this expression that recalls the political terminology or perhaps the famous debate within the German Socialdemocracy at the end of 19th century? According to Neuber (p. 23-43), Kant's transcendental philosophy was subject – in the late 19th century and at the very beginning of the 20th century – to a process of revision aimed at avoiding both its simply "immunisation" vis-a-vis the new issues of mathematics and physical science (non-Euclidean geometry, theory of relativity) and its rejection. The strategy of revision has two sides: the relativization of a priori endorsed by Cassirer, on the one hand, and Schlick's attempt to conceive Kant's thing in itself as knowable on the other.

To be sure, Schlick's relationship to Cassirer and to neo-Kantian thought is nothing new for scholars working on Logical Empiricism. As Christian Möckel states precisely, Cassirer was deeply interested in Schlick's, and more generally, in the Vienna Circle's typical way of conceiving the philosophical relevance of

---

Wahrheit und Wissenschaft. Schlicks weites philosophisches Spektrum", in: *Grazer Philosophische Studien* 80, 2010, pp. 263-285.

6 See Iven, *Moritz Schlick. Die frühen Jahre (1882–1907)*, *op. cit.*, pp. 125-169 and Mathias Iven, " ,... er ist nur da ganz Mensch, wo er spielt'. Moritz Schlick liest Friedrich Schiller", in: Engler/Iven (Eds.), *Moritz Schlick. Ursprünge und Entwicklungen seines Denkens*, *op. cit.*, pp. 129-148. Iven has the great merit of having introduced, also on the basis of unpublished sources, the critical edition of both Schlick's ethical most important contributions, i.e. *Lebensweisheit* (1908) and *Fragen der Ethik* (1930): see Schlick, *Kritische Gesamtausgabe*, vol. 3, *op. cit.*, pp. 1-40, 335-344. Regarding Schlick and Vienna I allow myself to refer to Massimo Ferrari, "1922: Moritz Schlick in Wien", in: Friedrich Stadler / Hans Jürgen Wendel (Eds.) *Stationen. Dem Philosophen und Physiker Moritz Schlick zum 125. Geburtstag*, Wien–New York: Springer 2009 (*Moritz Schlick Studien*, vol. I),, pp. 17-62.

7 Friedrich Stadler/Hans Jürgen Wendel, "Vorwort", in: Stadler/Wendel (Eds.), *Stationen. Dem Philosophen und Physiker Moritz Schlick zum 125. Geburtstag*, *op. cit.*, p. 7.

natural science so as to consider philosophy not as a subjective point of view, but rather as a science.[8] Nevertheless, Neuber has chosen an illuminating and original perspective by approaching both Schlick's and Cassirer's analyses of the problem of space within the wider frame of their respective philosophical agendas. Neuber offers a very detailed analysis of Schlick's early elaboration of a conception of space, from the period in Zurich and Rostock, when his work was focussed on the psychological problem of space, to the following phase in which Schlick adopted the psychological and epistemological insight of space already presented in the *General Theory of Knowledge* and made a crucial contribution to the philosophical understanding of Einstein's theory of relativity.[9] The decisive background of this elaboration is reflected by the critical realism Schlick claims represents the necessary alternative both to positivism *à la* Mach and to Kant's refusal of things in themselves as objects of possible knowledge. This made it possible, for Schlick, to elaborate the so-called method of coincidences and to deny any Kantian attempt to "constitute" independent reality (which is the object of science) through the (supposed) transcendental activity of thought. But in the same way as for Cassirer, Schlick maintains that the scientific objectivity resting on similar realistic presuppositions has to be fundamentally conceived as a network of relational *structures*.[10]

By contrast, Cassirer's scientific idealism suggests a revision of the Kantian philosophy by endorsing the relativization of the a priori. In this context Cassirer's theory of the logical invariants of experience assumes a transcendental status and – as Karl-Norbert Ihmig has argued – it is considered by Cassirer to be a historical, a priori but changeable set of principles and categories enabling the constitution of scientific experience.[11] Surprisingly enough, Neuber describes all of this as the "scandal" of Cassirer's epistemology[12] and tries to show that the consequence of such an idealistic view is the wrong interpretation of Riemann's and Einstein's

8 Christian Möckel, "Moritz Schlick und Ernst Cassirers Auseinandersetzung mit dem ‚Wiener Kreis'", in: Engler/Iven (Eds.), *Moritz Schlick: Ursprünge und Entwicklungen seines Denkens*, *op. cit*., p. 220. See also J.M. Krois, "Ernst Cassirer und der Wiener Kreis", in: Friedrich Stadler (Ed.), *Zur Interaktion von Philosophie, Geschichte und Theorie der Wissenschaften*, Wien–New York: Springer 2000, pp. 105-121. A stimulating picture is offered by Tobias Breidenmoser, "Der logische Aufbau der Welt und seine Baumeister. Rudolf Carnap zwischen Moritz Schlick und Ernst Cassirer", in: Engler/Iven (Eds.), *Moritz Schlick. Leben, Werk und Wirkung*, *op. cit*., pp. 193-214.

9 To this topic is also devoted the contribution of Fynn Ole Engler, "Über das erkenntniskritische Raumproblem bei Moritz Schlick, Wilhelm Wundt und Albert Einstein", in: Stadler/Wendel (Eds.), *Stationen. Dem Philosophen und Physiker Moritz Schlick zum 125. Geburtstag*, *op. cit*., pp. 107-145.

10 Neuber, *Die Grenzen des Revisionismus*, *op. cit*., pp. 119-120.

11 *Ibid*., p. 159. See also Karl-Norbert Ihmig, *Cassirers Invariantentheorie der Erfahrung und seine Rezeption des „Erlanger Programms"*, Hamburg: Meiner 1997.

12 Neuber, *Die Grenzen des Revisionismus*, *op. cit*., p. 155.

physical space as a kind of topological (or a kind of Leibnizian) a priori space resting on an idealistic foundation.[13]

It is not difficult to understand why Neuber, starting from this evaluation of Cassirer, and in contrast with scholars such as Michael Friedman, Thomas Ryckman, and, *si parva licet*, myself, denies that Schlick's famous discussion of Cassirer's interpretation of the theory of relativity is based upon a fundamental misunderstanding of Cassirer's neo-Kantian view. To be sure, Neuber quotes a letter to Schlick written in October 1920, in which Cassirer clearly defends his project as a *revision* of Kant and, in particular, of his theory of a priori in the spirit of a dynamical, historical transformation of Kant's overly "rigid" schemes of intellect.[14] Neuber is still convinced that Schlick was right in denying any form of a priori principles of scientific knowledge and in refusing the synthetic a priori, even though in the sophisticated form proposed by Cassirer. Cassirer – Neuber claims – has failed by identifying reality with its conditions of possibility;[15] Schlick, on the contrary, has perfectly understood that scientific relations and, more precisely, metric relations in the theory of relativity are the core of a realistic interpretation of physical objectivity.[16] But this is, at the same time, the reason for Neuber's doubts about the solidity of Schlick's realism: by admitting the conventional status of scientific principles, and first of all of geometrical axioms, Schlick is not able, according to Neuber, to solve the essential tension between conventionalism (that is something belonging to "human mind") and the transcendence of the physical world.[17] So both Cassirer and Schlick have left behind the "transcendental revisionism" and the Kantian heritage they aspired to reform. It is, however, doubtful that Schlick's realism is a better answer to this bet, whereas Cassirer's idealism has to be interpreted as a kind of return not to Kant, but to a pre-Kantian view of knowledge.[18]

Neuber's important contribution nevertheless shows once again what recent scholarship has fully recognized already, namely that Schlick's thought and achievements simply must be examined in the wider context of the scientific and cultural landscape at the beginning of 20th century. This context has two faces. On the one hand, it seems very illuminating to consider the tradition of scientific philosophy and its first beginnings within the German-speaking world. Björn

13 *Ibid.*, p. 178.

14 Neuber wrongly affirms that up to now this letter has never been quoted (*Ibid.*, p. 161). Contrary to Neuber's opinion, this letter is avaible in Ernst Cassirer, *Nachgelassene Manuskripte und Texte*, vol. 18, *Ausgewählter wissenschaftlicher Briefwechsel*, ed. by J.M. Krois, Hamburg: Meiner 2009, pp. 50-51. For extensive quotations from this letter and a commentary to the discussion between Cassirer and Schlick on this topic see also Massimo Ferrari, *Ernst Cassirer. Stationen einer philosophischen Biographie*, Hamburg: Meiner 2003, p. 128.

15 Neuber, *Die Grenzen des Revisionismus*, *op. cit.*, p. 200.

16 *Ibid.*, p. 212.

17 *Ibid.*, p. 215.

18 Ibid., pp. 204, 222.

Henning has, for instance, carefully analyzed the *Vierteljahrsschrift für wissenschaftliche Philosophie* founded in 1877 by Richard Avenarius: a journal which aimed, according to Avenarius's quite original empiricism, to constitute a tribune for both philosophy and science in their mutually relationship.[19] The basic insight characterizing the *Vierteljahrsschrift* was to conceive of philosophy as a *Begriffswissenschaft* (a kind of definition remembering the old Herbartian one), that is, to assign to philosophy the role of a science of the sciences grounded on the empirical data of special sciences.[20] Although the review was originally projected by Avenarius as a philosophical alternative to Neo-Kantianism, a Neo-Kantian philosopher such as Alois Riehl became later co-editor giving a partially new address to the long history of the journal. So the *Vierteljahrsschrift* knew at the turn of the century a significant transformation by offering a synthesis between Positivism and Neo-Kantianism, in the spirit of a scientific philosophy able to elaborate a scientific worldview starting from the achievements of various scientific disciplines.[21] And it is exactly in this perspective that we can understand Schlick's early collaboration to the *Vierteljahrsschrift*, where he published in 1910 both the paper about "Die Grenzen der naturwissenschaftlichen und philosophischen Begriffsbildung" and the very important essay "Das Wesen der Wahrheit nach der modernen Logik", as well as, in the following years, a considerable amount of reviews regarding contemporary philosophy and epistemology.[22]

On the other hand, Schlick's philosophical beginnings clearly show that his early work can to some extent explain later developments the standard-view usually considers as typical expression of his late period in Vienna. Schlick, for example, was one the first German-speaking philosophers (together with Wilhelm Jerusalem) to seriously consider William James and his conception of truth.[23] As emerges from his very early, unpublished notices on Pragmatism, Schlick was

19 Björn Henning, "Richard Avenarius' Konzeption der wissenschaftlichen Philosophie und die Herausbildung der wissenschaftlichen Philosophie in der zweiten Hälfte des 19. Jahrhunderts", in Engler/Iven (Eds.), *Moritz Schlick. Ursprünge und Entwicklungen seines Denkens*, *op. cit.*, pp. 149-170, On this topic see also Klaus Christian Köhnke, *Entstehung und Aufstieg des Neukantianismus. Die deutsche Universitätsphilosophie zwischen Idealismus und Positivismus*, Frankfurt am Main: Suhrkamp 1986, pp. 388-404.

20 Henning, "Richard Avenarius' Konzeption der wissenschaftlichen Philosophie und die Herausbildung der wissenschaftlichen Philosophie in der zweiten Hälfte des 19. Jahrhunderts", *loc. cit.*, pp. 165-166.

21 *Ibid.*, pp. 170-171.

22 I allow my self to refer to Massimo Ferrari, "An Unknown Side of Moritz Schlick's Intellectual Biography: the Reviews for the "Vierteljahrsschrift für wissenschaftliche Philosophie und Soziologie" (1911–1916)", in: Friedrich Stadler (Ed.), *The Vienna Circle and Logical Empiricism. Re-evaluation and Future Perspectives*, Dordrecht–Boston–London: Kluwer Academic Publishers 2003, pp. 63-77.

23 Michael Pohl, "Moritz Schlicks Kritik des pragmatischen Wahrheitsbegriffs", in: Engler/Iven (Eds.), *Moritz Schlick: Ursprünge und Entwicklungen seines Denkens*, *op. cit.*, pp. 171-189.

deeply interested in the new "Yankee" philosophy.[24] Although Schlick, similarly to other German philosophers, elaborated in short time an incisive critique of James' insights about truth, he was able to comprehend how important the central role of verification was for James.[25] Michael Pohl, however, points out that in his early works, and especially in his essay about the "Essence of truth", Schlick was still committed to an atomistic view of verification, whereas James defended an holistic and functionalistic one, which is no less important from an epistemological point of view as the clear distinction James endorsed between truth and criterion of truth.[26]

The case study of Schlick's relationship to pragmatism is a good example (obviously one among many others) of historical approach and contextualisation of philosophical arguments. Recent scholarship on Schlick has also established new standards of research seeking to explore on the basis of Schlick's manuscripts and correspondence not only the very early beginnings of his intellectual biography, but also the golden period of his later philosophical enterprise in Vienna. Thobias Breidenmoser can undoubtedly be credited with applying a similar methodology to acquire a more detailed account of the debate between Schlick and Carnap about physicalism, verification and foundation of knowledge.[27] The main sources here are the letters Schlick and Carnap exchanged from 1932 to 1934, which allow the convergences and divergences characterizing this central issue within the Circle to be reconstructed step by step. One of the main questions is whether Schlick's conception of *Konstatierungen* can be compatible with Carnap's protocol state-

24 Pohl seems to ignore a not irrelevant detail. In his philosophical apprenticeship Schlick was indeed favourable to the pragmatistic concept of truth (it was no coincidence that he had read with enthusiasm the works of Nietzsche and Georg Simmel). Schlick, in particular, endorsed truth as "utility of judgement", although he was not directly informed about Pragmatism and its main works (see Schlick's autobiographical notes held in Wiener-Kreis-Archiv im Rijksarchief in Nord-Holland/Amsterdam, Inv. Nr. 082C2b, pp. 15-16). According to Schlick, his "pragmatistic" notes go back to the years 1904–1905. It is noteworthy that Schlick in the following period modified his opinions, probably under the influence of Gustav Störring, who was his teacher in philosophy in Zurich from 1907 to 1909. Moreover, Störring took part in the International Congress of Philosophy in Heidelberg (September 1908) and he made a contribution to the heated discussion about Pragmatism, in particular criticizing Royce's lecture and emphasizing the necessity of making a clear distinction between the psychological question and the logical one. Interesting enough, in this occasion Störring encouraged Royce to be aware of the difference between truth as logical concept and the criterion establishing what is true (see Theodor Elsenhans (Ed.), *Bericht über den III. Internationalen Kongress für Philosophie zu Heidelberg. 1. bis 5. September 1908,* Heidelberg: Winter's Universitätsbuchhandlung 1909, pp. 92, 729).

25 Pohl, "Moritz Schlicks Kritik des pragmatischen Wahrheitsbegriffs", *loc. cit*., p. 177.

26 *Ibid*., pp. 184-185.

27 Thobias Breidenmoser, "Physikalismus, Verifikation und Fundamente im Lichte des Briefwechsels zwischen Moritz Schlick und Rudolf Carnap", in: Engler/Iven (Eds.), *Moritz Schlick: Ursprünge und Entwicklungen seines Denkens*, *op. cit*., pp. 277-302.

ments, and to what extent. Interesting enough, the discussion between Schlick and Carnap reflects a kind of fluctuating development, until the final conclusion is reached that the initial agreement was indeed only a temporary one. In his paper on *Konstatierungen* Schlick was no longer able to conceive all the statements of our languages as a hypothesis (as Carnap did) and, at the same time, to consider the simply statement of lived experience as the foundation of knowledge. In this way, as Breidenmoser rightly suggests, Schlick's solution amounts to denying one of the central thesis of Logical Empiricism endorsed by Carnap (and Neurath), that is, the universality of physicalistic language and the consequent project of the Encyclopaedia of Unified Science.[28]

The uniqueness of Schlick's position within the Circle is a well-known aspect of the history of Logical Empiricism, especially with regard to the great influence Ludwig Wittgenstein exerted on him. This topic is at the core of several investigations about Schlick and is documented now also by Brian McGuinness's short essay, "Von der Mathematik zur Psychologie".[29] The main thesis here is that Schlick's *Allgemeine Erkenntnislehre* (in the second edition of 1925 too) is "radically" different from Wittgenstein's point of view. There are obviously also some affinities, in particular because Schlick considers typical problems of philosophy to be pseudo-problems. But, first and foremost, Schlick's *opus magnum* is quite different from the aphorisms of the *Tractatus* for the reason that Schlick ignores the theory of tautologies as well as the Fregean idea of a "third reign" – objective, not real – which is indeed indispensable in order to comprehend Wittgenstein.[30] This explains the absence, in the second edition of Schlick's *Allgemeine Erkenntnislehre*, of any reference to Wittgenstein. Moreover, and contrary to Herbert Feigl's opinion, is highly probable that Schlick at this time had no direct knowledge of Russell's logical atomism.[31] And finally the "turn" in philosophy announced by Schlick in 1930 is not identifiable simply with a Wittgensteinian impetus, or – more precisely – with a loyalty towards the *Tractatus*, inasmuch as at this time Wittgenstein was changing, and in no marginal way, the views he had previously endorsed in his short, enigmatic book which Schlick greatly admired.[32] Following McGuinness's suggestions, we still have to carefully unravel the long

28 *Ibid.*, p. 295.

29 Brian F. McGuinness, "Von der Mathematik zur Psychologie", in: Brian F. McGuinness, *Wittgenstein und Schlick. Mit einer »Erwiderung« von Mathias Iven*, *op. cit.*, pp. 9-21. McGuinness's book offers also a vivid portrait of Friedrich Waismann ("Die Irrfahrten des Friedrich Waismann", pp. 41-53). The English version of this paper is available as "Waismann: the Wandering Scholar" in Brian F. McGuinness (Ed.), *Friedrich Waismann – Causality and Logical Positivism*, Wien–Dordrecht–Heidelberg–London–New York: Springer 2011, pp. 9-16. About Schlick, Wittgenstein, and the Vienna Circle see at least Schmitz, *Le Cercle de Vienne*, *op. cit.*, pp. 235-284.

30 McGuinnes, "Von der Mathematik zur Psychologie", *loc. cit.*, pp. 13-14.

31 *Ibid.*, p. 15.

32 *Ibid.*, p. 17.

and complicated history developed in Vienna at the time of Schlick's enthusiastic encounter with the author of the *Tractatus*.

In this review we have only considered some new contributions to a deeper understanding of Schlick's philosophy. Yet in the books we have briefly discussed are available other quite remarkable papers dealing with his thought. This is the case, for example, of Olaf Müller's long essay discussing Schlick's considerations in *Meaning and Verification* about the possibility of applying the principle of verification to the possibility to see our body after death.[33] Flavia Padovani shows in a very convincing way Schlick's shift from a Kantian point of view of causality in 1920 to the interpretation in the 1930s of the causal principle as a "directive" in the spirit of Wittgenstein's terminology.[34] Jan G. Michel discusses the theory of identity in the Schlick's and Feigl's version from the standpoint of contemporary debates.[35] For his part, Fynn Ole Engler is the author of a very stimulating paper, in which he attempts to give an account of the relationship between Schlick and Ludwik Fleck. The topic is not only of biographical and cultural interest, it is an illuminating consideration of both Schlick's conception of objectivity (in the sense of foundation of knowledge) and Fleck's elaboration of sociological frames summarized by the expression *Denkkollektiv* based on his consideration of origins and developments of "scientific facts".[36]

These contributions, and others that we apologize for not having discussed in greater detail, focus on several aspects of Schlick's theoretical philosophy and testify the highly specialized level of contemporary scholarship about the eminent father of the Vienna Circle. To be sure, the critical editions of his works promoted by a team of scholars working intensely on his books, papers, manuscripts, and letters are the main presupposition of this renewed interest, within the frame of the new historiography concerning the Logical Empiricism, in Schlick's philosophical adventures. This field of research is still open to further developments both in historical and in systematic perspective. Moritz Schlick, one of the leading figures

33 Olaf Müller, "Warte, bis du stirbst. Dualistische Hausaufgaben von Moritz Schlick", in: Engler/Iven (Eds.), *Moritz Schlick: Ursprünge und Entwicklungen seines Denkens*, *op. cit*., pp. 11-71.

34 Flavia Padovani, "Statistical or Dynamical Lawfulness? Reichenbach and Schlick on the Laws of Nature", in: Engler/Iven (Eds.), *Moritz Schlick: Ursprünge und Entwicklungen seines Denkens*, *op. cit*., pp. 225-255. A quite similar subject is dealt with by Tobias Fox, "Die letzte Gesetzlichkeit – Schlicks Kommentare zur Quantenphysik", in: Stadler/Wendel (Eds.), *Stationen. Dem Philosophen und Physiker Moritz Schlick zum 125. Geburtstag*, *op. cit*., pp. 212-258.

35 Jan G. Michel, "Die Schlick-Feigl-Variante der Identitätstheorie aus heutiger Sicht", in: Engler/Iven (Eds.), *Moritz Schlick: Ursprünge und Entwicklungen seines Denkens*, *op. cit*., pp. 303-338.

36 Fynn Ole Engler, "Koinzidenzen, Gestalten und Denkstile. Moritz Schlick und Ludwik Fleck über die Objektivität am Fundament der Erkenntnis und die Bedeutung der Wissenschaftsgeschichte", in: Engler/Iven (Eds.), *Moritz Schlick: Ursprünge und Entwicklungen seines Denkens*, *op. cit*., pp. 339-386,

of scientific philosophy and of philosophy in general in 20th century, has finally received the attention he deserved ever since his tragic death on the stairs of the Vienna University in 1936.

Department of Philosophy
University of Torino
Palazzo delle Facoltà umanistiche (Palazzo Nuovo)
Via Sant'Ottavio
20-Torino
Italy
massimo.ferrari@unito.it

JAN FAYE, *After Postmodernism. A Naturalistic Reconstruction of the Humanities.* Basingstoke: Palgrave-Macmillan, 2012, pp. vi + 224.

Jan Faye has written a timely and important book. Its aim is the overcoming of the separation of the natural from the human sciences and the provision of a unified understanding of scientific knowledge. One might think that enough such attempts have been mounted in the past for another to one to be superfluous, but this is not so: rather, the task has only become more urgent. As is well-known, the separation in question was initiated in the 19th century by the dichotomies of supposedly mutually exclusive cognitive processes (explanation/understanding) and types of sciences (nomothetic/idiographic), with interpretation the distinctive key notion of the human sciences. With that conclusion radicalised and the domain of interpretation universalised in the later 20th century by so-called postmodernists – for whom all of the sciences could not but aim only for interpretations of the world: an imperialist anti-scientism, if you like, that assimilates natural science to cultural studies – the very idea of objective scientific knowledge was called in question. It is this challenge that requires renewed engagement, Faye argues, by going back to the beginning and reconceptualising the relation between explanation and understanding such that neither of them gets assigned exclusively to the natural or the human sciences. Instead, their interaction is to be explored in manifold ways that respect the distinctive research traditions of the different sciences.

Faye's self-styled "neomodernism" aims out to provide a "reasonable defense of scientific objectivity that goes between the extremes of narrow empiricism and the inevitable scepticism fostered by postmodernist relativism" (2) by means of a novel strategy of naturalization. Here it is important to note that Faye distinguishes between the natural sciences, the social sciences and the humanities: under the latter heading he comprises "archaeology, classics, history, languages, linguistics, philosophy, cultural studies, rhetoric, literature, art history, music, film and media studies, communication, etc." (3). Even though the distinction of these humanities subjects from the social sciences is said to be inessential to Faye's thesis, what is essential to reading his book aright is noting that he is concerned with the naturalization of the humanities. With a rich menu of examples and broadly in the style of Dagfinn Føllesdal's by now classic proposal – albeit radically decoupled from the hypothetico-deductive method, as we shall see – Faye urges us "to think of hermeneutics and other methods in the humanities as nothing but the application of empirical methods to obtain knowledge of meaning or intentional objects" (30).

What is also essential is that Faye proposes a non-reductive naturalism. Central to his project is the claim that notions like 'intention' are not only central to any philosophical understanding of the humanities, that they are no less natural than notions like 'atom' or 'brain-state', but also that they need not be reduced to physical or biological notions: there is no need to replace the intentional discourse about ourselves and others. (All the while it is held that mental states and brain states are "structurally identical" (37). Faye's naturalism thus crucially turns on

M.C. Galavotti et al. (eds.), *European Philosophy of Science – Philosophy of Science in Europe and the Viennese Heritage*, Vienna Circle Institute Yearbook 17,
DOI 10.1007/978-3-319-01899-7, © Springer International Publishing Switzerland 2014

his argument that everything that can be said from a first-person perspective can be said from a third-person perspective. It is the potential rendition of all intentional discourse into the third-person perspective that allows for suitable standards of intersubjective agreement about the statements of the humanistic sciences and so allows for their objectivity (24).

Naturalization requires, of course, that cultural phenomena are explained with respect to their origin: psychologically and socially/culturally determined intentions. This means, among other things, that the norms, meanings and perspectives that enter into the intentions behind human actions also come into the purview of the naturalised humanities. Faye's naturalism does not require them to be reduced either, only that they be rendered intersubjectively intelligible in the third-person perspective. Importantly, Faye holds that "talk of norms and intentions does not mainly refer to the individual psychological states of our experience" (54). The attribution of experiences, norms and intentions to others is treated like the postulation of theoretical entities, invoked to explain behaviour and its results – but likewise our introspectively experienced mental episodes are said to have become understandable only by reference to expressions of the public common language. This wholesale (rather than piece-meal) naturalization of intentionality is compared to Darwin's effort to "naturalize biology without relying on other than biological explanations" (37) for the explanation of which we are referred to other works by the author.

The particular point of the naturalization pursued in this volume lies in restoring objectivity to interpretation. For Faye, the humanities aim to provide a systematic understanding of human history, language and culture. Establishing objectivity in these disciplines therefore means to reclaim reasoned self-reflection from the spectre of relativism. To this end Faye re-engages in the debate between Gadamer on the one side and Emilio Betti and E. D. Hirsch on the other, siding with the latter and assembling new arguments in favour of Hirsch's distinction between the textual meaning and the broader significance of a work of art (118-119). While the former is tied to the intentions of its author – and defended as such against contemporary opposition – only the latter is granted to be the free-for-all that all interpretation is for postmodernists. In the course of Faye's argumentative tour de force readers are presented with arguments not only against the likes of Foucault (156-157) and Eco (139-142) but also analytic philosophers like Searle (157-158).

As can be gathered, Faye does not shy from controversy. His "pragmatic-rhetorical" theory of explanation holds that explanation is a discursive response "used directly to address an explanation-seeking question": nothing counts as an explanation in isolation. Unlike other pragmatic theories of explanation (and especially non-pragmatic ones), however, Faye's "permits false answers to act as explanations when they are consonant with the background knowledge of the explainer and the explainee" (59). The reason is to be found in the theory's grounding not in syntax or semantics but rhetoric. "What will count as a remedy to the inquirer's ignorance will depend on his or her background knowledge, beliefs and assertions, and cognitive interests" (60). These elements form the "problem context" which itself is part of a larger context, the "explanatory situation" which includes the background knowledge, beliefs, assertions and interest of the speaker as well. In turn, the speaker's explanatory situation is characterised (following the rhetorical theory of Lloyd Blitzer) by the elements of exigence (the audience's lack of knowledge), audience and constraints (the circumstances under which the audience is to be addressed, including its problem context). Now, as a speech act an explanation is successful if it fulfils not only a "perlocutionary purpose" by being performed with the intention to impart information so that the audience can reach a better understanding, but also if it fulfils a "perlocutionary effect", that is, if the answer given "has an informative effect" on the intended audience "by changing his or her cognitive state" (62). In sum: "*S* explains *q* to *A* by uttering *u* (rather than an alternative statement *v*) if and only if *S* utters *u* with the intention that his utterance of *u* renders *q* understandable in the problem context *P*" (63). And such an explanation is successful if and only if it also achieves the intended perlocutionary effect as well.

It is, of course, the caveat that it is not a requirement that, while *q* is assumed to be true, *u* has to be true as well "as long as *u* renders *q* understandable in the specific problem context" (63), that will bring many readers up short. There seems to be no room for the notion of a pseudo-explanation. If the answer "Mars square Uranus in the birth-chart" satisfies my interlocutor who asked about a third person's short temper, then that would make it a successful explanation. Faye points out the fallacy of thinking that there always is one and just one way of correctly describing the world and stresses that our practice of providing explanations is fallible. But neither of these points addresses the concern just aired. Some readers may wonder whether Faye's strategy for smoothing the passage from explanation to understanding – the decoupling of the notions of explanation and understanding from that of knowledge (which retains its truth condition component) – is worth the price of allowing back into scientific discourse myth and superstition in a supposedly explanatory capacity. Further discussion of this point can be expected.

Another aspect of the naturalization here attempted that needs further discussion is Faye's rejection of the reduction of intentional to causal descriptions as "not mandatory for an explanation to be considered naturalistic" (55). Faye sympathises with the logical connection argument (actions and intentions are not logically independent as causes and effects must be) and finds Davidson's response unsatisfactory (but doesn't say why), yet he also concedes that intentional explanations "have a causal element attached to them" (68). Accordingly he states: "The criterion for an intention being regarded as a cause is the evidence we have for the intention. The evidence referred to is the behaviour which does not just follow from the intention but, qua evidence, must be caused by it." (69) This circularity, Faye seems to think, is inescapable because of the problem posed by deviant causal chains. For many readers this will not settle the matter. The problem of the reduction of intentional explanations as a class to causal explanations does not touch the issue of whether individual reason explanations are such that, if true, they can count as causal explanations – and what may count as evidence for that. Thus readers may applaud the general anti-reductionism of Faye's account and yet remain puzzled as to his position on the causal dimension of individual intentional explanations.

Given Faye's notion of explanation, of course, the project of naturalization as such remains unaffected. Concerning issues particularly prominent in the humanities as defined by him, Faye does sterling work in detailing, in Chapter 4, the nature of interpretational problems (when we are confronted with a problem of identifying the reference or the relevance of a set of representations); in Chapter 5, on defending the idea that in the interpretation of artistic products the intention of the producer matters; and, in Chapter 6, on how the problem of evidential underdetermination plays out in interpretation (importantly, only as far as the works in question communicate a cognitive content). Faye's neomodernism, contrasted in Chapter 7 to postmodernism, takes on board the advances afforded by the historical and social turns within the philosophy of science since the demise of orthodox logical positivism. So educated, it presents to us attractive options on how to embrace, within the umbrella of a "unified science" that seeks unity in diversity, a collection of highly distinctive disciplinary perspectives. Faye's efforts to remedy the postmodern condition deserve the widest critical support.

Thomas Uebel (Manchester)

Paul Feyerabend, *The Tyranny of Science*. Edited by Eric Oberheim. Cambridge: Polity Press 2011. Pp. xii + 153. ISBN 978-0-7456-5189-7 (hardcover) €54,00, ISBN 978-0-7456-5190-3 (paperback) €15,60.

A red depiction of the Doomsday Clock, signalling only three minutes to midnight, glows on the book cover's black background. Bomb rockets are seen falling from the sky across the Clock's face. The skyline is in flames; destroyed buildings and sketchy ruins are visible. Mushroom clouds rise at a distance. Beneath the Clock, the reader is greeted with the book's title: *THE TYRANNY OF SCIENCE*, whose whitish letters are shaded with washed-out red splatters of blood. Underneath, at the bottom of the cover, the author's name, *PAUL FEYERABEND*, finally appears, the letters now uniformly filled with the same washed-out red colour.

Judging from the cover, the intended audience will be deeply disappointed by the book's content: The author envisions no alarmist end-of-the-world scenarios in it and he doesn't argue anywhere that science will lead to doomsday. The book is an edited transcription of four lectures on the arguably more moderate, but no less complex, topic 'What is knowledge? What is science?' given by Feyerabend in May 1992 at the University of Trento, Italy. (I will refer to them simply as "lectures" from now on.) While it is easy to tell what this book isn't about, it isn't as straightforward to tell what this book is actually about. Through the course of his lectures, divided into four chapters, Feyerabend tackles a seemingly chaotic wealth of topics, ranging from Clinton's then President-elect economic policies to a detailed analysis of Homeric poems. It is therefore excusable, although not helpful, that the chapter titles summarize the content only tangentially. Each chapter is followed by a Q&A session with the audience, where Feyerabend engages with questions related to his lectures, as well to his previous works. The content of the lectures themselves overlaps heavily with the likewise posthumously published collage *Conquest of Abundance*[1] – an unsurprising fact given that Feyerabend had been working on several articles included in it at the time he gave the lectures. To the avid reader of Feyerabend's works, then, I can promise no new major insight by reading *The Tyranny of Science.*

* * *

The main recurring theme of the book is that in order to understand science we have to look beyond science, both topically and temporally. Feyerabend tries to make the case that science is a specific, although not uniform, blend of practices found also in other human activities; that it started, one might say, as an historical accident in ancient history; and that practical knowledge is key to the inner working of science. To develop these claims Feyerabend adopts a particular historicist vantage point, which could be aptly dubbed a state of "artificial naivety" in the vein of Ernst Mach,[2] rejecting the ready-made and evident outlook of the phenomenon under scrutiny. The specific phenomenon under scrutiny here is science: Feyerabend asks why science could develop in the first place and gain such a wide acceptance. He rejects the ready-made answer – "because of its success" – with two counterarguments: First he argues that the track-record of science neither explains nor warrants the unquestioned epistemic and social acceptance, even social prestige, we witness today. Secondly, he argues that the particular form of rationalism underlying scientific knowledge was developed and gained recognition at a time when its success was neither established nor foreseeable. Science becomes a genuine puzzle, which is in need of explanation.

The first counterargument is developed in the second chapter by introducing the distinction between "a performance and the ideology that encourages the

1 *Conquest of Abundance: A Tale of Abstraction Versus the Richness of Being.* Ed. by Bert Terpstra. Chicago: University of Chicago Press 1999.

2 "Künstliche Naivität", cf. Ernst Mach, *Erkenntnis und Irrtum: Skizzen Zur Psychologie Der Forschung.* Leipzig: Barth 1905, p. 15. Transl. *Knowledge and Error: Sketches on the Psychology of Enquiry.* Dordrecht: Reidel 1976.

performance or seems to be supported by it" (37). Despite what the book's title might suggest, Feyerabend does not criticize science as an epistemic enterprise, nor does he deny that it is epistemically successful; he is genuinely fascinated by it. He criticizes the "ideology" of science, i.e. the pretension that the success of science warrants the extension of a "scientific world view" (36) across most or all domains of human thought and agency. In Feyerabend's vocabulary the "scientific world view" does not denote scientific rationality; it denotes a misleading self-conception of scientific rationality, ironically quite like what Otto Neurath criticized with his notion of "pseudorationalism".[3] Science as an epistemic enterprise works despite the "scientific world view", not because of it. The expansion of the "scientific world view" is, however, instrumental to the expansion of science *as a social enterprise*, in that it legitimizes the privileged position of science in the social world on a global scale. Then, the author argues, it may become a threatening force, in that it restricts the variety of conceptions of rationality in other domains of human agency. Here one finds again his notorious conclusion that "you can criticize science without becoming a scientist yourself. More especially, you can criticize scientific demands for more money, more power, greater influence in politics and especially in education – without becoming a scientist. A democratic criticism of science not only is not an absurdity – it belongs to the nature of knowledge" (36). It's worth noting that his previous strictly anti-interventionist policy aimed at the preservation of closed traditions and cultures – a cornerstone of his previous relativist standpoint – is definitely abandoned in favour of a collaborative model: "What we need therefore, is not an increasingly aggressive application of science that treats locals as if they were idiots; what we need is a closer collaboration between experts and the people whose surroundings the experts want to judge, change, improve" (48-9).

The second counterargument opens up a fascinating window onto the history of ideas, with a particular focus on ancient Greek philosophy, in which Feyerabend tracks the evolution of different conceptions of apt reasoning connected to different notions of knowledge. Why an analysis of ancient Greek philosophy and not, say, of philosophy in the early modern period as a source of the scientific revolution? Since the mid 1960s Feyerabend had been increasingly interested in what he called "the rise of rationalism" in the ancient history of ideas. Through the course of a decade this interest developed into a full-blown research project, planning as much as seven volumes (!) devoted to a detailed chronological study of the rise and evolution of "Naturphilosophie" (natural philosophy), beginning with cave paintings and ending with quantum theory. As one might suspect, the

3 "Pseudorationalismus", cf. e.g. Otto Neurath, "Die Verirrten des Cartesius und das Auxiliarmotiv: Zur Psychologie des Entschlusses", in: *Jahrbuch der Philosophischen Gesellschaft an der Universitat zu Wien*, 1913, pp. 45-59. Transl. "The Lost Wanderers of Descartes and the Auxiliary Motive (On the Psychology of Decision)", in: *Philosophical Papers 1913-1946*. Ed. by Robert S. Cohen and Marie Neurath. Dordrecht: Reidel 1983, pp. 1-12.

project was never completed. The available typescript of the first volume remained unpublished and the whole project was abandoned in the late 1970s[4] – unfortunately, I might add, because *Against Method*'s main claim that the rationality of science is incompatible with its "rational reconstruction" leaves open at least two options: The failure of standard accounts of rationality to account for science and its progress "means either that there cannot be any discoveries unless one leaves the house of reason, or that the house of reason is very different from what philosophers and other idea-mongers make it out to be" (112). Unfortunately, the sole publication of *Against Method* lent credibility to the idea that Feyerabend himself went with the first option, instead of hinting at his true intentions of giving a long and detailed account of which way(s) "the house of reason" might differ from formal philosophical accounts.

Even if the monumental project was abandoned, the topic of the rise of rationalism and natural philosophy continued to permeate Feyerabend's research and publications ever since, as witnessed also in the present book, particularly in the third chapter. Feyerabend diagnoses a widespread belief of the primacy of theory over practice in science and tries to show how this belief came about. The belief "which is taken for granted by many scientists and philosophers has very old origins. Actually, it has a double origin. It has an *intellectual origin* in the groups that started Western rationalism and a '*primitive*' origin in the myths that preceded it." (68) Through Homeric poems and the writings of pre-Socratic and later philosophers, Feyerabend tries to convey a picture of how different notions of rationality were 'at work' in the activities of politicians, rhapsodists, carpenters, rhetoricians, priestesses and philosophers in classical antiquity. Accordingly, different styles of reasoning and differing conceptions of what a "compelling argument" might be competed with each other, establishing boundaries between different domains and occasionally crossing them. From here Feyerabend tracks the formation of a canon of rationality tied to a specific self-conception of expertise, later to enter science and its "ideology". Alongside the 'local' expertise of the craftsman rooted into practice (different crafts require different expertise), a kind of 'general' expertise developed, in which an abstract, and therefore generalizable, notion of knowledge could gain currency, purportedly applicable across different domains. Feyerabend diagnoses the birth of a universal entitlement put forward by the new self-proclaimed "experts", which eventually would give rise to the "ideology" of science. He is openly critical of this early development, claiming that different ideals of knowledge are tied to different social and political outcomes. He doesn't shy away from professing his own sympathies: "I am on the side of practice because it seems to be more democratic, but I realize that theory can improve practice, only, this is a complicated matter and not easily understood" (120). His main point, as I understand it, is that an abstract notion of knowledge, namely theoretical knowledge,

4 The German typescript has been published in the meantime: *Naturphilosophie*. Ed. by Helmut Heit and Eric Oberheim, Frankfurt/Main: Suhrkamp Verlag 2009. An English translation is forthcoming.

is necessary to science, but not sufficient, as science only succeeds embedded in specific practices.

Particularly in the fourth chapter Feyerabend invites the reader to rethink the relationship between the empirical and the theoretical dimensions of human agency, science being one such activity among many. One claim is that the theoretical dimension is but a kind of particular practice. In the particular case of science "the opposition between theory and experiment is not an opposition between theory – understood as Platonic ideas – and a moving and partly subjective practice; it is an opposition between two kinds of (moving and subjective) practice, the one applied to things, the other applied to formulae" (124). On the other hand, Feyerabend re-evaluates the domain of practical knowledge: It is not just a domain of mechanical procedures, but also a realm of practical abilities and proper, albeit tacit, knowledge. Science without such abilities would be wishful thinking, a void project at best: "The knowledge we claim to possess, the very general knowledge provided by modern physical theory included, is an intricate web of theoretical principles and practical, almost bodily abilities and it cannot be understood by looking at theories exclusively." (108) Science is, therefore, best understood as a theoretical and practical craft.

* * *

The editorial history of Feyerabend's lectures is somewhat convoluted: A transcription of the recorded lectures had been made available to Feyerabend, who prepared an edited typescript for publication. This typescript first appeared in a posthumously edited Italian translation, which was followed by an unauthorized but tolerated German translation.[5] Eric Oberheim, the editor, is to be lauded for taking up the job of finally making available the original typescript to the English reading audience.

That the lectures are now available under the misleading title *The Tyranny of Science* says more about the publisher's marketing intent to cash-out on Feyerabend's anti-science image than about Feyerabend's authorial intent. Given that Oberheim is the author of a comprehensive book on Feyerabend's philosophy[6] devoted not least to demystify Feyerabend's anti-science image, it is most probably the publisher that should be blamed for these unfortunate marketing decisions. The book is prefaced by an introduction by the editor, giving a good overview of Feyerabend's complex philosophical career and informing the reader of the newest developments in the Feyerabend-scholarship, while being partial to the editor's interpretation of Feyerabend as a pluralistic thinker struggling against conceptual conservatism. The book also contains a helpful index at the end. Given Feyerabend's erudition and the fact that the original typescript lacked any references, the editor did a good job in providing footnotes with additional information on

5 *Ambiguità e armonia: Lezioni trentine*. Roma: Laterza 1996; *Widerstreit und Harmonie: Trentiner Vorlesungen*. Hrsg. von Peter Engelmann. Wien: Passagen Verlag 1998.

6 Eric Oberheim, *Feyerabend's Philosophy*. Berlin: De Gruyter 2006.

public figures mentioned and world events alluded to. What is missing, however, are references informing the reader about explicit and implicit allusions to ancient and current literature disseminated throughout the book. There are literally dozens of implicit and explicit references, which the reader might want to explore further, but for which no bibliographic information has been provided.

A final thought on the very publication of Feyerabend's lectures in print: Particularly in the first chapter Feyerabend is keen to suggest an historical approach as a way to resolve the puzzle of science's predominance in today's world. This judgment should not come as a surprise, given that Feyerabend is among those credited with bringing about the historical turn in the philosophy of science. However, in this book he distances himself from any means of systematic presentation within historical analysis itself: "Strictly speaking my lectures will be fairytales woven around events that are vaguely historical" (13); "we have to go into history and, with that, storytelling" (113). As should be clear from Feyerabend's attempt at writing a colossal work on the history of ideas, this was not always his understanding of historiography. Considering that the content of the lectures overlaps with already available books by the author and that Feyerabend's live performances was one of his strongest feats, there might have been a good case to be made for publishing the original tapes, rather than an edited transcription.[7] It is within the setting of a live performance that "history as storytelling" might actually make sense and persuade us.

Daniel B. Kuby (Wien)

PAOLO PARRINI, *Il valore della verità.* Milano: Guerini e Associati, 2011, pp. 253.

This volume by Paolo Parrini collects eight independent essays covering a long period of studies from the 1970s to more recent years and now partially revised. The contributions inquire into some of the major themes concerning the theory of knowledge and can be grouped as follows: Kant's perspective, focusing in particular on the meaning of the synthetic a priori and its manifold interpretations (I, IV, VIII); the main approaches to epistemology like conventionalism, holism, metaphysical realism and radical relativism (II, III, VI, VII); more general and meta-philosophical reflections on the risks deriving from both absolutism and nihilism (V). Parrini's essays share the common purpose of outlining an alternative

7 A short passage of these lectures is available in a published collection of original recordings, mostly in German. Cf. *Wissenschaftstheoretische Plaudereien: Originaltonaufnahmen 1971-1992*. Hrsg von Klaus Sander. Audio-CD, 60 Minuten. Begleitbooklet, 24 Seiten. Berlin: Suppose Verlag 2000.

concept of truth and a consequent theory of knowledge able to reject both defeatist relativism and coercive foundationalism.

As the title suggests, Parrini defends the value of the concept of truth, in contrast to all those conceptions that debase it by implying that truth cannot be a guiding-concept for our experience (scientific or not). As specified in Essay I, the expression "to be true" can be considered neither a mere descriptive term nor a performative one (according to Strawson's distinction): in the first case, a predicate asserting the truth of a claim would not actually add anything to that claim; in the second, the truth would be exclusively used to perform an action, thus defending a deflationist view. Instead, Parrini prefers to refer to the Kantian and Neokantian traditions. He interprets Kant's conception of truth, as presented in his theory of knowledge, as an intermediate notion between correspondentism and coherentism, empiricism and rationalism. In Kant's view, truth concerning empirical objects depends not only on material elements but also on subjective formal assumptions: truth thus becomes something not inherent in the objects themselves, but in our judgments about them. Formal assumptions constitute a negative (necessary but not sufficient) general formal criterion (Kriterium) ensuring the truth of any knowledge and including both logical-analytical principles (non-contradiction) and logical-transcendental conditions (space, time, categories). Kant claimed that the Aristotelian formula "truth is the adequation of things and intellect" is indeed only a nominal definition (Namenerklärung), as it does not guarantee any certain rule through which truth can be established (similar to the Tarskian paradigm in Parrini's view). In Kant's approach, no specific content is thus ascribed to the notion of truth, but its value is preserved.

Truth, objectivity, reality and knowledge are closely interlinked notions. Once again in Kant's footprints, Parrini endorses the idea that objectivity cannot result from a passive reflection of a reality alleged to be known. But unlike Kant, he claims that truth should not be sought in the relation between our thoughts and things, since objective truth exists only as a regulative ideal guiding the synthesis of our experience (cf. p. 198). According to Parrini, objectivity does not correspond to an effective reality, but rather to an achievement produced by a set of theoretical and empirical processes. Knowledge itself becomes a process aiming not at capturing an objective world, which could not exist in the way we conceive it, but at unifying the disordered and manifold material of experience. In this case too, Kant's influence is apparent.

In outlining his perspective, Parrini reconstructs and criticizes a wide range of conceptions of truth. Essay II deals with Popperian epistemology, criticizing both its deductivism and falsifiability and the notion of verisimilitude (truthlikeness). Parrini demonstrates how deductivism is not better grounded than inductivism at all, as in the end deductivism itself requires inductivism. He claims the formal notion of verisimilitude is not sufficiently articulated either, since it is unable to establish a hierarchy among theories on the basis of the higher or lower content of their truth (or falsehood), a limit later admitted by Popper himself. Popper had paradoxically suggested an informal and intuitive notion of verisimilitude to suffice for his purpose, falling into the same trap arising from Humean inductivism.

The last contribution in the book is concerned with the concept of truth in Heidegger's works. More specifically, Parrini rejects Heidegger's interpretation of Kant's truth as mere correspondentism, since he seems to ignore Kant's attempt to find a Kriterium to establish what is worth accepting as true. Moreover, he objects to Heidegger's ontological interpretation of truth as disclosure (aletheia), as such a concept proves completely unable to supply any precise criteria to establish the judgments' truth in our ontic context.

Despite not ascribing a specific content to truth, Parrini strongly refuses to interpret it as arbitrary. Truth has a value in guiding both scientific and general knowledge and thus represents a primitive and undefinable notion, an empty regulative ideal, aiming at directing our cognitive processes: "[truth] is a guiding-ideal which supports and inspires the synthesis of current and possible experiences" (p. 156). Consequently, it also has to be distinguished from mere opinion and justification: truth requires an agreement among all judging subjects that mere opinions do not require, and beliefs could be reasonably justified without being actually true (cf. p. 15). To some extent Parrini draws on both realism and relativism. Like realists, he believes in the importance of maintaining the significance of the notion of truth, but without filling it with a precise content. At the same time, he agrees with relativists in so far as they refuse any kind of absolutism, but without sharing the defeatist idea of the impossibility of ultimately gaining a precise notion of truth.

Parrini is also concerned with the different approaches to epistemology: by analyzing the opposite perspectives of metaphysical realism and radical relativism, he defends a third way, able to avoid any ontological commitment and also the conviction that, in Nietzsche's words, "it is precisely facts that do not exist, only interpretations". If realism treats the world as existing independently from our beliefs, radical relativism (paradoxically) assumes that no theory can ever capture the world as it is. Instead, Parrini adopts a "moderate epistemological relativism" that consists in keeping separate the subject and object of knowledge (thus adopting a kind of realism), but recognizing that truth and objectivity, although mediated by a linguistic, theoretical, and methodological structure, are always relative to the coordinate system taken into account. In this way, he avoids any commitment to the ontological existence of the world, also defining such a position as a "non-metaphysical objectivism" (p. 179).

Parrini constantly refers to two principles characterizing contemporary philosophy of science: the "empirical underdetermination of scientific theories" and the "theoretical overdetermination of experience". Empirical underdetermination recognizes that facts and characteristics of reality can only be pointed out within an interpretative scheme, i.e. a structured complex of epistemic conditions allowing for knowledge. Conversely, theoretical overdetermination leads us to accept that our knowledge cannot transcend experience, which only allows us to verify the empirical adequacy of assertions, theories and hypotheses.

Parrini does not completely reject the Kantian a priori/a posteriori dichotomy, as the neoempirists did in the wake of the debate on non-Euclidean geometries characterizing the early 20th century. Nevertheless, Parrini recognizes the need to rethink the dichotomy at the core of Kant's theory of knowledge. He argues for a "relativization and contextualization version of a priori", establishing a parallel with both the Duhem-Quine holistic thesis, and the Kuhnean notion of "disciplinary matrix". According to Parrini, a holistic dependence of theories on experience can only be admitted by pinpointing the roles of the various elements: some assertions are synthetic a posteriori, as they can be directly evaluated through experience; others are linked only indirectly to experience and can be properly called a priori. Assertions are analytic when they concern the linguistic component, synthetic when they concern the theoretical part of the framework. Hence, Parrini's epistemic relativism (unlike Carnap's view) rejects any semantic characterization, as "no convention concerning language could close a non-linguistic gap" (p. 111n). Parrini's proposal finally leads him to interpret the relation between hypotheses and empirical observations as threefold, as a reference to a background of assertions taken for granted is always required. In this way, conventionalist perspectives like Poincaré's can also be reasonably avoided: if cognitive processes are inevitably bound by a set of assumptions, this entails neither that that assumptions are entirely arbitrary nor that truth is a mere convention without any effective value.

Parrini's outstanding work thus combines studies on historical debates with the development of a personal perspective aimed at the elaboration of an innovative theory of truth and knowledge, merging the best aspects of different approaches. However, it is to be hoped that Parrini will elaborate such a theory in greater detail in a specific work, since it promises to open up a whole new approach not fully explored in the present volume.

Beatrice Collina (Bologna)

András Máté, Miklós Rédei and Friedrich Stadler (Eds.), *Der Wiener Kreis in Ungarn/The Vienna Circle in Hungary*. Veröffentlichungen des Instituts Wiener Kreis, Bd. 16. Wien: Springer 2011. 300 p.

This volume brings the latest assessments of the influence of the Vienna Circle on Hungarian philosophy and science both before and after World War II. With the exception of a few publications, the research that has been done in this field has not received much consideration until recently. This is mainly due to the asymmetrical relationship between the German and Hungarian language and the ideological condemnation of logical empiricism during the communist regime in Hungary. The editors of the volume grouped the eight papers written in English and the three in German under three headings: History of Philosophy, Physics and Logic. M. Rédei, together with F. Stadler also contributed an introductory essay in which they try to identify evidence of Austria-Hungary linkages and co-operation in philosophy and science.

The results of this research, which are further reflected in different ways in the individual essays in the volume, can be summarized in the following general historical overview. In Hungary, before 1914, there was already reception of works of Franz Brentano and Ernst Mach and also of the new logic. After World War I scholars emigrated from Hungary to the West, via Vienna, and until 1933/34 mutual contacts and joint projects evolved in the context of the Vienna Circle and Logical Empiricism. From 1933/34 these trends had already come to resemble a 'science in exile' and after 1938 they were, in a few isolated cases, carried on in the USA. In the post-World War II period, following the contacts with the 'Kraft Circle' in Vienna, featuring, among others, Paul Feyerabend and Béla Juhos, Imre Lakatos and Karl Popper communicated on scientific issues in England. During the Cold War scientists from both nations were forced to break off scientific exchanges.

The volume´s first section titled History of Philosophy begins with K. Kókai's essay which is devoted to the work of Anton Fischer, a physician educated in Vienna (1919–1926) and a solitary proponent of scientific philosophy in post-war Hungary, who developed a systematic reflection on philosophical foundations of scientific knowledge. In the second essay, G. Á. Zemplén reassesses the philosophical views of another Hungarian scientist, the physicist Győző Zemplén (1879–1916), showing his conventionalism in both Hungarian as well as wider European intellectual context. Two further contributions in the section are dedicated to Béla Juhos (1901–1971) who was born into a Hungarian gentry family. He spent his childhood in Budapest and later he studied (a dissertation 'Inwieweit ist Schopenhauer der Kantschen Ethik gerecht geworden?' defended under Moritz Schlick in 1926) and worked (a habilitation thesis 'Die Erkenntnis und ihre Leistung' defended under Viktor Kraft in 1948) in Vienna where he spent the rest of his life. Juhos and Kraft were the only two members of the Vienna Circle who rode out the era of National Socialism in Vienna and then had a stint at the University of Vienna after the war. In the first contribution, based on his study of various archive materials, W. L. Reiter conveys more biographical data about Juhos. In the second one, G. Ambrus discusses Juhos' antiphysicalist views on the mind-body problem, contrasting them with the accounts given by other members of the Circle, particularly Carnap and Neurath.

The middle section of the volume consists of four papers concerning Physics. In the first one, L. Kvasz interprets epistemologically the development of classical mechanics, from Newton to Lagrange, as changes of the pictorial form in the sense of Tractatus. In the course of this reconstruction he distinguishes perspectivistic, projective, coordinative, compositive and interpretative forms of representation by means of language as peculiar to individual historical stages. In the second one, J. X. Madarász and G. Székely compare relativistic and Newtonian dynamics by their axiomatization in first-order logic and illustrate how the assumption that "the center-lines of a system of point masses viewed from two different reference frames are related exactly by the coordinate transformation between them", though being natural and harmless in Newtonian dynamics, leads to a contradiction in relativistic dynamics. Then G. Székely, in his paper, offers both a definition and criteria for a comparison of acceptable answers to general why-questions in relativistic physics. Finally, L. E. Szabó shows, by means of logico-empiricist analysis, that Lorentz-type theories and new special relativity are completely identical as theories about space and time and that they provide an equal description of the behavior of moving physical objects.

The volume´s last section Logic is introduced by A. Máté who analyses, in a historical perspective, the reception of the new symbolic logic and philosophy of mathematics in Hungary. A. Máté presents the work of the most important Hungarian logicians and scientists of the 20th century, including, among others, Lászlo Kalmár (1905–1976), Rózsa Péter (1905–1977), Béla Bencsik (1905–1943), Ferenz Lehner (1911–1988) and Imre Ruzsa (1921–2008). He also reflects on the influence of the Marxist-Leninist dialectical logic and its protagonists, e.g., Béla Fogarasi (1891–1958). The central part of the section is devoted to the research findings (i.e., three axiom systems) of the "Logic and Relativity Theory" school, made up of a group of algebraic logicians, namely H. Andréka, J. X. Madarász, I. Németi, P. Németi and G. Székely, from the Renyi Institute of Mathematics of the Hungarian Academy of Sciences. This school, explicitly referring back to principles and results of logical empiricism, and particularly to Reichenbach's work, evolves the logical analysis of the theory of relativity by means of the first order logic. The objective of the study is to make the theory of relativity more transparent and understandable by founding it upon unambiguous axioms and showing which of them are needed for which of the novel relativistic predictions, such as the twin paradox. The logical section, and with it the entire volume, ends with the translation of László Kalmár's lecture "The Development of Mathematical Rigor from Intuition to Axiomatic Method" which was first published in 1942. Kalmár here describes, in an accessible and non-technical way, the path that he himself had to traverse in trying to reach a rigorous system of concepts and theorems within mathematics.

The volume as a whole can be seen as evidence that the legacy of the scientific philosophy of the Vienna Circle has endured the adversities of communist ideology and is still alive and well within Hungarian intellectual circles. On the one hand, it has become clear that apart from the famous, fascinating historical cases of mutual scientific interchange, such as, for instance, the debate between John von Neumann and Kurt Gödel about the consequences of the second incompleteness theorem for the Hilbert's program, there are still some neglected issues deserving historiographical study. On the other hand, it has been proven that the ideas and methods of logical positivism, refined and developed in different, intricate ways, still have the power to draw the attention of current and active Hungarian scientists such as the logicians of the Renyi Institute of Mathematics with their program of logical axiomatization of relativistic physics.

Radek Schuster (Pilsen)

JACQUES LE RIDER, *Fritz Mauthner. Scepticisme linguistique et modernité. Une biographie intellectuelle*. Éditions Bartillat: Paris 2012

JACQUES LE RIDER, *Fritz Mauthner. Le langage*. Translation of "Die Sprache" from German and foreword by Jacques Le Rider, Éditions Bartillat: Paris 2012

The recent lull in the (re)discovery of Fritz Mauthner (1849–1923) may lead one to believe that all has been said and written about this intriguing proponent of a radical linguistic skepticism. Two recently published books by the French scholar Jacques Le Rider who has made himself a name as an expert on fin-de-siècle Vienna are certainly proof that this is not the case. His highly engaging, well-structured intellectual biography of Fritz Mauthner is the first work to introduce Mauthner's extensive work to the French reader. As in one of his earlier books (Modernity and Crises of Identity: Culture and Society in Fin-de-Siècle Vienna, Vienna 1993), the central theme of Le Rider's latest book are crises of identity – in this case the crisis of Jewish identity and the crisis of linguistic identity.

Born into a German-speaking, assimilated Jewish family living in a small Bohemian town, Mauthner was sensitized to the issue of language from his childhood on. Even though he was later able to find professional acclaim working as a journalist and novelist in Berlin, Mauthner was, as Le Rider argues persuasively, never entirely at ease with his identity. Le Rider sees Mauthner's experience of the battle of languages in Prague as a strong influence in his critique of language. In his adamant defense of German against Czech Mauthner first developed a nationalism – which stood in stark contrast to the "Habsburgian myth" of a harmonious coexistence of different nationalities, ethnic groups in the Austrian-Hungarian monarchy. In 1866, at the age of 17, he developed an unbridled enthusiasm for Bismarck's Prussia. From then on he felt that he lacked a "real" mother tongue and was left with the dead remnants of the languages of his childhood. Le Rider sees this sense of crisis of language – heightened by Mauthner's calling into question the knowledge of classical languages as the basis of a neo-humanist culture – as one of the main roots of his critique of language.

Le Rider identifies the crisis of Jewish identity as another crucial force in his critique of language, devoting an entire chapter to this facet of Mauther's life, which has not been explored so thoroughly before. Even if Mauthner's family belonged to the assimilated Jewish population, Rider shows that he was steeped in Jewish ideas from an early age on. Le Rider draws a line leading from the religious mysticism of the messianic movement of Jacob Frank that Mauthner's grandfather associated with to Mauthner's later interest in "Atheism and its History in the West" (76-79). He also sees the development of Mauthner's critique of language as an echo of the Berlin controversy on anti-Semitism (1879–1881). In his novel "Der neue Ahasver" (The New Wandering Jew) Mauthner had taken a stance in this controversy in which anti-Semites opposed the intellectual anti-anti-Semites. To further corroborate Mauthner's search for his lost Jewish identity Le Rider cites an article written in 1924 and published in English in The Menorah Journal, where he explores the possibility of a direct link between Jewishness (Judenschaft) and skepticism.

In the second half of his biography Le Rider not only offers a detailed reconstruction of Mauthner's main ideas, he also explores numerous philosophical sources of his radical linguistic skepticism. One important one was, of course, Friedrich Nietzsche's Truth and Lie in an Extra-moral Sense – words being the origin of translations, transpositions, metaphors ("lies in an extramoral sense"). Herein also lies one of Mauthner's main arguments – language does not describe the world but can only express it in a metaphorical sense. Le Rider's reconstruction of Mauthner's thought through a number of different philosophical discussions of his time is intellectual history at its best. He is able to uncover a wealth of fascinating historical details (e.g., how the writer Gerhart Hauptmann directed Mauthner's attention to a book on translation by the philosopher Otto Friedrich Gruppe), while at the same time clearly identifying the most important players and their arguments.

As others who have written on the critique of language, Le Rider reflects on how Fritz Mauthner's linguistic skepticism can be positioned within the Austrian tradition of philosophy. Le Rider carefully delineates this tradition, taking into account studies that trace its historical development from Occam to the Austrian school of analytical philosophy and the semantic turn (Brentano, Wittgenstein, Vienna Circle) (238). As Le Rider notes, Mauthner was clearly influenced by the "realist" Johann Friedrich Herbart and the "empiricist" Ernst Mach. But he also shows how Mauthner broke with the Austrian tradition in several important points. Mauthner was convinced that it would never be possible to construct a formalized language in which all concepts and rules of deduction could be defined without any ambiguity. He adopted Mach's anti-metaphysical orientation but he would not accept the possibility of scientific knowledge as the latter had. As Le Rider convincingly argues, Mauthner's linguistic skepticism was also strongly inspired by Schopenhauer and Nietzsche. Le Rider is also good at uncovering certain paradoxes and inconsistencies in Mauthner's thought. Mauthner saw his critique of language as filling a lacuna in Kantian thought, as going beyond Kant. As Le Rider correctly observes, Mauthner can be seen both as an anti-Kantian but also as a representative of a neo-Kantian linguistic turn. Le Rider also identifies an alliance between Kantian and Nietzschean thought in Mauthner's radical skepticism – as is also illustrated by the latter's association with Hans Vaihinger. And even if Mauthner opposed Aristotelian logic, he entertained ideas that could be likened with those of the neo-Aristotelian Friedrich Adolf Trendelenburg who was also interested in the historical geneaology of concepts (253). Yet at the end of the day, the only philosophy Mauthner could really identify with was "skepticism and mysticism".

In his translation of one smaller work by Fritz Mauthner, "Die Sprache" from the year 1907, Le Rider introduces the reader to some of the main ideas in a very informative foreword. In his first major work, *Contributions to a Critique of Language*, Mauthner had argued that language is first and foremost an individual phenomenon, even if he also viewed language as a social reality, with communication being by necessity imperfect and subject to misunderstandings. In his short monograph on language he now turned to the social and psychological dimensions of language. Jacques Le Rider gives a lucid account of Mauthner's own version of linguistic psycho-sociology in which he borrowed heavily from the "Völkerpsychologie" (folk psychology) propagated by Wilhelm Wundt. Wundt, whose first volume of Völkerpsychologie was also titled "Die Sprache", had distinguished between individual psychology and collective psychology. Le Rider identifies another important influence on Mauthner in his attempt to grasp both the individual and social side of language: Gabriel Tarde (*Les Lois de l'imitation. Étude sociologique*, Paris 1890) who was also cited by Wundt: "Society is the imitation of imitation, a sort of somnambulism" (26). Le Rider shows that some of the main questions Mauthner wrestled with not only reflected his Humboldtian leanings but also anticipated Ferdinand de Saussure's distinction between langage and langue, the latter being for Mauthner a sort of sensorium commune.

Mauthner also explored the possibility of a sort of international *Volksprache*, a universal language that would unite all people which, as he concluded, is simply a futile utopia. Still clinging adamantly to German nationalism, Mauthner also rejected the idea of a multinational, multicultural nation (breaking once again with the Habsburg dream).

Le Rider also presents one of Mauthner's most interesting ideas: the importance of translation. Even if he rejected the idea of an artificial universal language, he saw the growing number of translations in his time creating a sort of common bond between the peoples of the world. Translation in his view also had the effect of making the distinction between mother tongue, national language and foreign languages become increasingly blurred in loan translations, calque, words "borrowed" from one language, with each word constituting a monograph of intercultural history. Anticipating his *Dictionary of Philosophy* (1910/11), Fritz Mauthner presented several "conceptual histories", one of which addressed the object/subject dichotomy (which in his analysis prove to be interchangeable words). The fact that there are two words in German for object – the older word 'Objekt' and the newer one 'Gegenstand' – is certainly a challenge for the translator. Here Le Rider shows great sensitivity in maneuvering the nuances of language as can also be seen in his translation of 'Sprache' as either 'langue' or 'langage' depending on the context.

Relating Mauthner's ideas to contemporary thinkers such as Georg Steiner and Michel Foucault, the biographer sees radical linguistic skepticism as part of a critical undercurrent of modernity that has anticipated later intellectual developments. As both works under review here show, there still remains 'tierra incognita' to be discovered in Mauthner's writings. The bibliography at the end of Le Rider's intellectual biography is a veritable treasure trove, including not only an exhaustive bio-bibliographic chronology but also a list of all published correspondence and archival sources. A publication of both works in English can only be encouraged.

Camilla R. Nielsen (Wien)

Activities of the Institute Vienna Circle

Activities 2012

International Conferences and Workshops

12th Vienna International Summer University
Scientific World Conceptions (VISU/SWC)
Main topic: "Applied Science. Historical, Epistemological, and Institutional Characteristics"
Main Lecturers: Martin Carrier (University of Bielefeld), Rose-Mary Sargent (Merrimack College), Peter Weingart (University of Bielefed)
Venue: Campus der Universität Wien
Date: July 2–13, 2012
Together with: University of Vienna, Department of Philosophy and Department of Contemporary History
www.univie.ac.at/ivc/VISU

Philosophie, Wissenschaft, Religion / Philosophy, Science and Religion
International Conference
Together with: Universität Tübingen and Helsinki Collegium for Advanced Studies
Date: February 12–13, 2012
Venue: Forum Scientiarum, Universität Tübingen
www.forum-scientiarum.uni-tuebingen.de/
www.helsinki.fi/collegium/

New Directions in the Philosophy of Science
Closing plenary conference of the Europan Science Foundation (ESF) Research Networking Programme
Date: October 17–21, 2012
Venue: Bertinoro Conference Centre
(University of Bologna)
www.pse-esf.org

M.C. Galavotti et al. (eds.), *European Philosophy of Science – Philosophy of Science in Europe and the Viennese Heritage*, Vienna Circle Institute Yearbook 17,
DOI 10.1007/978-3-319-01899-7,

Philosophy and the Sciences – Old Visions, New Directions
International Symposium
Together with: the ESF-Programme „The Philosophy of Science in a European Perspective"
Date: November 30 – December 1, 2012
Venue: University of Cambridge

Lectures and Lecture Series

20. Wiener Kreis Vorlesung / 20th Vienna Circle Lecture
Don Howard (University of Notre Dame)
Einstein and the Vienna Circle
Date: December 13, 2012

Ian Hacking (Collège de France / University of Toronto)
The Lure of Pythagoras
Together with: DK+ Doctoral Programme "The Sciences in Philosophical, Cultural and Historical Context" of the University of Vienna
Date: April 17, 2012
Venue: Kapelle, Campus of the University of Vienna

Philosophy of Science
Lecture series in cooperation with the Department of Philosophy at the University of Vienna
http://wissenschaftstheorie.univie.ac.at/vortragsreihe/

Wissenschaftsphilosophisches Kolloquium
Weekly lectures on the philosophy and theory of science given by scholars from Austria and abroad.
www.univie.ac.at/ivc/koll/

Publications

Jubiläumsausgabe/Anniversary Edition
Wissenschaftliche Weltauffassung. Der Wiener Kreis
(Wien: Artur Wolf Verlag 1929)
Reprint of the first edition with translations into English, French, Spanish and Italian
Ed. by Friedrich Stadler and Thomas Uebel

Moritz Schlick Gesamtausgabe, hrsg. von Friedrich Stadler und Hans Jürgen Wendel. Wien–New York: Springer Verlag:
www.springer.com/series/7287
Abt. I, Band 4: *Zürich – Berlin – Rostock. Aufsätze, Beiträge, Rezensionen, 1907–1916.* Hrsg. von F. O. Engler. 2012
Abt. II, Band 6: *Erkenntnistheoretische Schriften 1925–1936*
Hrsg. von Johannes Friedl. 2012

Ernst Mach Studienausgabe in 9 Bänden. Berlin: xenomoi Verlag
Band 3: *Die Mechanik.* Hrsg. von Gereon Wolters und Giora Hon. Berlin 2012
www.xenomoi.de

Forschungsprojekte / Research projects

Forschungs- und Editions-Projekt: Moritz Schlick (1882–1936)
Kritische Gesamtausgabe und intellektuelle Biographie
Phase 3 (2011f.): Schriften aus dem Nachlass (Abteilung II). Gemeinsam mit der Hamburgischen Akademie der Wissenschaften, Moritz Schlick Forschungsstelle der Universität Rostock
www.univie.ac.at/ivc/Schlick-Projekt/

Running Research Projects: Austrian Science Fund (FWF)

Christian Damböck: Wilhelm Dilthey und Rudolf Carnap

Richard Dawid: Realismus und Stringtheorie

Günther Sandner: Die Politik des Logischen Empirismus

Georg Schiemer: Zwischen Logizismus und Metalogik

Activities 2013

International Conferences and Workshops

13th Vienna International Summer University
Scientific World Conceptions (VISU/SWC)
Main topic: "Climate Studies"
Main Lecturers: James Fleming (Colby College), Roman Frigg (London School of Economics), Wendy Parker (Ohio University)

Guest Lecturer: Angela Kallhoff (University of Vienna)
Date: July 1–12, 2013
Venue: Kapelle and Department of Contemporary History, University Campus, Court 1
Together with: University of Vienna, Department of Philosophy and Department of Contemporary History
Part of the DK+ Programme „The Sciences in Philosophical, Cultural and Historical Context" of the University of Vienna

A Tribute to Otto Neurath – Otto Neuraths Bildsprache
In connection with the exhibition *Zeit(lose) Zeichen. Gegenwartskunst in Referenz zu Otto Neurath*
December 13, 2012 to February 17, 2013, Künstlerhaus, Wien

International Workshop
Isotype
Date: January 26, 2013
Venue: Institut Wiener Kreis

Internationaler Workshop
Wilhelm Dilthey als Wissenschaftsphilosoph / Wilhelm Dilthey as a philosopher of science
Date: 3. und 4. Juni 2013
Venue: Institut Wiener Kreis

Logischer Empirismus und Pragmatismus – Logical Empiricism and Pragmatism
Together with: Universität Tübingen and Helsinki Collegium for Advanced Studies (im Rahmen eines Netzwerkes zwischen Helsinki, Tübingen und Wien)
Date: November 8–9, 2013
Venue: Campus der Universität Wien

Lectures and lecture series

21. Wiener Kreis Vorlesung / 21st Vienna Circle Lecture
Ilkka Niiniluoto (University of Helsinki)
Eino Kaila and the Vienna Circle
Date: November 7, 2013, 5 p.m.
Venue: University of Vienna

Philosophy of Science
Lecture series in cooperation with the Department of Philosophy at the

University of Vienna
http://wissenschaftstheorie.univie.ac.at/vortragsreihe/

Wissenschaftsphilosophisches Kolloquium
Weekly lectures on the philosophy and theory of science given by scholars from Austria and abroad.
www.univie.ac.at/ivc/koll/

Publications

Vienna Circle Institute Yearbook 17:
Maria Carla Galavotti, Elisabeth Nemeth, Friedrich Stadler (eds.)
European Philosophy of Science – Philosophy of Science in Europe and the Viennese Heritage. Dordrecht: Springer 2013

Veröffentlichungen des Instituts Wiener Kreis, Band 18:
Elisabeth Nemeth und Friedrich Stadler (Hrsg.)
Die Europäische Wissenschaftsphilosophie und das Wiener Erbe
Wien–New York: Springer 2013

J. Allan Hobson
Dream Consciousness
The William James Lectures
Foreword by Wolf Singer and commented by renowned scholars of the cognitive and life sciences
Edited by Nicholas Tranquillo
Dordrecht: Springer 2013

Ernst Mach Studienausgabe in 9 Bänden. Berlin: xenomoi Verlag:
Populärwissenschaftliche Vorlesungen
Hrsg. von Elisabeth Nemeth und Friedrich Stadler, Berlin 2013

Library/Documentation/Archives

Estate and library of Arthur Pap (1921–1959)
www.univie.ac.at/ivc/pap.pdf

Preview 2014

14th Vienna International Summer University
Scientific World Conceptions (VISU/SWC)
"Humans/Animals. A Contested Boundary"
July 7–18, 2014
University Campus
www.univie.ac.at/ivc/VISU

5th Conference on Integrated History and Philosophy of Science (&HPS)
June 26–29, 2014
University Campus

Preview 2015

XII. Internationaler Kant-Kongress: Natur und Freiheit
Focus: Kant und der Wiener Kreis / Kant and the Vienna Circle
June 2015
University of Vienna
http://kant2015.univie.ac.at/

Index of Names

Not included are: Footnotes, Tables, References and Figures

M.C. Galavotti et al. (eds.), *European Philosophy of Science – Philosophy of Science in Europe and the Viennese Heritage*, Vienna Circle Institute Yearbook 17,
DOI 10.1007/978-3-319-01899-7, © Springer International Publishing Switzerland 2014

Abstracts

Cristina Chimisso

A matter of substance? Gaston Bachelard on chemistry's philosophical lessons

Many historians of philosophy of science have pointed out that traditionally philosophers of science have paid selective attention to some sciences rather than others. Physics has been the model for a large part of mainstream of philosophy of science. Critics have argued that this has created a philosophy of science that has often been claimed to be general but has in fact reflected the particular science taken as model. On the other hand, some philosophers, as for instance Bernadette Bensaude-Vincent, have argued that the distinctiveness of an important part of French philosophy of science comes from its focus on chemistry. In this paper I shall focus on the significance of chemistry for some crucial aspects of the philosophy of Gaston Bachelard. He believed that 'philosophy should follow science' and that 'the major lesson that the philosopher should learn from the evolution of science is that philosophy itself should be improved'. Here I will look specifically at some of the lessons that for Bachelard philosophy should learn from chemistry, including lessons about the role of analysis and synthesis in knowledge, the concept of substance and of the scientific object.

Richard Creath

(Anti-)Metaphysics in the Thirties: Why Should Anyone Care Now?

We live in a metaphysical age. So some philosophers are now baffled (and others outraged) that members of the Vienna Circle in the 1930s attacked metaphysics as empty at best and utterly unintelligible at worst. This paper examines this rejection of metaphysics to understand both what metaphysics was then taken to be and why the rejection took the form that it did. The answer, it turns out, is as much political – in the widest sense – as it is scientific. Once we see that, we can begin to see why that discussion of metaphysics more than 80 years ago is relevant to our own time.

Claude Debru

On the Relationship between Neuroscience and Philosophy: the Case of Sleep and Dreaming

In this paper, I will try to examine from both philosophical and epistemological points of view the major developments of neurophysiological research on sleep and dreaming during the last sixty years. In the nineteen fifties and sixties, neuro-

M.C. Galavotti et al. (eds.), *European Philosophy of Science – Philosophy of Science in Europe and the Viennese Heritage*, Vienna Circle Institute Yearbook 17,
DOI 10.1007/978-3-319-01899-7, © Springer International Publishing Switzerland 2014

physiologists and psychophysiologists had great hopes regarding the contribution of sleep and dreaming studies to the understanding of consciousness mechanisms. These hopes were not entirely fulfilled. Progress in the field of sleep research was achieved thanks to developments in disciplines like neurophysiology properly said, neurochemistry, clinical medicine, etc. Today, more modern EEG Imaging techniques bring new and interesting results on cerebral correlates of conscious experience. The purpose of the following analysis is to show how some problems could be solved while others remain open for future investigation, be it experimental or (hopefully) philosophical. I will start with a sketch of the history of the field since the nineteen fifties.Then I will present some epistemological remarks on experimental approaches in neurophysiology, and I will go on with discussions on both psychophysiological parallelism and physiological functions.

Michael Esfeld

Reductionism Today

The paper outlines the main argument for ontological reductionism in today's discussion, claims that ontological and epistemological reductionism (theory reduction) stand or fall together and finally sketches out how today's most widespread form of reduction, namely functional reduction, can be developed into a fully-fledged theory reduction, thus taking up the programme of the Vienna circle in today's philosophy.

Jan Faye

Does the Unity of Science Have a Future?

The main idea behinds the program of logical positivism was that of the unity of science. The idea carries the belief that all the sciences including the social sciences and the humanities ought to share some common features if these disciplines were to be considered genuine sciences (Wissenschaften). In the end, according to the standard view, the unity of science program failed because it entailed unrealistic commitments to anti-metaphysics, behaviourism, reductionism, nomothetic descriptions, objectivity, universalism, value-freeness, and verificationism. Today no philosopher of science would even think of these commitments to hold unrestrictedly for the natural sciences.

The criticism raised against the logical positivist view of the unity of science program was that neither the social sciences nor the humanities were able to meet most of these requirements which were given a strong positivistic interpretation. The opponents pointed out that both the social sciences and the humanities shared an orientation towards idiographic descriptions. Hence Carl Hempel's covering law model of explanation was useless as an account of explanation in history, psychology, sociology or anthropology. In fact, humanistic disciplines were more interested in understanding than explanation. Also the critics emphasized that research within the social and humanistic fields was not value-free, and the humanities at least were interested in subjectivity as much as objectivity. Moreover, they stressed that our understanding of social and humanistic phenomena depends on the cultural and historical context of our inquiry.

In my talk I shall nevertheless argue for a unity of all sciences in spite of all the differences. I point to some of the mistakes made by the positivists but claim that these mistakes do not falsify the main idea. To carry such an argument through requires some revision of some philosophers' understanding of what it takes to be an interpretation, an explanation, a law, a model, and a theory.

Maria Carla Galavotti

Probabilistic Epistemology: A European Tradition

Probabilistic epistemology holds that probability is an essential ingredient of science and human knowledge at large, and that induction is a necessary constituent of the scientific method. Developed in some detail by a number of authors including Patrick Suppes, Richard Jeffrey and Brian Skyrms, this view has been embraced by so many, as to gradually become predominant. While probabilistic epistemology has been growing, awareness of its origins was somehow left behind. Probabilistic epistemology is usually seen as a product of the encounter of logical empiricism with American pragmatism. Without denying the impact of such an encounter, it is argued that the probabilistic approach was already part of the European scenario before the dissolution of logical empiricism in the late 1930s. Notably, such an approach is not linked to a particular interpretation of probability, having been endorsed by upholders of different probability notions, including Janina Hosiasson, Frank Plumpton Ramsey, Bruno de Finetti, Harold Jeffreys, and Hans Reichenbach, to mention only those authors whose contribution is addressed here. The list could extend to the French probabilists Émile Borel and Maurice Fréchet, and others not considered in this survey. The work of so many authors operating at the four corners of Europe in the first decades of the 19th century reflects a European tradition in probabilistic epistemology, somewhat overshadowed in the literature by the attention paid to logical empiricism.

Veronika Hofer and Michael Stöltzner

Vienna Circle Historiographies

During the almost 20 year both of us have been interested in the Vienna Circle, its historical perception has changed in many ways. While back then a large number of historical studies and philosophical analyses set out with the declared intention to overcome the 'received view' and go back to the sources themselves, today there exists a broad historical scholarship about the movement's core members and their embedding into Austrian and German intellectual traditions. While this scholarship could initially be subdivided into a strand that took the Vienna Circle as part of an Austrian Philosophy and a contrasting one that analyzed the movements' neo-Kantian roots, Otto Neurath and Rudolf Carnap being the main protagonists respectively, today we witness a stunning plurality both as regards the themes and the methodologies applied. Albeit less scrutinized, the same holds true for the broader movement of Logical Empiricism. Commencing from the recently published Cambridge Companion, our paper provides a provisional field guide on methodology and investigates the lessons of the surprising diversity on the interactions of philosophy and history of science.

On a first level of classification, one finds standard historiographic methods, among them network history, institutional history, intellectual history, history of ideas, cultural history, applied to the Vienna Circle alongside the history of philosophy. On a second level, these methods interact with diverging intellectual and philosophical agendas, ranging from upholding the legacy of Red Vienna or the late Enlightenment to specific programs in contemporary philosophy of science, ranging from Neurath's economics to Carnap's Aufbau and quasianalysis. It is interesting, Again historicizing, on a third level, these justificatory employments we find that they were started by the movement itself, both the form of programmatic writings and intellectual autobiographies, a fact which influenced the critics' appraisal of the Circle. On a fourth level, the more recent debates about what constitutes, both historically and thematically, the history of philosophy of science can be understood as an attempt to redefine the Vienna Circle's historiographic narrative.

Eckehart Köhler

Gödel and Carnap· Platonism versus Conventionalism?

Kurt Gödel (1955) seemed to successfully refuted the core of Carnap's "Syntax-Program", namely that one cannot do or even just formulate mathematics without content. (I.e. one needs significant portions of Hilbert's "intuitive" mathematics.) Gödel omitted an explication of Conventions – other then that they allegedly exclude intuition –, but he also omitted an explication of Intuition itself (the faculty of observation which makes out the content of mathematics). Surprisingly, Gödel himself contributed to Carnap's famous "Principle of Tolerance" (Logical Syntax §17), the core of Carnap's Conventionalism – although of course Gödel would never have agreed to the formulation there. But Gödel's main thesis about Conventionalism used in his refutation of the "syntax program", namely that Carnap resorted to conventions in order to "eliminate" intuition, is misleading. For it can be shown that, unbeknownst to Carnap (or Gödel), conventions in fact "reveal" intuitions in any case. Moreover it can also be shown that Gödel's definition of Platonism, which makes essential reference to intuition, actually makes Platonism compatible with Conventionalism! Intuition simply needs to be explicated as normative value judgment – Gödel came to within a hair's breadth to this idea. With reference to Hume's Law (fact/value dichotomy), Platonism is just the belief in objectively valid norms. Hume's dichotomy is presupposed in (Bayesian) Decision Theory, and using it we may easily distinguish the real from the ideal (Platonic) world. We also quickly see that Conventionalism is compatible with Platonism – so long as conventions are sufficiently well established. In retrospect we may view Carnap as decidedly Platonistic, at least for certain areas: namely in his Inductive Logic.

Ladislav Kvasz

Mathematics and Experience

Mathematics was considered by the members of the Vienna Circle an *a priori* discipline consisting of analytic propositions. The aim of the present paper is to draw attention to the experiential dimension of mathematical knowledge. Following the interpretation of physical knowledge as knowledge constituted by the use of instruments, I am trying to interpret also mathematical knowledge as knowledge based on instrumental experience. This interpretation opens a new perspective on the place of the logicist program in philosophy of mathematics.

The aim of the present paper is to argue that mathematics is based on experience and that mathematical experience is instrumental. The difference between mathematics and physics is that in the case of mathematics the instruments are not measuring devices but *tools of symbolic representation*. In order to distinguish the mathematical experience from the physical one (i.e. from the experimental experience), I will call the instrumental experience in mathematics as *symbolic experience*. In analogy with the measuring devices of physics I will interpret the tools of symbolic representation as tools constituting symbolic experience.

Pablo Lorenzano

What is the Status of the Hardy-Weinberg Law within Population Genetics?

The aim of this communication is to analyze the status of the Hardy-Weinberg law within (classical) population genetics. The analysis will be carried out with the notions of the structuralist view of theories, specially those of fundamental law or guide-principle, specialization, and special law, having as a background a rational reconstruction of (classical) population genetics – sketched in this communication – made within the framework of such a metatheory.

Matthias Neuber

Is Logical Empiricism Compatible With Scientific Realism?

Scientific realism is the view that the theoretical entities of science exist. Atoms, forces, electromagnetic fields, and so on, are not merely instruments for organizing observational data but are real and causally effective. This view seems to be hardly compatible with the logical empiricist agenda: As common wisdom has it, logical empiricism is mainly characterized by a strong verification criterion of meaning, i.e., by the project of defining the meaning of theoretical terms by virtue of the meaning of purely observational terms. However, it has been largely ignored by the historians of logical empiricism that there indeed existed a *realist faction* within the logical empiricist movement. Among the few authors who have recognized both the historical and the programmatic relevance of this realist faction is Stathis Psillos who, in two recent papers, attempts to emphasize the important role played in this connection by Herbert Feigl and by Hans Reichenbach. According to Psillos, it was these two thinkers who documented in their writings the *compatibility* of logical empiricism and scientific realism.

Like Psillos I am of the opinion that the realist faction within the logical empiricist movement deserves more attention than it has received so far. However, I will come to a different result than Psillos. According to the view I wish to defend, Feigl and Reichenbach (and with them Psillos) are still too optimistic about the ontological impact of language. In order to establish the intended realist account of logical empiricism, more metaphysics is needed than Feigl and Reichenbach (and with them Psillos) would allow. As will be shown, among the logical empiricists themselves it was Eino Kaila who came closest to this—less linguistic and more metaphysical—kind of approach.

John D. Norton

Infinite Idealizations

Idealizations are distinguished from approximations in that only idealizations involve novel reference. This difference is important when idealizations are created by taking infinite limits such as in statistical mechanics. These infinite limits may have strange properties, much odder than the discontinuities of phase transitions now widely acknowledged in the literature. The infinite limits may be indeterministic, or may not exist at all, so that the idealization of an infinite limit should not or cannot be used.

Herlinde Pauer-Studer

Kelsen's Legal Positivism and the Challenge of Nazi Law

In this paper I am going to examine Kelsen's legal positivism in the light of Nazi legal theory. My claim will be that Kelsen's thesis that law and morality constitute two distinct normative spheres is highly plausible, but that some of his metaethical assumptions are seriously flawed. Kelsen's insistence on the separation of law from morality seems much more persuasive if one takes into account the Nazi legal theorists' program of a 'unification of law and morality'. The critique that legal positivism was responsible for the compliance of the German judiciary with the Nazi's interpretation and use of law is not viable. I will, however, argue that Kelsen's relativist account of morality made his position vulnerable to the post-war objections that legal positivism provided no safeguard against the Nazi perversion of law.

The paper is structured in the following way: Section II explains briefly why Nazi law is thought to pose a special problem for legal positivism, Section III outlines the main assumptions of Kelsen's legal positivism and his relativist understanding of morality. In section IV, I depict the anti-positivist view of some prominent Nazi legal theorists. Section V reflects on the lessons we should draw from the arguments of the Nazi legal theorists. I conclude by arguing that the weak point of Kelsen's position is not the separation between law and morality, but his skepticism towards an intersubjectively valid morality.

Jeanne Peijnenburg and David Atkinson

Biased Coins. A model for higher-order probabilities

Is it coherent to speak of the probability of a probability, and the probability of a probability of a probability, and so on? We show that it is, in the sense that a regress of higher-order probabilities can lead to convergent sequences that determine all these probabilities. By constructing an implementable model which is based on coin-making machines, we demonstrate the consistency of our regress.

Tomasz Placek

Vienna Circle on Determinism

Members of Vienna Circle explicated determinism in terms of predictability in principle, or calculability. This paper attempts to uncover the rationale for this explication. It argues that the explication was an attempt to escape trivialization arguments; another important factor was the Circle's views on meaning as testability.

Wlodek Rabinowicz and Lina Eriksson

Betting Interpretation and the Problem of Interference

The paper concerns the historically influential betting interpretation of subjective probabilities due to Ramsey and de Finetti. While there are several classical and well-known objections to this interpretation, the paper focuses on just one fundamental problem: There is a sense in which degrees of belief cannot be interpreted as betting rates. The reasons differ in different cases, but all these cases have in common one crucial feature: The agent's degree of belief in a proposition *A* need not coincide with her degree of belief in a conditional that *A* would be the case if she were to bet on *A*, which itself is conditioned on the supposition that she will have an occasion to make such a bet. Even though the two degrees of belief sometimes can coincide (they will coincide in those cases when the bet has no expected causal bearings on the proposition *A* and the opportunity to bet have no evidential bearings on that proposition), it is the latter belief rather than the former that guides the agent's rational betting behaviour. Or so, at least, will be argued. The reason is that this latter belief takes into consideration potential interferences that bet opportunities and betting itself might create with regard to the proposition to be betted on. It is because of this /interference problem/ that the agent's degree of belief in *A* cannot be interpreted as her betting rate for *A*. This suggestion will be developed in the talk.

Günther Sandner

Political Polyphony. Otto Neurath and Politics reconsidered

Among the Vienna Circle's members, it was certainly Otto Neurath who was most active and ambitious in matters of politics. His particular self-image as a "social engineer" influenced his understanding of politics to various degrees throughout his intellectual life. A biographical overview shows, however, that politics always played a special and sometimes the most important role in Neurath's intellectual life. The orientation on social reform issues in his youth, his leading role in the Bavarian socialisation debate, his pedagogical relevance in Red Vienna and, not least of all, his contributions and interventions in the discussions on National Socialism as an émigré are perhaps the most outstanding examples.

Although political questions were never a central theme in research on Neurath (at least not to an extent comparable to very frequently addressed fields such as the philosophy of science and visual education), a number of essays and studies during the last three or four decades have addressed political issues directly.

The paper first addresses these subsequent interpretations of Otto Neurath's political writings and activities, and discusses their different foci and perspectives as well as accordance and possible contradictions among them. Second, it re-evaluates the main lines of argumentation by looking in detail at exemplary biographical periods of special political interest. And third, it raises the question of how political elements influenced Neurath's philosophy of science – and vice versa.

Friedrich Stadler

From the Vienna Circle to the Institute Vienna Circle: On the Viennese Heritage in Contemporary Philosophy of Science

The Vienna Circle as part of the intellectual movement of Central European philosophy of science is certainly one of the most important currents for the emergence of modern philosophy of science. Independent from this uncontested historical fact there remains the question of the direct and indirect influence, reception and topicality of this scientific community in contemporary general philosophy of science as well as in the philosophy of the individual sciences, including the social sciences and humanities.

In my account I will focus on the "present situation in the philosophy of science" by identifying some relevant impacts, results, and unfinished projects since the classical Vienna Circle, by dealing with specific European features of this globalized philosophical tradition up to the present, and by exemplifying some future perspectives after the linguistic, historical and pragmatic turns. This reconstruction is partly linked to the history of the Institute Vienna Circle which was established in 1991 in Vienna, and which was a supporting institution of the ESF Research Network Program the "Philosophy of Science in a European Perspective" (PSE) from 2008 to 2013.

Thomas Uebel

Carnap's *Aufbau* and Physicalism: What Does the "Mutual Reducibility" of Psychological and Physical Objects Amount to?

The present paper is part of a larger project of investigating how far puzzling questions about Carnap's philosophical deflationism – as expressed most prominently in "Empricism, Semantics and Ontology" – can be answered by reference to his own preferred position in areas upon which this meta-philosophical position can be expected to have a bearing. For that project the explorations below provide a starting point; on the present occasion they will, I hope, be found to be of independent interest. At issue is the question of what makes Carnap's *The Logical Structure of the World* (the *Aufbau*) a *pre*-physicalist work.

Jan Woleński

Kazimierz Twardowski and the Development of Philosophy of Science in Poland

Kazimierz Twardowski studied with Brentano and followed his style of doing philosophy, in particular, the thesis that the method of philosophy is the same as the method of natural science. Hence, philosophy of science (Polish philosophers preferred the name "methodology of sciences") became of a crucial importance for philosophy as well as science. Twardowski did not contribute to philosophy of science in its contemporary setting. Perhaps his definition of reasoning as based on the concept of logical consequence and his ideas related to the division between a priori and a posteriori science should be remembered.

Twardowski established so-called Lvov-Warsaw School at the beginning of the 20th century. The history of this school can be divided into two periods: before 1918 and 1918–1939. Several members of this school students intensively worked in logic and philosophy of science and achieved remarkable results, particularly in the second period when philosophy of science was influenced by logic. The (at least) following points are to be mentioned:

— the problem of induction (Jan Łukasiewicz, Janina Hossiasson-Lindenbaum);
— classifications of reasoning (Łukasiewicz, Kazimierz Ajdukiewicz);
— radical conventionalism (Ajdukiewicz);
— applications of many-valued logic to science (Zygmunt Zawirski);
— styles of thought (Ludwik Fleck – outside of Twardowski's school);
— operationalism (Edward Poznański, Alexander Wundheiler);
— uses of formal semantics in analysis of science (Alfred Tarski, Maria Kokoszyńska);

Polish philosophy of science was closely related to Vienna, not only via Twardowski himself, but also by close contacts and affinities with the Vienna Circle.

Gereon Wolters

Is There a European Philosophy Science? A Wake-up Call

In this paper I would like to give a sober and unadorned analysis of the situation, and at the same time make some proposals how to improve it. An improvement of the present situation is, indeed, badly needed, for the answer to the question in the title is not an unequivocal "yes", but rather "yes and no". In a geographical sense there is certainly a European Philosophy of Science. Our network, EPSA and its journal are ample proof of it. So the real question is whether there exists a European philosophy of science in a qualitative sense. Here the answer is no, with some qualifications, however. Globalization in the world of learning has led to an ever increasing use of the English language also in the humanities, and here particularly in disciplines like logic and philosophy of science. So, when I speak of "European" in my talk it is at the exclusion of the British Isles. The linguistic preponderance of English means also and most importantly that the agenda in philosophy of science is set in the Anglophone world, particularly in the US. As a consequence European approaches are largely marginalized and not even taken notice of in other European countries, since the "relay station" is the recognition and discussion of such approaches by major Anglophone figures. As some examples show that the use of the English language is, however, only a necessary condition for a European to be heard. Sufficient would be at best the complete immersion in existing Anglophone networks. But this is, of course, not a guarantee, for promoting specific European approaches. One has to have them first. In closing Vienna Logical Empiricism with its deep enlightenment intentions ("double approach") is described as a typical European way to conduct philosophy of science. It went lost, when the logical empiricists sought refuge in the US.

The corresponding volume to this is *Die europäische Wissenschaftsphilosophie und das Wiener Erbe Band 18, Nemeth, Elisabeth; Stadler, Friedrich (Hrsg.)* Further volumes can be found at: http://www.springer.com/series/3410 and http://www.springer.com/series/6669

MIX
Papier aus verantwortungsvollen Quellen
Paper from responsible sources
FSC® C105338

If you have any concerns about our products, you can contact us on
ProductSafety@springernature.com

In case Publisher is established outside the EU, the EU authorized representative is:
Springer Nature Customer Service Center GmbH
Europaplatz 3, 69115 Heidelberg, Germany

Printed by Libri Plureos GmbH
in Hamburg, Germany